# SOILS: BASIC CONCEPTS AND FUTURE CHALLENGES



This book was born as an international tribute to Fiorenzo C. Ugolini, an outstanding soil scientist who recently retired from university teaching and research. It is a fully up-to-date synthesis of the present knowledge of soils, their genesis, functions and management. It includes contributions from leading soil scientists and the result is a book that provides the basic concepts as well as the latest data and practical examples from across the discipline, including many issues that are overlooked in other treatments. The book also discusses the increasingly important role of soils in enabling the preservation of life.

*Soils: Basic Concepts and Future Challenges* provides the necessary keys to name soils and soil horizons. It contains a rare attempt to cross-harmonize the Reference Soil Groups of the World Reference Base of Soil Resources with the Soil Orders of the Soil Taxonomy, and presents a novel analysis of the various soil-forming factors. The book also quantifies the global extent of human-impacted soils, and the possible existence of extraterrestrial soils based on the findings from the last space missions.

This volume will be a valuable resource for researchers and students of soil science, soil conservation, geography and landscape ecology.

The Editors of this book, Giacomo Certini and Riccardo Scalenghe, are researchers at the Universities of Florence and Palermo respectively. Authors of numerous papers in international journals dealing with soil science and ecology, they have carried out studies in various European countries including Italy, Norway, Poland, Spain, Switzerland and the UK. They teach topics related to soil, water and the environment.

# SOILS: BASIC CONCEPTS AND FUTURE CHALLENGES

GIACOMO CERTINI
*University of Florence*

RICCARDO SCALENGHE
*University of Palermo*

CAMBRIDGE
UNIVERSITY PRESS

CAMBRIDGE UNIVERSITY PRESS
Cambridge, New York, Melbourne, Madrid, Cape Town,
Singapore, São Paulo, Delhi, Mexico City

Cambridge University Press
The Edinburgh Building, Cambridge CB2 8RU, UK

Published in the United States of America by Cambridge University Press, New York

www.cambridge.org
Information on this title: www.cambridge.org/9781107406438



First published 2006
First paperback edition 2012

*A catalogue record for this publication is available from the British Library*

ISBN 978-0-521-85173-2 Hardback
ISBN 978-1-107-40643-8 Paperback

'We might say that the earth has the spirit of growth; that its flesh is the soil.'
*Leonardo da Vinci*

This book pays homage to Professor Fiorenzo C. Ugolini, who has recently retired from his long career as a university professor of soil science. All the authors of this book had spirited interactions with him. He is an enthusiastic and inspirational teacher and scientist, a tremendous mentor and friend, and a talented Renaissance man. We all benefited greatly from his valuable contributions through which he enriched us and our discipline, soil science. For this reason, and perhaps also for the vital energy Professor Ugolini showed on every occasion, the authors accepted with enthusiasm the invitation to contribute to this book dedicated to his career.

We hope that this text represents a useful resource for preparing future soil scientists.

# Contents


*List of contributors* — *page* xi

*Preface* — xiii

*Acknowledgements* — xvi

1 **Concepts of soils** — 1
*Richard W. Arnold*
- 1.1 Some Greek and Roman concepts — 2
- 1.2 The transition — 4
- 1.3 The awakening — 4
- 1.4 Genetic supremacy — 5
- 1.5 Sampling volumes — 7
- 1.6 Landscape systems — 9
- 1.7 The new millennium — 9

2 **Pedogenic processes and pathways of horizon differentiation** — 11
*Stanley W. Buol*
- 2.1 Horizonation processes — 11
- 2.2 Studies of soil genesis — 12
- 2.3 Surface horizons — 14
- 2.4 Subsurface horizons — 15
- 2.5 Formation of structural features in soil — 21

3 **Soil phases: the inorganic solid phase** — 23
*G. Jock Churchman*
- 3.1 Description — 23
- 3.2 Future prospects — 44

4 **Soil phases: the organic solid phase** — 45
*Claire Chenu*
- 4.1 Soil organic matter complex composition — 46

4.2 Organomineral associations 51
4.3 Soil organic matter dynamics 54

5 **Soil phases: the liquid phase** 57
*Randy A. Dahlgren*
5.1 The liquid phase of soils 59
5.2 Methods of soil solution characterization 63
5.3 Application of soil solution studies to pedogenesis 66
5.4 Conclusions 73

6 **Soil phases: the gaseous phase** 75
*Andrey V. Smagin*
6.1 Gaseous components of soil 75
6.2 Sources, sinks and transport of gases in the soil 77
6.3 Agroecological evaluation of the soil air 84
6.4 Gases emissions and global ecological functions of the soil 85

7 **Soil phases: the living phase** 91
*Oliver Dilly, Eva-Maria Pfeiffer and Ulrich Irmler*
7.1 Physiological capabilities of soil organisms 92
7.2 The role of organisms for soil functions 95
7.3 Aerobic and anaerobic metabolisms in soil 96
7.4 The living phase indicates soil quality 98
7.5 Modification of biotic communities during soil degradation 100

8 **The State Factor theory of soil formation** 103
*Ronald Amundson*
8.1 The soil system 105
8.2 State factors 108
8.3 Importance of State Factor theory 111

9 **Factors of soil formation: parent material. As exemplified by a comparison of granitic and basaltic soils** 113
*Michael J. Wilson*
9.1 Mineralogical properties 114
9.2 Physical properties 116
9.3 Chemical properties 119
9.4 Conclusions 127

10 **Factors of soil formation: climate. As exemplified by volcanic ash soils** 131
*Sadao Shoji, Masami Nanzyo and Tadashi Takahashi*
10.1 Global climate and soil formation 132

10.2 Influences of climatic factors on soil formation based on the studies on volcanic ash soils 137

11 **Factors of soil formation: topography** 151

*Robert C. Graham*

11.1 Topographic elements of landscapes 151
11.2 External factors mediated by topography 155
11.3 Pedogenic processes linked to topography 156
11.4 Topography-based models of soil distribution 162

12 **Factors of soil formation: biota. As exemplified by case studies on the direct imprint of trees on trace metal concentrations in soils** 165

*François Courchesne*

12.1 Approach 168
12.2 Case study 1: Trace metal distribution at the soil–root interface 169
12.3 Case study 2: Trace metal patterns in organic horizons 175
12.4 In conclusion 179

13 **Factors of soil formation: time** 181

*Ewart A. FitzPatrick*

13.1 Time for horizon differentiation 182
13.2 Soil development 183
13.3 Holocene soil formation 185
13.4 Soil age and progressive change 185
13.5 Time and soil classification 190

14 **Soil formation on Earth and beyond: the role of additional soil-forming factors** 193

*Giacomo Certini and Riccardo Scalenghe*

14.1 The anthropogenic factor 194
14.2 Other factors of pedogenesis 205
14.3 Extraterrestrial soils 208

15 **Soil functions and land use** 211

*Johan Bouma*

15.1 How to deal with future demands on our soils 212
15.2 To characterize soil functions better 215
15.3 Storylines: what can the soil tell us when we listen? 219
15.4 In conclusion 221

16 **Physical degradation of soils** 223

*Michael J. Singer*

16.1 Soil compaction 224
16.2 Sealing and crusting 227
16.3 Physical soil management 229

16.4 Secondary effects 231
16.5 Conclusions 232

17 **Chemical degradation of soils** 235
*Peter Blaser*
17.1 Chemical soil degradation processes 236
17.2 Our duty 253

18 **The future of soil research** 255
*Anthony C. Edwards*
18.1 Soils and their buffering capacities 257
18.2 The soil resource 258
18.3 Soil phosphorus 258
18.4 Soil processes 260
18.5 Nitrogen cycling 261
18.6 The continued investigation of soil processes 263

*Appendix: Naming soils and soil horizons* 265
*Stanley W. Buol, Giacomo Certini and Riccardo Scalenghe*

*References* 277

*Index* 303

# List of contributors

Ronald Amundson
Division of Ecosystem Sciences, University of California, Berkeley, USA

Richard W. Arnold
Fairfax, Virginia, USA

Peter Blaser
Swiss Federal Institute for Forest, Snow and Landscape, Birmensdorf, Switzerland

Johan Bouma
Wageningen Unioversity and Research Centre, Wageningen, The Netherlands

Stanley W. Buol
Department of Soil Science, North Carolina State University, Raleigh, USA

Giacomo Certini
Dipartimento di Scienza del Suolo e Nutrizione della Pianta, Universitá degli Studi di Firenze, Firenze, Italy

Claire Chenu
Département AGER, UMR BIOEMCO, Thiverval Grignon, France

G. Jock Churchman
School of Earth and Environmental Sciences, University of Adelaide, Adelaide, Australia

François Courchesne
Département de Géographie, Université de Montréal, Montréal, Canada

Randy A. Dahlgren
Department of Land, Air and Water Resources, University of California, Davis, USA

Oliver Dilly
Lehrstuhl für Bodenschutz und Rekultivierung, Brandenburgische Technische Universität, Cottbus, Germany

Anthony C. Edwards
Peterhead, Scotland, UK

Ewart A. Fitzpatrick
Department of Plant and Soil Science, University of Aberdeen, Scotland, UK

Robert C. Graham
Department of Environmental Sciences, University of California, Riverside, USA

Ulrich Irmler
Ökologie-Zentrum, University of Kiel, Kiel, Germany

Masami Nanzyo
Graduate School of Agricultural Science, Tohoku University, Sendai, Japan

Eva-Maria Pfeiffer
Institute of Soil Science, University of Hamburg, Hamburg, Germany

Riccardo Scalenghe
Dipartimento di Agronomia Ambientale e Territoriale, Universitá degli Studi di Palermo, Palermo Italy

Sadao Shoji
Sendai, Japan

Michael J. Singer
Department of Land, Air and Water Resources, University of California, Davis, USA

Andrey V. Smagin
Faculty of Soil Science, Moscow State University, Moscow, Russia

Tadashi Takahashi
Graduate School of Agricultural Science, Tohoku University, Sendai, Japan

Michael J. Wilson
The Macaulay Institute, Aberdeen, Scotland, UK

# Preface

Soil is a dynamic natural body occurring in the upper few metres of the Earth's surface at the interface between the atmosphere, biosphere, hydrosphere and geosphere. A soil is both an ecosystem in itself, and a critical part of the larger terrestrial ecosystem. From the earliest perceptions of soils as the organic enriched surface layer to today's pedologic horizonation of profiles, there is a rich history of beliefs and understanding of this vital life-sustaining resource.

In Chapter 1 changes in perceptions of soils and their classification are explored. Chapter 2 describes some of the specific reactions that are components of the soil-forming processes that transform geologic materials into recognizable pedologic features and horizons. Solids, along with the liquids and gases that fill pore spaces between the solids, compose the three-phase soil system.

Chapter 3 treats the inorganic fraction of the solid phase, examining differences between primary minerals, derived directly from rocks, and secondary minerals, formed by pedogenic processes. Soil organic matter is discussed in Chapter 4. It is often a minor fraction of soil in quantitative terms, but exerts a major control on soil properties. Soil organic matter is complex, being a mixture of a multitude of different components. Organic matter may be tightly bound to clay surfaces by adsorption or physically protected by entrapment within aggregates. These associations modify the physicochemical and physical properties of the mineral phase and affect organic matter biodegradation rates.

The liquid phase of soil is an aqueous solution of solids and gases. It is dynamic and highly sensitive to changes occurring in the soil ecosystem. As shown in Chapter 5, studies examining soil solution chemistry can be a powerful approach to elucidate pedogenic processes, equilibrium and kinetic factors, solute transport, soil fertility, nutrient cycling, and the fate and transport of environmental contaminants. Chapter 6 provides an overview of the composition and dynamics of the soil gaseous phase. This phase has received considerable attention in recent years due to the realization that soils act as a global source, sink and reservoir of gaseous substances that control the atmospheric composition and thus affect the global climate.

Soil biota, the biologically active powerhouse of soil, includes an incredible diversity of organisms. It has been reckoned that there may be greater than 4 trillion organisms per kilogram of soil and more than 10 000 different species in a single gram of soil! Chapter 7 examines how soil biota plays a tremendous role in a number of soil properties and processes.

Genetic soil science (pedology), espoused by Dokuchaev and colleagues in Russia in the late 1880s, described soils as independent natural bodies resulting essentially from interaction of five environmental factors: parent material, climate, topography, biota, and time. Chapter 8 outlines how Hans Jenny rigorously defined or redefined the meaning of these factors, and more importantly, added the new concept of the *soil system*, which when combined with these factors provides a powerful conceptual framework to study and understand soils.

The influence of parent material as a soil-forming factor is an inverse function of time, making it especially important in young soils. Chapter 9 focuses on the impact of dissimilar parent materials – granite and basalt, the two most widely occurring igneous rocks on the Earth's surface – on the physical and chemical properties of soils in the context of different weathering intensities.

Climate, often the predominant soil-forming factor when considering soil development over the long term, is treated in Chapter 10. Temperature and precipitation are the most important components of climate. Temperature strongly influences the rates of chemical and biological reactions while soil moisture contributes to the dissolution, neoformation and transport of materials. Climate also determines the type and productivity of vegetation that, in turn, affect soil formation.

Topography, referring to the configuration of the land's surface, can have a major control on soil genesis. Chapter 11 examines the influence of topography on the disposition of energy and matter experienced by soils on the landscape. Slope, aspect, elevation and position modify the regional climate, causing soils to intercept more or less water and solar energy. Fine-scale topographic features may also influence pedogenesis by trapping aeolian dust, altering water infiltration patterns, modifying localized thermal regimes, and providing niches for biological activity.

Chapter 12 deals with the effects of biota on soil formation. Two field studies that illustrate the direct impact of trees on the spatial distribution of trace metal concentrations in uncontaminated forest soils are described.

The length of time needed to convert geological material into a soil varies, depending on the nature of the material and its interaction with climate, topography and living organisms. A given period of time may produce large changes in one soil and have little effect on another soil. Some horizons differentiate before others, especially those at the surface which may take only a few decades to form in unconsolidated deposits. Middle horizons differentiate more slowly, particularly when a considerable amount of translocation of material or weathering is necessary, some taking several millennia to develop. Chapter 13 examines the evolution of soil properties over time.

Pedogenesis is also possible in the absence of biota, as documented in some ice-free areas of the Arctic and Antarctic regions. On this basis, the physically and chemically weathered substrata of the Moon and Mars must be considered soils. Chapter 14 discusses extraterrestrial soils, as well as a variety of factors that can affect soil genesis on Earth, in addition to the five soil-forming factors first proposed by Dokuchaev.

Soil functions and land uses are described in Chapter 15. Several ideas are provided that will allow soil scientists to be better prepared for collaboration in the interdisciplinary arena. The pressure of a constantly growing population along with its demands and activities increasingly threaten the soil as a slowly renewable resource. Major problems arise from the cumulative use of land for living space, infrastructure, food and industrial production. Chapter 16 examines the most common issues due to physical degradation of soil, while Chapter 17 discusses the various forms of chemical degradation and their causes. The non-linearity of many soil processes and the spatial and temporal variability associated with their kinetics is particularly worthy of further investigation. Some of the questions that more urgently need an answer from soil science are discussed in Chapter 18.

Finally, this book includes an Appendix that provides: (a) the rudiments for naming genetic horizons, (b) a list of diagnostic horizons, properties and soil materials of the World Reference Base for Soil Resources (WRB) with their Soil Taxonomy (ST) equivalents, (c) description of the 32 WRB Reference Soil Groups, and (d) an approximate correlation of the WRB Reference Soil Groups with the 12 Soil Orders of Soil Taxonomy.

# Acknowledgements

G. J. Churchman thanks P. Rengasamy, R. C. Graham and M. J. Wilson, as reviewers, for useful suggestions regarding various drafts of this manuscript.

A. V. Smagin thanks the Russian Science Support Foundation.

R. C. Graham thanks K. Kendrick for advice on and drafting of the figures, and D. H. Yaalon and M. J. Wilson for reviewing the chapter.

O. Dilly, E.-M. Pfeiffer and U. Irmler thank the Ecology-Centre of the University of Kiel and the Institute of Soil Science of the University of Hamburg for their support in the preparation of this chapter.

S. Shoji, M. Nanzyo and T. Takahashi thank K. Minami, T. Makino, Y. Shirato, and R. J. Engel for their valuable information and suggestions.

F. Courchesne thanks N. Kruyts, P. Legrand, S. Manna and V. Séguin because the data presented in his chapter are part of the work accomplished by these graduate students or post-doctoral fellows. Data were also contributed by R. R. Martin, S. J. Naftel, S. Macfie and W. M. Skinner. B. Cloutier-Hurteau, N. Gingras, H. Lalande and J. Turgeon are sincerely thanked for their help with field and laboratory work. A special thanks to M.-C. Turmel, for managing the information originating from all of the above. Financial support for the researchers cited was provided by the Fonds Québécois de la Recherche sur la Nature et les Technologies (FQRNT), the Metals in the Environment Research Network (MITE-RN) and the National Science and Engineering Research Council of Canada (NSERC).

G. Certini and R. Scalenghe thank R. Amundson, R. A. Dahlgren, A. C. Edwards and B. Sundquist for critically reviewing the manuscript and T. Osterkamp for providing useful information.

P. Blaser thanks I. Brunner, B. Frey, F. Hagedorn, J. Innes, J. Luster, W. Shotyk, and R. A. Dahlgren for fruitful discussions and critically reviewing the manuscript. Figures 17.1 and 17.2 were provided by I. Brunner, while Figure 17.3 was provided by B. Frey and C. Sperisen (Swiss Federal Institute for Forest, Snow and Landscape Research, WSL).

S. W. Buol, G. Certini and R. Scalenghe thank R. J. Engel for critically reviewing the Appendix.

The Editors especially thank D. H. Yaalon, S. Francis, E. J. Pearce and J. Robertson

# 1

# Concepts of soils

*Richard W. Arnold*

In this chapter we will explore some changes of people's perceptions of soils and their classification as background for the dominant concepts of today.

Close your eyes for a moment and imagine that when you open them you are at the beginning of human time, a hunter and gatherer somewhere in the world, isolated, with barest of necessities, and you are hungry. By trial and error and stories passed on to you, you now know which plants and berries are okay to eat and how to stalk and kill animals and how to fish for your survival. As a keen observer you detect the location of specific plants and the common habitats and behaviour of the animals that become your food. You can't go far from where you are because your source of protein is here – not somewhere else. One or two million years pass by almost unnoticed.

Close your eyes again and when you open them imagine that you look beyond the bank of a river to small plots of irrigated land where grain is growing. Fish are still an important protein source but now with harvestable and storable grains you can easily carry protein with you. The world around you opens up to exploration and conquest. Ideas and technology are transferable to faraway places. It is known locally that some lands are better than others for producing grains and are easier to prepare and manage. Your observations reveal many new relationships – for instance, that the effort expended and the yields returned are geographic for the most part.

Throughout the Holocene, starting 11–12 000 years BP, there has been evidence of increasing use of land for cultivated grains and fruits. From the early habitats at the edge of sloping uplands to the later migrations into the lower lying river and lake plains, there arose complex systems of irrigation enabling the blossoming of early civilizations. Egypt, the Middle East, India, China, and

*Soils: Basic Concepts and Future Challenges*, ed. Giacomo Certini and Riccardo Scalenghe.
Published by Cambridge University Press. 

subtropical America – each with a remarkable history of use of land for agriculture and other needs of society (Stremski, 1975; Krupenikov, 1992). The concept of land (and soil) during the global expansion of people seems to have been twofold: one was the suitability for growing specific plants, and the other was the energy required to prepare and use the land. In general, sandy soils were much easier to prepare but the yields were more difficult to maintain, whereas clayier soils were hard to prepare but the yields were much better. Thus properties of soils, functions of soils, and classification of soils have been around a long, long time.

Soil, as we understand it today, is a concept of the human mind. From the earliest perceptions of soils as the organic enriched surficial layer to today's pedologic horizonation of profiles there is a rich history of beliefs and understanding of the vital life-sustaining resource. The earthy material is real, it exists, you can touch it, feel it, stand on it, and dig in it, but defining it is far more complex because it can be what you want it to be. The Mother of life, a healer of sickness, a home of spirits, a geomembrane that sustains ecosystems of which we are a part – yes, soil surely is all of these and likely much more depending on your cultural background and heritage, education and training, and your personal experiences.

An interesting aspect of thinking about soils is the uncertainty expressed as dichotomies, which have been present throughout humankind's involvement with this surficial layer of 'dirt' that somehow is vital to our existence and survival. From sacred to profane, from beautiful to filthy, from productive to unresponsive – all are human perceptions of soils – brave and bold and highly subjective.

Several references (Boulaine, 1989; Yaalon and Berkowicz, 1997) may help you get started in your search of vignettes of what the 'ancients' did. There is a natural tendency to look for the initiator – the first – the beginning – and then follow the paths of evolution, the birth and death of ideas. Why? Perhaps because we are also cyclic.

## 1.1 Some Greek and Roman concepts

By the time of Greek civilization there had already been several millennia of records of humankind's achievements and failures to control soils scattered among the languages of the Earth's inhabitants. Let us pick up the story with Aristotle. He said that there were four elements formed and shaped from the same amorphous matter by a spirit endowed with reason. Fire, air, water and earth were in opposition to ether, the fifth element, which could not be perceived by the senses. These four elements were carriers of both active and passive qualifiers. Earth was characterized by opposing qualities, such as warm and cold, dry and wet, heavy and light, and hard and soft.

One of Aristotle's students, Theophrastos (371–286 BC) gave soil the name of 'edaphos' to contrast it with earth (terrae) as a cosmic body. Edaphos was a layered system; a surface stratum of variable humus content, a fatty subsoil layer that supplied nutrients to grass and herb roots, a substratum that provided juices to the tree roots, and below was the dark realm of Tartarus. He described numerous relationships of soils and plants, and indicated six groups of lands suitable for different crops. The Greeks paid special attention to grapevines and Theophrastos even noted that an important way to increase productivity on stony soils was by transplantation of soil.

Herodotus (*c.* 485–425 BC), an experienced traveller, considered soil as an important element in characterizing a place, noting for example that Egyptian soil was black and friable, and consisted of silt brought by the Nile from Ethiopia.

In summary, the Greek intelligentsia concluded that soil was something special and important, had a profiled (layered) structure, fertility was its main quality, soil was spatially variable, plants were both wild and cultivated, and plant selection and cultivation were highly dependent on the properties of the soils.

In ancient Rome, the problems of agronomy including technology and organization of agriculture, and better land utilization were important. The nature of Italy is diverse and these features created a complicated mosaic of soil cover; thus it was necessary for Roman farmers to determine 'which land likes what'.

Cato, the senior (234–149 BC), was a government official, a big landlord, and traveled on assignment for the Senate. One of his major works, *De Agricultura*, appeared about 160 BC. In addition to knowing 'which land likes what', he also admonished that careful ploughing and application of dung and use of green manure crops was necessary to create those conditions that are best for plant development. Cato dealt at length with the problem of dung manure. He developed a classification of arable soils based on farming utility with nine major groups that were subdivided into 21 classes.

Varro (116–27 BC), an encyclopedia specialist, was assigned by Julius Caesar the task of organizing a public library in Rome and may have been the first to recognize the independent status of farming as a science. He also observed that it teaches us what should be sown on which field so that the earth will constantly produce the highest yields. Varro devised a classification recognizing as many as 300 types of soil using soil properties such as moisture, fattiness (texture), stoniness, colour and compaction. Maintaining productivity by rotating crops was important advice to farmers.

Twelve volumes about agriculture by Columella (first century AD) covered the gamut of agronomy of the Mediterranean region. With regard to declining soil fertility he said that the guilt lies with people who deal with agriculture like a hangman with a prisoner, the lowliest among slaves. He developed a classification based on combinations of properties yet conceded that no one can know '*in toto*' the whole diversity of soils. He conducted many field experiments, noting that science shows the learner the correct path.

Stremski (1975) summarized Roman heritage by noting that Cato emphasized the suitability of soils for farming and their quantitative productive potential, Varro was concerned mainly with physical composition of soils, Columella emphasized physical properties, and Pliny the Elder focused on rocks and minerals as soil-forming materials. It is obvious that ancient knowledge of soils was extensive; however, agricultural soil science stagnated with the downfall of Rome only to be revitalized in the eighteenth and nineteenth centuries.

## 1.2 The transition

Close your eyes again and when you open them imagine that the Renaissance and the Age of Enlightenment have just finished. Here in the nineteenth century there abound a myriad of discipline-oriented concepts of soil based on the background and interests of scientists in different disciplines. The geologists refer to the straight-line function of rocks to soils; thus there were granite soils, limestone soils, shale soils, and so forth. Geomorphologists recognized upland soils, river valley alluvial soils, colluvial soils, mountain soils, steppe soils, desert soils and so forth. Botanists associated plant communities with soils; thus there were oak soils, prairie soils, pine soils, desert shrub soils, taiga soils, etc. Chemists denoted alkali soils, carbonaceous soils, base saturated soils, acid soils, and so on. Agriculturists referred to maize soils, wheat soils, pasture soils, fertile and infertile soils and many others. People concerned with mechanical behaviour recognized sticky soils, clayey soils, push soils, silty soils, one, two and three water buffalo soils, stony soils and so forth.

Throughout this period there was no general agreement on how to recognize and refer to soils. One cultural attitude still prevailed – that of the lowly status of those who tended the fields. Serfs, peasants and slaves were associated with the menial, filthy aspects of preparing, tilling and harvesting produce from the earth. By association soil was not generally worthy of serious consideration.

## 1.3 The awakening

As you open your eyes once more you suddenly stand in a gently waving sea of prairie grass looking across a seemingly infinite expanse of open landscape – the home of the famous Russian Chernozem. Severe droughts in 1873 and 1875 in this region caused untold misery and economic loss. In 1877 the Free Economic Society instituted the 'Chernozem Commission' and funded V. V. Dokuchaev, a geologist at the University in St Petersburg, to conduct geologic-geographic investigations of the Chernozem. In the report of the second year of work he described soil generally as a mineral-organic formation of unique structure lying on the surface and continuously being formed as a result of the constant interaction of living and dead organisms, parent rock, climate and relief of the locality. He also stated that 'soil exists as an independent body with a specific physiognomy, has its own special origin, and properties unique to it alone' (Krupenikov, 1992, p. 161). Dokuchaev's classic monograph, '*Russian Chernozem*' published in 1883, was the final report to the Free Economic Society about the Chernozem problem and it was defended as his doctorate dissertation.

In 1882 the Nizhi Novgorod province requested Dokuchaev to conduct geological and soil investigations for a rational assessment of land. The project continued from 1882 to 1886, was published in 14 volumes, and laid the foundation for the new school of genetic soil science. According to Dokuchaev the main aim of pedology was to study soils 'as they are' and to understand the regularities of their genesis, interrelations with the factors of soil formation, and geographical distribution. This was the principal difference from the prevailing notions of soil as just an object of agricultural activities. What did this really mean? It brought together many of the ideas about soil, restructured them into a set of integrated causal relationships, and provided a framework for research and understanding of soil as an independent science.

## 1.4 Genetic supremacy

After blinking your eyes again, you realize that another hundred years has passed and that pedology has been constantly evolving (Bockheim *et al.*, 2005). Genetic soil science has been accepted around the world and soil surveys have been underway in many countries for a number of years, associated mainly with agriculture and forestry. Pedologic and geologic concepts and terminology of soil horizons, solum, profile and weathering layers have come into existence and been adapted to meet both scientific and societal needs (Tandarich *et al.*, 2002).

The conservation of soils has usually been stimulated by catastrophic events related to their degradation. For example, the Dust Bowl in the USA in the 1930s spurred government action to create a Soil Conservation Service in the Department of Agriculture. Along with practices to mediate water and wind erosion, there were attendant actions to better manage water resources and maintain fertility. Advice about protecting soils was based on knowledge of the soil resources; consequently an expanded programme of soil survey was undertaken. The need for basic units of classification and for mapping was evident and pedological concepts prevailed.

Based on the 'neo-Dokuchaev paradigm of pedology' that relates factors → processes → properties (Gerasimov as referenced by Sokolov, 1996, p. 253) there arose two major pedological concepts of soils. One is represented by the pedon, or arbitrary volume; the other by the polypedon, or small landscape unit (Fig. 1.1). The literature contains many terms for both small arbitrary volumes of soils and the spatial entities identified by named and defined kinds of soils (Arnold, 1983).

A major Russian textbook based on Dokuchaev's concepts stated that 'the moisture and thermal regimes determine the dynamics of all phenomena in soils, i.e. they are fundamental in soil formation as a whole' (Gerasimov and Glazovskaya, 1965, p. 147); however, neither soil temperature nor soil moisture state were considered or defined as 'soil properties'. Climatic regimes,

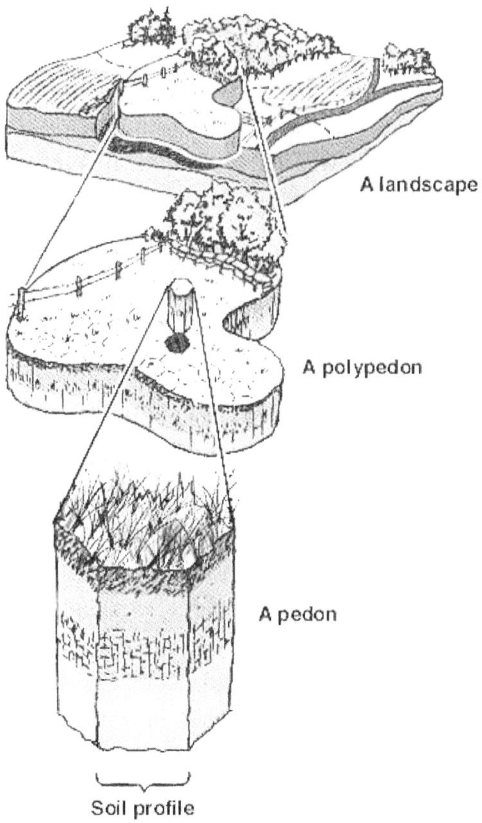


Fig. 1.1. A hierarchy of spatial relations of soil bodies commonly used in pedology. The profile, a thin rectangular section is the basis for soil description; the pedon represents a small sampling volume for property characterization; the polypedon represents a body having a set of similar properties in a landscape; and the landscape represents a portion of the pedosphere containing bodies of several kinds of soils. Adapted from Fig. 3.1 in Brady and Weil (1996).

by contrast, were commonly identified as important environmental conditions for specific kinds of soils in Russia. The United States Department of Agriculture's (USDA) '7th Approximation' and subsequent editions of *Soil Taxonomy* (Soil Survey Staff, 1999) have described and defined soil temperature and soil moisture state as soil properties. Proper placement of soils in Soil Taxonomy has been achieved by defining patterns of soil temperature and soil moisture as regimes and using them to define specific taxonomic classes. This revolutionary deviation from other taxonomies recognized soils as dynamic entities in addition to being historical records of soil evolution by quantifying these properties.

## 1.5 Sampling volumes

Arbitrary small volumes of soils are the basic source of information about the genesis and properties of soils (Holmgren, 1988). This information abstracts the central concept of a soil and the properties characterize a soil mainly for purposes of classification and correlation.

Because soil-forming factors occupy space and their influence is over time, there is a concept of soil as a geographic entity whose recognition and distribution depend on limits associated with defined kinds of soils and external features associated with the processes and properties of the dominant soil.

The USDA Soil Survey Manuals of 1951, 1962 and 1993 highlighted a number of concepts guiding soil surveys in many parts of the world. Soil was thought of as the collection of natural bodies on the Earth's surface, in places modified or even made by humans of earthy materials, containing living matter and supporting or capable of supporting plants out of doors. The upper and lower limits with non-soil were discussed but not quantified.

Natural soil bodies were still considered to be the result of climate and living organisms acting on parent material, with topography or local relief exerting a modifying influence and with time required for soil-forming processes to act. Some confusion still existed about whether soil referred to the broader concept of a resource, or to its component members; that is, the soil, or kinds of soils.

In the first version of a new Russian soil classification scheme (Shishov *et al.*, 2001) soil was considered to be a system of interrelated horizons composing a genetic profile, which derived from the transformation of the uppermost layer of the lithosphere by the integration of soil-forming agents.

A pedon was regarded as the smallest body of one kind of soil large enough to represent the nature and arrangement of horizons and variability in other properties that are preserved in samples (Soil Survey Staff, 1999). It had a minimal horizontal area of 1 square metre but ranged to 10 square metres depending on the variability in the soil. In the USA, the pedon was originally considered to be a sampling unit within a polypedon that was a unit of classification, a soil body homogeneous at the series level, and big enough to exhibit all the soil characteristics considered in the description and classification of soils (Fig. 1.1). Because of the difficulty in fitting boundaries on the ground and the circular nature of the concept, the polypedon seldom served as the real thing to be classified.

In the 1998 *Keys to Soil Taxonomy* (Soil Survey Staff, 1998) soil was referred to as a natural body that comprises solids (mineral and organic), liquid and gases that occur on the land surface, occupies space, and is characterized by one or both of the following: horizons, or layers, that are distinguishable from the initial material as a result of additions, losses, transfers and transformations of energy and matter; or the ability to support rooted plants in a natural environment. This expanded definition of soil was meant to include soils of Antarctica where pedogenesis occurred but where the climate was too harsh to support the higher plant forms. For purposes of classification the lower limit of soil was set at 200 cm. The deposition, alteration, and layering of sediments helped explain discontinuities of parent materials. Recognition of *in situ* alteration such as weathering, hydrothermal influence or contamination expanded the concept of factor interactions.

For the most part profile features were combined into models of soil formation involving the processes and events of geomorphology that had influenced and helped to shape the hypothesized features.

As portrayed by the National Cooperative Soil Survey in the United States, a soil series was a group of soils or polypedons that had horizons similar in arrangement and in differentiating characteristics. The soil series had a relatively narrow range in sets of properties. Map unit delineations had commonly been identified as phases of the taxonomic soil series. This process attempted to bridge the gap between classification and geography as classification became more quantitative; however, the resolution by refining soil series definitions to fit the limits imposed by the hierarchical Soil Taxonomy resulted in the loss of much landscape information.

After several decades of study the Food and Agriculture Organization's (FAO) legend for the map of World Soil Resources was accepted at the 1998 World Congress of Soil Science as the basis for developing a World Reference Base to correlate soil classification systems with the intent to provide an updated legend, map and database for global soil resources. Although no real definition of soil was reiterated (FAO/ISRIC/ISSS, 1998), Reference Soil Groups were defined by a vertical combination of horizons within a defined depth and by the lateral organization of the soil horizons, or by the lack of them, at a scale reflecting the relief of a land unit. Soil horizons and properties were intended to reflect the expression of genetic processes that are widely recognized as occurring in soils (Bockheim and Gennadiyev, 2000).

## 1.6 Landscape systems

Several other approaches to describe and define geographic bodies of soils were summarized by Fridland (1976). Most of these concepts arose where detailed soil mapping was not the major soil survey activity, but where exploratory and other small-scale studies were being undertaken. The French school of pedology has refined and implemented many concepts related to soil landscape mapping (Jamagne and King, 2003). Soils were believed to result from transformations that affect the material of the Earth's crust, and that successive climates and biological and human activities had been the agencies directly responsible. Their effect depended not only on the nature of the rock and their derived formations that have resulted from them, but also on landscape relief and the migration of matter in solution or in suspension in water. The overall result was that the original arrangement of geological material disappeared, leaving an entirely new arrangement of pedological origin.

They maintained that the genetic conditions of soils resulted in a double differentiation: in their vertical arrangement and in their spatial distribution. The former corresponded to the common notion of soil profiles that are vertical sections through the nearly horizontal layers of altered parent materials (horizons) (Fig. 1.1), and the latter corresponded with the lateral arrangement of different types of horizons within the landscape, thus allowing for the definition of soil systems in space.

The concept of soilscape or 'pedolandscape' was defined as the soil cover, or part of the cover, whose spatial arrangement resulted from the integration of a group of arranged soil horizons and other landscape elements. A soil system was a type of soilscape, a toposequence, where the differentiation was linked with a functioning process. A reference relief unit was a catchment or watershed area and the analysis of lateral transfers on, in and through the soils (vertically and laterally) had to be considered to understand the functioning of the landscape units. The systems could be open or closed relative to the flow of water and energy. Soil systems provided a framework to describe the process dynamics of the evolution of a landscape and its associated soils.

## 1.7 The new millennium

Now as you open your eyes once more, a new millennium has begun. It is one full of uncertainty, especially concerning the extent to which humans have irreversibly altered their global habitat. It has been postulated that the achievements of the Industrial Revolution and the rapid changes of the Information Age are characteristic of the Anthropocene, the current geological period.

In the 1997 edition of the Russian classification (Shishov *et al.*, 2001) special attention was given to agrogenically and technogenically transformed soils. These soils were considered to be the result of soil evolution under the impact of human activities. When considering natural soil, the anthropogenically transformed soils formed an evolutionary sequence grading finally to non-soil surface formations. Recognition was based on morphology and did not include direct impacts on soil fertility. The initial proposal recognized Agrozems as soils whose profiles had been agrogenically modified providing a homogeneous topsoil more than 25 cm deep over diagnostic subsoils or parent material. Agrobrazems lacked a surface diagnostic horizon due to erosion, deflation or mechanical cutting but had a specific surface horizon formed from subsoil or parent material. Abrazems, although similar to Agrobrazems, were recognized by the presence of a subsoil horizon or transition to parent material and were not suitable for cropping. In addition degraded soils due to chemical impacts, Chemdegrazems, could be recognized in any other class of soil. A special group of artificially constructed materials (non-soils) called Fabricats were suggested for trial use. They included Quasizems that had a humus-enriched surface layer placed over chaotic mixtures, Naturfabricats that lacked an organic matter enriched surface layer but consisted of human transported or mixed materials, Artifabricats that had substrates whose materials are absent in nature, and Toxifabricats that consisted of toxic, chemically active materials unsuitable for agriculture or forestry.

Man as an important soil-forming factor is also reflected in the concepts of soils as functional entities now within the realm of the noosphere where man and nature are considered to be co-evolutionary factors of the biogeosphere. Ecological and environmental soil functions as described are human values associated with actual and potential behaviour of soil landscapes and their relevance to society. Although not far removed from the concepts attributed to use of the soil by ancients, the details are more focused and even global in scope.

One perspective describes the function of soils in the pedosphere as they interact with associated spheres, namely the atmosphere, biosphere, hydrosphere and the lithosphere (Arnold *et al.*, 1990). These concepts portray the pedosphere as the active geomembrane interface that mediates energy fluxes and enables terrestrial life to exist (Ugolini and Spaltenstein, 1992).

Another perspective describes major soil functions as: biomass producer and transformer; filter, buffer and reactor; habitat for macrobiota and microbiota; direct utilization as raw material and infrastructure support; and cultural and heritage aspects (GACGC, 1995). Both perspectives are significant to our understanding of soil quality, soil health, ecosystem sustainability, and the world of human-influenced soils. The chapters that follow describe major aspects of pedology and lead us to consider the challenges and changes of the concepts that will guide the future.

# 2

# Pedogenic processes and pathways of horizon differentiation

*Stanley W. Buol*

Soils acquire and maintain their characteristics and composition while undergoing simultaneous alteration by an almost infinite number of biogeochemical reactions. The possible number of pedogenic events and combinations and interactions among them in soils is staggering. Although laboratory experiments can demonstrate that specific processes can produce specific soil features, the actual course of events within undisturbed soil will probably never be fully known because the cumulative impact of soil-forming processes spans such long periods of time relative to the lives of humans who observe those impacts.

## 2.1 Horizonation processes

The entire volume of material defined as soil is but one layer within a larger context of the lithosphere. Soil is a layer of the lithosphere where minerals formed at high temperatures in the absence of water during the cooling of the Earth's magma are being decomposed by water, and new minerals (secondary minerals) are being formed at lower temperatures. In soil, organic compounds formed in plants primarily from carbon taken from the air are mixed into the mineral material of the lithosphere. Soil can be conceptualized as an open system where material can be added, transformed, translocated and removed. Generalized processes responsible for the presence of identifiable horizons and other features within soil are outlined in Fig. 2.1.

These processes include:

1 Energy exchange as the soil surface is daily heated by the sun and cooled by radiation to space each night.
2 Water exchange as soil is periodically wetted by precipitation and dried as water evaporates and/or taken from the soil by plant roots and transpired through the plant leaves.

*Soils: Basic Concepts and Future Challenges*, ed. Giacomo Certini and Riccardo Scalenghe.
Published by Cambridge University Press. 

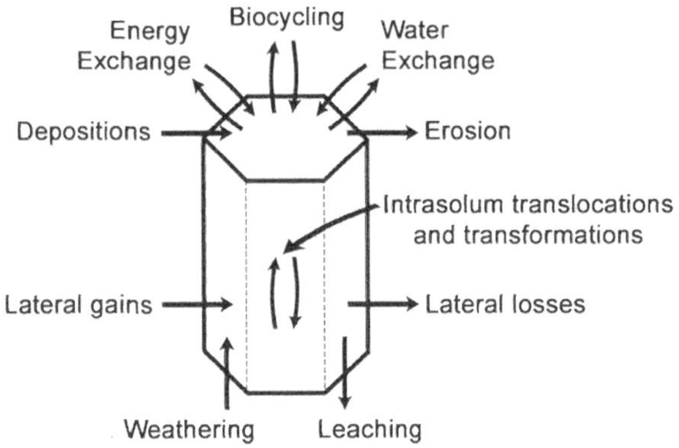


Fig. 2.1. Schematic representation of the generalized processes that actively create horizons and other features in soil.

3 Biocycling depicts essential plant nutrients being taken from the soil, combined with carbon, hydrogen and oxygen, temporarily stored in vegetation and concentrated on or near the soil surface as organic compounds as the vegetation dies. Human activities that remove vegetation for use as food and fibre disrupt this cycle.
4 Erosion and deposition are processes engendered by the movement of wind and water that physically remove material from the surface of some soils and deposit soil material on the surface of other soils.
5 Weathering processes alter minerals that are unstable in the current soil environment. Some elemental components of the primary minerals are removed and some are restructured into secondary minerals as the soil wets and dries.
6 Leaching processes remove soluble organic and inorganic compounds from the soil as water percolates beyond the rooting depth of the vegetation present, negating the biocycling capability of the vegetation.
7 Lateral transfer of soluble and suspendable material takes place in the flow of groundwater between some adjacent soils.
8 Intrasolum translocation processes represent the movement of mineral and organic substances and particles within the soil. Most downward translocation is via the downward movement of water with subsoil accumulations developing as that movement is attenuated within the soil. Fauna and floral activity, shrink and swell movement of wetting and drying, freezing, earth tremors, violent shaking of rooted trees and human activities also physically move material within some soils.

## 2.2 Studies of soil genesis

Soil-forming processes are assemblages of reactions occurring simultaneously or in sequence that create soil horizons and other morphological features. For a soil property to be present, it must be compatible with the existing environment within the soil. This may be because current processes promote formation of the present soil property or because the soil property formed by past environments is stable enough to persist under present conditions. Generally, sequences of processes occur when the result of one process triggers the initiation of a subsequent process. The results of a given process may tend to maintain the soil in its current condition, or may tend to change the soil.

No person has ever seen a mature soil form *in toto*. Pedogenic processes include gains and losses of materials from a soil body in accordance with the degradational, aggradational, or intermediate geomorphic character of the site, as well as translocations within the soil body. While appearing permanent in the short timeframe of human observations the volume of soil in the lithosphere is transient in space–time. Soil material dissolves, erodes or is buried over geologic timescales. All soils move vertically in space over time. Downward movement of soil occurs as erosion and dissolution remove material and in some locations upward movement occurs as material is deposited by flooding water, dust or volcanic depositions. Although the soil surface may appear stable during the short periods of human examination, the space–time dimension of soil is necessary to fully understand processes of soil formation (Buol *et al.*, 2003).

Two scientific approaches termed static pedology and dynamic pedology have been used in studies of soil genesis. The static approach proceeds by obtaining data from field observations and laboratory analysis of samples (NRCS, 1996) and then inferring what processes could have been capable of producing the observed soil properties. The dynamic approach is to monitor processes *in situ*, using apparatus such as suction-plate lysimeters that extract samples of percolate in host horizons (Ugolini, 2005). The dynamic approach also employs laboratory simulations using leaching columns and other devices. Whereas the dynamic approach measures some of the current processes, these processes may not accurately reflect the impacts of long-term processes and may miss entirely those processes that operated only sporadically in the past. A combination of these two approaches provides the most useful information about present, past and sporadic soil-forming processes.

Most identifiable soil layers or horizons are formed in response to the movement of water into and out of the soil. Some soluble and suspendable components of soil eluviate in percolating water and accumulate to form distinct and contrasting horizons as downward water movement ceases and water is extracted via evapotranspiration. Some horizons form by the physical mixing of various organic and inorganic components. Soil horizons persist only if their components are stable or in steady state within ambient conditions. Horizons with unique quantities of elements, salts, organic or mineral compounds develop only if the necessary components are present in the media within which the soil is formed or can be formed and added to the soil from other components of the ecosystem.

Two overlapping trends in soil development are horizonation and haploidization. Simply stated, these are processes that tend to form layers or horizons within soil or to mix soil, respectively. Both trends are present in all soils. The degree to which certain soil properties are present represents the relative intensity of these contrasting forces. Mixing processes are more prevalent near the soil surface and produce surface horizons while translocation processes are more active in forming distinctive subsoil horizons.

## 2.3 Surface horizons

Most surface horizons, called epipedons in Soil Taxonomy (Soil Survey Staff, 2006), result from the mixing of organic material and mineral material. Organic materials are formed as plants capture carbon as carbon dioxide from the air and combine that carbon with hydrogen, oxygen and the other life essential elements captured from the soil by their root systems. When plants and other organisms die the organic materials mix with mineral material of geologic origin in the soil. This process is known as biocycling. The result is a concentration of inorganic elements essential for plant growth and organic carbon compounds in surface horizons. Organic carbon compounds in soil are transient as soil microbes oxidize the carbon, returning it to the air as carbon dioxide. Organic carbon content in soil at any moment of time reflects a steady state condition between rate of organic additions and rate of organic decomposition reactions.

Distinctive surface horizons or epipedons[1] include:

- Mollic epipedons (mollic and voronic horizons) are thick, friable, dark coloured surface horizons containing more than 1% organic matter and having a base saturation of 50% or more of the cation exchange capacity (CEC) determined at pH 7. They form in calcium-rich mineral material usually under grass vegetation. The seasonal decomposition of the fibrous grass roots well below the soil surface assures that extremely high surface temperatures do not facilitate the rapid microbial decomposition of the organic carbon compounds.
- Umbric epipedons (umbric horizons) are similar to mollic epipedons in organic matter content but have base saturations less than 50% of the CEC as determined at pH 7. They are most often formed in acid geologic material under several types of vegetation. Stability of organic matter content is favoured in soils where organic compounds are not subjected to high temperatures. Cool climatic conditions and landscape position that shield the soil surface from direct solar radiation favour their

[1] Surface horizons are identified as epipedons in Soil Taxonomy (Soil Survey Staff, 2006). The approximate equivalents in the World Reference Base (IUSS Working Group WRB, 2006) are in parentheses.

formation. Umbric epipedons are often present in poorly drained soils. Water has a high heat capacity; thus soils with a high water content experience lower maximum daytime temperatures and have a slower rate of organic matter oxidation than equivalent drier soils.

- Ochric epipedons have organic matter contents too small or are too thin to meet the specific thickness requirements of umbric or mollic epipedons. Their low organic matter content can be related to low production of organic residues because of the slow growth of plants in moisture or temperature limited climates. They are commonly formed under tree vegetation where most organic additions are from the surface deposition of plant residue and organic additions below the surface are minimal.
- Histic epipedons (histic horizons) contain more than 20% to 30% organic matter. They are almost entirely present where the soil is saturated with water much of the year. Saturated conditions deprive the decomposing micro-organisms in the soil of the oxygen they need to completely decompose the organic residues deposited on the surface, and the high heat capacity of the water buffers the soil from high daily maximum soil temperatures, further slowing microbial decomposition.
- Plaggen epipedons (plaggic and terric horizons) have been formed by intensive human additions of manure and organic residues to form thick surface horizons with high organic matter content. Other distinctive surface horizons created by a shorter-term human activity in Soil Taxonomy are the anthropic epipedons.

The World Reference Base (WRB) recognizes specific kinds of human-created surface horizons as:

- Anthric, formed by long-term ploughing, liming, fertilization, etc.;
- Irragric, formed by long-term application of muddy irrigation water;
- Hortic, formed by deeper than normal cultivation with intense fertilizer or manure application;
- Anthraquic, formed over a 'puddled' slowly permeable plough pan created by many years of cultivating crops, primarily rice, in flooded fields.

## 2.4 Subsurface horizons

Preferential losses, accumulations and mineral transformations of specific soil components are responsible for most subsoil horizons. Water and the physical movement of water is the primary agent for these intrasolum processes. Water movements in soil are dynamic and sporadic events largely controlled by weather events. Spatial differences result as relief (slope) of the land and affect infiltration of water. Most water enters the soil through the surface. Within the soil water attains its greatest downward velocity in the larger pores. Fast-moving water suspends small clay and organic particles and soluble components of both inorganic minerals and organic compounds. After infiltration ceases, the water from the larger pores is distributed via capillary action into smaller pores and some suspended particles are filtered and deposited on the walls of the larger pores.

The water from most rain events does not penetrate to a great depth and terminates in a subsoil layer also occupied by plant roots. Between rainfall events plants extract water from those pores between about 0.01 and 0.0002 mm in diameter. As plants extract water the concentration of those ions dissolved in the water and not ingested into the roots is increased and they precipitate into solid crystal structures. The type of crystal formed depends on the ions present. For example, if there is an abundance of $Ca^{2+}$ and $H_2CO_3$ present in the soil, $CaCO_3$ will form in the subsoil as the vegetation extracts water. As Si and Al ions dissolved from primary silicate minerals are concentrated in the soil solution they form aluminium silicate clays, either 1:1 clays if in equal proportion or 2:1 clays if Si ions are more abundant than Al ions.

The upper part of the soil is subjected to more downward movement of water and tends to lose soluble and suspendable material (eluviation processes) while at somewhat greater depths soil ('subsoil') tends to accumulate suspended and soluble materials (illuviation processes). Excepting soil in the most arid climates some water sporadically moves below the rooting depth of plants, causing some chemical alteration of minerals present and loss from the entire soil. Plants do not remove water at the same rate throughout the year and in temperate latitudes more water leaches below the rooting depth in the winter when vegetation is not actively transpiring. However, whenever temperatures are sufficient for active transpiration, plants daily remove water from the soil. Rainfall events are spasmodic assuring some drying, and therefore concentration of suspended and soluble constituents in the soil solution takes place within the rooting volume of the plants in almost all but the most continuously saturated soils.

Specific subsoil horizons are identified by the properties they acquire from the relative intensity of the eluviation, illuviation and leaching processes that have been and continue to be active at a specific depth. Dissolution, suspension and subsequent deposition of a particular element or compound depend on the abundance of that material in the soil, i.e. the parent material. If the parent material does not contain an adequate supply of the element needed to form a secondary mineral that mineral will not form. Also, pedoturbation processes such as creep movement of soil material on steep slopes, physical disruptions by soil fauna or extensive expansion and contraction upon wetting and drying may so rapidly mix soil material that subsoils of substantial illuvial accumulations will not form.

Subsoil horizons with specific properties and characteristics are defined in modern soil classification systems.

- Argillic horizons (argic horizons) have 1.2 times as much clay as horizons above and are formed by the illuviation of clay and its filtering on the walls of large pores as the percolating water is drawn into smaller pores by capillarity. This process is also known as lessivage with the formation of clay skins, also known as clay films or tonhatchen (Fig. 2.2). This process appears most active when new clay is being formed. Clay that has been present in the soil for a long period of time is often coated with iron oxide and appears less subject to illuviation (Rebertus and Buol, 1985).
- Kandic horizons (some ferralic horizons) are like argillic horizons but contain primarily kaolinitic clay and have apparent CEC less than 16 $cmol_c^+ kg^{-1}$ clay at pH 7. The lessivage process is slow in soils with kandic horizons due to the lack of weatherable primary minerals from which new clay can form and low suspendability of clay that is coated with iron oxides. Clay skins are rare, apparently due to slow formation and their homogenizing into the soil matrix by pedoturbation processes.
- Albic horizons (albic horizons) are light coloured horizons below dark coloured epipedons and above several of the other subsoil horizons. Albic horizons are most often described as E horizons but also meet specific colour criteria. They form as percolating water suspends clay and organic particles and transports them to deeper horizons. They are usually formed under tree vegetation and far enough below the soil surface that little organic material is mixed into the E horizon.
- Calcic horizons (calcic horizons) are accumulations of calcium carbonate. Carbonates are dissolved as bicarbonate from mineral sources during the percolation of water and precipitate as carbonate-rich subsoil horizons as plant roots or evaporation-extract soil water. They are common in arid climates with little leaching when carbonate minerals

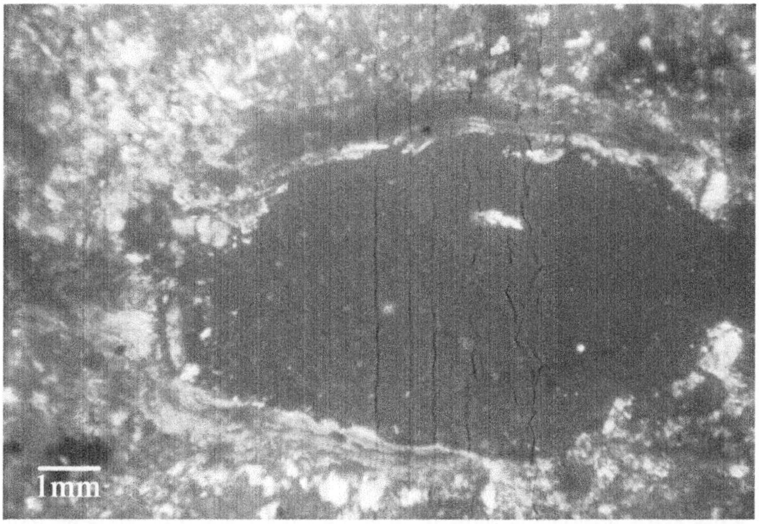


Fig. 2.2. Photomicrograph of clay skin surrounding a large void in the argillic horizon in Wisconsin, USA.

are present in the parent material or from dust. Hard calcic horizons usually containing more silicon are known as petrocalcic horizons (petrocalcic horizons).

- Spodic horizons (spodic horizons) are formed by the aluminium and iron eluviated from O, A, and E horizons and the immobilization of these metals in short-range-order complexes with organic matter, and in some cases silica, in the B horizon (Mokma and Evans, 2000). The process is driven largely by the production of organic acids from the decomposition of plant materials deposited on the soil surface. Ugolini *et al.* (1977) found direct evidence from lysimeter studies of the migration of organic matter particles (0.5–1.5 μm in diameter) in the solum and of mineral particles (2–22 μm in diameter) below that. In addition to aluminium and iron, monovalent cations and most $Ca^{2+}$ and $Mg^{2+}$ ions are leached down into underlying horizons or to groundwater, along with silica (Singer and Ugolini, 1974; Zabowski and Ugolini, 1990). The translocation of silica provides a mechanism by which allophane or imogolite forms in the spodic horizon (Dahlgren and Ugolini, 1989a, 1991). The capacity of some spodic horizons to sorb dissolved organic carbon may be due to the presence of imogolite (Dahlgren and Marrett, 1991).
- Cambic horizons (cambic horizons) are horizons that have some alteration of the primary minerals present in the parent material or the replacement of original structure in the parent material with pedogenic soil structure. The removal of iron, carbonate or gypsum when present in the parent material is the criterion for identifying some cambic horizons. Cambic horizons may have small accumulations of clay, iron or organic matter but such illuvial accumulations are present in insufficient quantities to qualify for argillic or spodic horizons.
- Gypsic horizons (gypsic horizons) contain 5% or more gypsum that accumulates as percolating water dissolves primary gypsum minerals which then recrystalize as plants extract water from the subsoil. Dense gypsic horizons where the material does not slake in water are known as petrogypsic horizons (petrogypsic horizons).
- Natric horizons (natric horizons) contain more than 15% sodium or more exchangeable sodium plus magnesium than calcium on the cation exchange capacity. They form from the accumulation of $Na_2CO_3$ or hydrolysis of sodium minerals (Chadwick and Graham, 2000).
- Oxic horizons (some ferralic horizons) are sandy loam or finer textured and almost always present in subsoil material that has been previously exposed to weathering in other soils and during fluvial transport to its present site (Buol and Eswaran, 2000). Silica has been lost and 1:1 (kaolinite) clays and gibbsite predominate in the clay fraction. Apparent cation exchange capacity is less than 16 $cmol_c^+ kg^{-1}$ clay at pH 7. The sand fraction contains less than 10% primary minerals that can decompose and provide ions for the formation of new clay. The clay present in surface horizons of soils with oxic horizons is not easily dispersed in water and there is little or no lessivage and clay skin formation. In the absence of clay illuviation the clay content in the oxic horizon is nearly the same as in surface horizons. Most oxic horizons have a strong grade fine granular structure (Fig. 2.3). Iron oxides are considered responsible for stabilizing the granular structure.

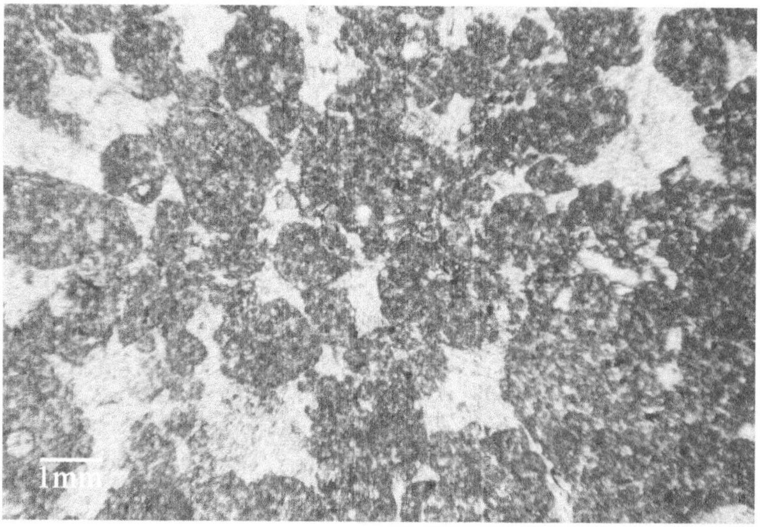


Fig. 2.3. Photomicrograph of granular structure in an oxic horizon in Brazil.

- Salic horizons (salic horizons) develop in arid climates where there is little or no leaching and soluble salts accumulate near the soil surface. The soil water accumulates soluble salts from the dissolution of geologic material rich in soluble minerals. Most salic horizons form in depressional areas of the landscape where the groundwater is near the soil surface or run-off water accumulates and evaporates. Salic horizons are frequently created by irrigation with salty water.
- Sulfuric horizons (thionic horizons) are formed as sulphates accumulated in some horizons that are saturated with brackish water (Fanning and Fanning, 1989).

### *2.4.1 Subsoil features related to chemical reduction*

Subsoil conditions resulting from the reduction and removal of iron oxides are known as aquic conditions in Soil Taxonomy and gleyic properties in the World Reference Base. As iron-bearing silicate minerals such as biotite decompose in soil the iron combines with oxygen to form iron oxides. The most abundant iron oxides are goethite and haematite. Mixtures of these oxides are responsible for the yellow and reddish colours present in many subsoil horizons (Scheinost and Schwertmann, 1999). When a subsoil horizon is saturated with water for prolonged periods of time the microbes in the soil that require oxygen for respiration first use all the oxygen dissolved in the water and then gain oxygen by reducing nitrate ($NO_3^-$), manganese oxides ($Mn_2O_3$), and iron oxides ($Fe_2O_3$). The iron in the iron oxides is reduced from ferric ions ($Fe^{3+}$) to ferrous ions ($Fe^{2+}$) that are soluble and readily leach from the soil. As a result the grey colour of the silicate minerals becomes the predominant colour of the subsoil.

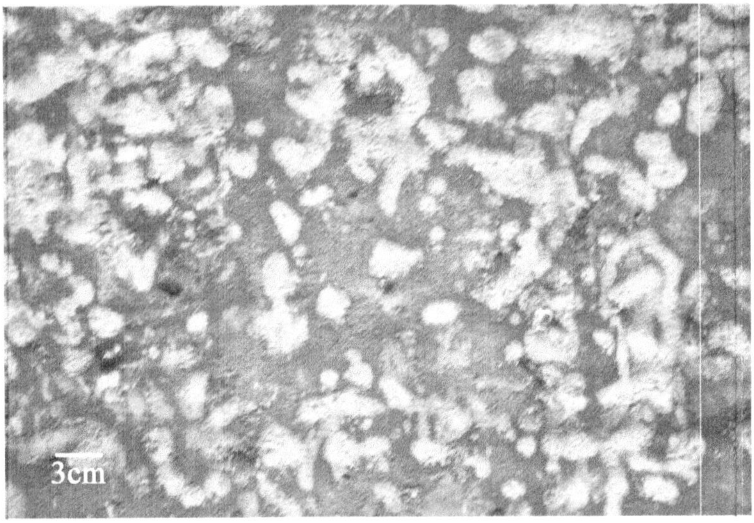


Fig. 2.4. Photograph of a mottled colour pattern in the subsoil horizon in North Carolina, USA.

To engender reduction in the subsoil it is necessary to have saturation that prevents air from entering the soil pores and providing oxygen to the actively respiring micro-organisms in the soil. Micro-organisms require a biologically digestible form of carbon and suitable temperatures for active respiration. A lack of warmth or a limited amount of digestible carbon can slow the process of reduction in saturated soil; thus the length of time that a soil must be saturated before iron is reduced is somewhat variable. Also, contents of nitrate and manganese oxides that reduce prior to the iron oxides contribute oxygen and a flow of oxygen containing water renders determination of the exact amount of time required for iron reduction problematic.

The processes of reduction in soil are known as gleization. Greyish colours of two or less chroma that result when iron oxides are removed from soil are often known as 'gley' (glei) colours. Gley colours may dominate a subsoil horizon but more frequently subsoil horizons undergo alternating periods of saturation and aeration as the water table rises and falls in response to rainfall events and transpiration demands. The dynamic shift from reducing to oxidizing conditions results in a spatial migration of iron forming a mottled pattern of gley areas where iron is removed and more yellowish to red coloured areas where iron oxides have accumulated within the horizon (Fig. 2.4). The presence of 'gley' colours is widely accepted as an indication of saturation during some period of time each year. When a horizon with a mottled colour pattern pictured in Fig. 2.4 is exposed and hardens upon repeated wetting and drying it is recognized as plinthite (plinthic horizon). If such a horizon is indurated in place within the soil it is identified as a petroferric contact (petroplinthic horizon) often referred to as ironstone or laterite.

## 2.5 Formation of structural features in soil

Structural development and expression in soil horizons often referred to as pedogenic structure involves a combination of processes and mechanisms. These processes are related to shrink–swell phenomena associated with wetting and drying cycles in soils with low shrink–swell potential (Southard and Buol, 1988). During periods of desiccation in confined subsoil horizons soil compression takes place because of capillary tension as water is withdrawn from the larger soil voids. The capillary tension causes cracks to form in the soil mass. The crack walls become incipient ped faces. As the soil material again becomes wetted it swells. The protruding points on the incipient ped faces make contact first and shearing forces are focused at these points of contact with coarser particles. Sand grains are forced away from the pressure zone and flat clay particles are concentrated and forced into a parallel orientation on the ped face. These modifications cause the soil to crack in the same place upon subsequent drying, thereby increasing the stability of the peds and resulting in recognizable blocky structure. These same shrink–swell processes can contribute to the conversion of rock fabric to soil fabric (Frazier and Graham, 2000).

Wetting and drying without the confining weight of overlying soil favours granular structure in surface horizons. As unconfined soil material dries the surface tension of water tends to form spherical aggregates. Most surface horizons are also well populated by fungi and other microbes associated with living and dead organic materials. Threadlike fungi associated with plant roots are known to secrete the sticky sugar protein glomalin that along with several other microbial exudates serve to temporally stabilize the granular aggregates. Physical disruption of surface horizons reduces the fungi activity and aggregate stability is decreased as the stabilizing organic compounds are decomposed.

# 3

# Soil phases: the inorganic solid phase

*G. Jock Churchman*

Inorganic solid phases in soils can generally be described as minerals. In soils, rocks provide the raw (i.e. 'parent') materials for minerals in soils. Minerals may derive directly from rocks, with little or no chemical or structural changes, although physical changes, e.g. comminution, commonly occur. In this case they are referred to as *primary* minerals. However, many of the minerals that are of most importance for soil properties are *secondary*. While these have formed from rock minerals under the influence of soil-forming processes, principally weathering, they usually comprise different phases from those present in the rocks.

## 3.1 Description

Table 3.1 comprises a compilation of (a) the characteristics and properties of inorganic solid phases that occur most commonly in soils and (b) the nature of their processes of formation and transformation and their occurrence in soils.

The information in Table 3.1 is extracted from Dixon and Weed (1989), Churchman and Burke (1991), Churchman *et al.* (1993, 1994), Churchman (2000), Olson *et al.* (2000), and Dixon and Schulze (2002), and is discussed as follows under the various categories in the table.

### *3.1.1 Characteristics and properties*

*Primary and secondary minerals*

Although there is little doubt that most occurrences of those designated as primary minerals in Table 3.1 have a direct origin as the residue of minerals that formerly constituted rocks, some – for example quartz, micas, calcite and the zeolite analcime – may also form pedogenically. In the case of those designated as

*Soils: Basic Concepts and Future Challenges*, ed. Giacomo Certini and Riccardo Scalenghe.
Published by Cambridge University Press. 

Table 3.1. *Common inorganic phases in soils*

| Name | Group name | Chemical formula[a] | Structural type | Related phases (or other names) | Usual particle size | Stability | Distinctive physical properties | Specific surface ($m^2g^{-1}$)[b] | CEC ($cmol^+$ $kg^{-1}$) | Soils of main occurrence |
|---|---|---|---|---|---|---|---|---|---|---|
| ***Primary minerals*** | | | | | | | | | | |
| Quartz | Silica | $SiO_2$ | Tectosilicate | Cristobalite, Tridymite, Opal-CT, Opal-A | Sand, silt | High | Hard, brittle | ~0 | 0 | Almost all, less in soils from basalt |
| Orthoclase | Feldspar | $KAlSi_3O_8$ | Tectosilicate | (K-feldspar), Sanidine | Silt, sand | K-feldspars usually | Hard, brittle | ~0 | 0 | Feldspars occur in many soils, but are |
| Microcline | Feldspar | $KAlSi_3O_8$ | Tectosilicate | (K-feldspar), Sanidine | Silt, sand | more stable than plagio- | Hard, brittle | ~0 | 0 | absent from highly weathered soils |
| Albite | Feldspar | $NaAlSi_3O_8$ | Tectosilicate | (Sodic plagioclase), Oligoclase, Andesine | Silt, sand | clases | Hard, brittle | ~0 | 0 | |
| Anorthite | Feldspar | $CaAl_2Si_2O_8$ | Tectosilicate | (Calcic plagioclase), Labradorite, Bytownite | Silt, sand | | Hard, brittle | ~0 | 0 | |
| Clinoptilolite | Zeolite | $Na_3K_3(Al_6Si_{30}O_{72})24H_2O$ | Tectosilicate | Erionite, Faujasite, Heulandite, Chabazite, | Silt, sand | Unstable to acid | Sorb water strongly | ~0 (I~800) | 100–300 | Zeolites rare; analcime formed |
| Analcime | Zeolite | $Na_{16}Al_{16}Si_{32}O_{96}\cdot16H_2O$ | Tectosilicate | Laumontite, Mordenite | Silt, sand | | | | | in high-pH saline soils |
| Muscovite | Mica | $KAl_2AlSi_3O_{10}(OH)_2$ | Phyllosilicate (Dioctahedral) | Paragonite, Margarite, Glauconite | Silt, sand | Quite high | Soft; white or yellow | ~0 | Low[c] | Widespread |
| Biotite | Mica | $K(Mg,Fe^{II})_3AlSi_3O_{10}(OH)_2$ | Phyllosilicate (Trioctahedral) | Phlogopite, Clintonite, Lepidolite | Silt, sand | Low | Soft, black | ~0 | Low[c] | Slightly weathered soils only |
| Chlorite | Chlorite | $(Fe,Mg,Al)_6(Si,Al)_4O_{10}(OH)_8$ | Phyllosilicate | (Chinochlore), Cookeite, Sudoite, Donbassite | Silt, sand | Very low | Soft | ~0 | Low[c] | Very slightly weathered ('raw') soils only |
| Vermiculite (Trioctahedral) | | $M^{II}_x(Mg, Fe)_3(Al_xSi_{4-x})O_{10}(OH)_2\cdot4H_2O$ | Phyllosilicate | | Sand, silt, clay | Low | Soft, exfoliates with heating | 50–150 | 100–210 | Mostly in temperate soils from micas |
| Chrysotile | Serpentine | $Mg_3Si_2O_5(OH)_4$ | Phyllosilicate | Antigorite, Lizardite, Amesite, Berthierine | Silt, clay | Low | Soft, fibrous | 50–150 | 100–210 | Rare |

| | | | | | | | | | |
|---|---|---|---|---|---|---|---|---|---|
| Pyrophyllite | | $Al_2Si_4O_{10}(OH)_2$ | Phyllosilicate | Ferripyrophyllite | Silt, clay | Moderate | Soft, flexible | <10 | <1 | Rare |
| Talc | | $Mg_3Si_4O_{10}(OH)_2$ | Phyllosilicate | Minnesotaite, Willemseite, Kerolite, Pimelite | Sand, silt, clay | Moderate | Soft, flexible, hydrophobic | ~0 | <1 | Rare |
| Hornblende | Amphibole | $(Ca,Na)_{2.3}(Mg,Fe,Al)_5(Si,Al)_8O_{22}(OH)_2$ | Inosilicate (Double chain) | Tremolite, Actinolite, Cummingtonite, Glaucophane, Riebecite, Anthophyllite | Sand, silt | Low | Moderately hard | ~0 | 0 | Quite widespread, but absent in highly weathered soils |
| Ice | | $H_2O$ | Hexagonal | | Sand | Very low | Soft | ~0 | 0 | Frozen soils |
| Augite | Pyroxene | $(Ca,Na)(Mg,Fe,Al)(Si,Al)_2O_6$ | Inosilicate (Single chain) | Enstatite, Hypersthene, Diopside, Pigeonite, Jadeite, Spodumene, Hedenbergerite | Sand | Very low | Moderately hard | ~0 | 0 | Relatively rare |
| Tourmaline | | $(Na,Ca)(Li,Mg,Al)(Al,Fe,Mn)_6(BO_3)_3Si_6O_{18}(OH)_4$ | Cyclosilicate | Beryl | Sand | Very high | Hard | ~0 | 0 | Rare |
| Epidote | | $Ca_2(Al,Fe)Al_2O(Si_2O_7)SiO_4(OH)_4$ | Sorosilicate | Zoisite | Sand | High | Hard | ~0 | 0 | Uncommon |
| Forsterite | Olivine | $Mg_2SiO_4$ | Nesosilicate | Fayalite, Tephroite, Monticellite | Sand | Very low | Hard | ~0 | 0 | Rare |
| Almandine | Garnet | $Fe_3Al_2(SiO_4)_3$ | Nesosilicate | | Sand | Very high | Hard | ~0 | 0 | Rare |
| Zircon | | $ZrSiO_4$ | Nesosilicate | Baddeleyite ($ZrO_2$) | Sand | Very high | Hard | ~0 | 0 | Widespread |
| Apatite | | $Ca_5(PO_4)_3(OH,F,Cl)$ | Insular, hexagonal | Variscite, Wavellite, Monazite | Clay | Very low | Quite hard | ~0 | 0 | Some in 'raw' soils |
| Rutile | Ti oxide | $TiO_2$ | Tetragonal | Brookite, Sphene ($CaTiSiO_5$) | Clay | Very high | Hard | ~0 | 0 | Widespread in small amounts |
| Ilmenite | | $FeTiO_3$ | Sheets | Pseudorutile, Spinel, Perovskite | Sand, silt | Low | Hard | ~0 | 0 | Widespread in small amounts |
| Magnetite | Iron oxide | $Fe_3O_4$ | Cubic | Titanomagnetite | Sand, silt | High | Hard, ferrimagnetic | ~0 | pH-variable | Rare in soils |
| Calcite | Carbonate | $CaCO_3$ | Rhombohedral | Aragonite, Siderite ($FeCO_3$) | Sand, silt, clay | Can be high; low in acid | Soft | ~0 | 0 | Common in soils in arid regions; some in others |
| Dolomite | Carbonate | $CaMg(CO_3)_2$ | Rhombohedral | Ankerite, Mg-calcite | Sand, silt, clay | Can be high; low in acid | Moderately hard | ~0 | 0 | From dolomitic rocks |
| Corundum | Al oxide | $Al_2O_3$ | Sheets of edge-shared octahedra | | Sand, silt | High | Very hard | ~0 | unknown, prob. low | Rare, may form in fires |

Table 3.1. (Cont.)

| Name | Group name | Chemical formula[a] | Structural type | Related phases | Usual particle size | Stability | Distinctive physical properties | Specific surface ($m^2g^{-1}$)[b] | CEC ($cmol^+ kg^{-1}$) | Soils of main occurrence |
|---|---|---|---|---|---|---|---|---|---|---|
| ***Secondary minerals*** | | | | | | | | | | |
| Kaolinite | Kaolin | $Al_2Si_2O_5(OH)_4$ | Phyllosilicate | Dickite, Nacrite | Clay | Quite high | Platy | 6–40 | 0–8 | Widespread; high in well weathered soils |
| Halloysite | Kaolin | $Al_2Si_2O_5(OH)_4{\cdot}2H_2O$ | Phyllosilicate | (Endellite; Meta-halloysite [~0 $H_2O$]) | Clay | Moderate | Mostly tubular or spheroidal | 20–60 | 5–10 | Where wet; espec. from volcanic ash |
| Illite | Mica | $K_{0.6}(Ca,Na)_{0.1}Si_{3.4}Al_2Fe^{III}Mg_{0.2}O_{10}(OH)_2$ | Phyllosilicate | | Clay | Moderate | Platy | 55–195 | 10–40 | Widespread; espec. weakly weathered soils |
| Montmo-rillonite | Smectite (Di-octahedral) | $M^{II}_{0.25}/M^{I}_{0.5}Si_4Al_{1.5}Mg_{0.5}O_{10}(OH)_2$ | Phyllosilicate | Stevensite, Hectorite (trioctahedral) | Clay | Low | Swell: extensively when M=Na | 15–160 | 45–160 | Mostly where drainage poor and pH high |
| Beidellite | Smectite (Di-octahedral) | $M^{II}_{0.25}/M^{I}_{0.5}Si_{3.5}Al_{2.5}O_{10}(OH)_2$ | Phyllosilicate | Saponite (trioctahedral) | Clay | Low | | (I~800) | | Not common except in acid leached horizons |
| Nontronite | Smectite (Di-octahedral) | $M^{II}_{0.25}/M^{I}_{0.5}Si_{3.5}Al_{0.5}Fe_2O_{10}(OH)_2$ | Phyllosilicate | (Hisingerite) | Clay | Low | | | | Rare |
| Vermiculite dioctahedral | | $K_{0.2}Ca_{0.1}Si_{3.2}Al_{0.8}(Al_{1.6}Fe_{0.2}Mg_{0.2})(Al_{1.5}[OH]_4)O_{10}(OH)_2$ | Phyllosilicate | Pedogenic chlorite, HIV, 2:1–2:2 intergrade; Chloritized vermiculite | Clay | Low-moderate | | unknown | pH-variable | Leached, mildly acid soils |
| Illite-smectite | Interstratified | Variable, intermediate between components | Phyllosilicate | Mica-smectite, Illite (Mica)-vermiculite, (Hydro-biotite [-mica]), Allevardite, Rectorite | Clay | Low | | unknown | unknown, probably moderate | From diagenesis and early stages of weathering |
| Chlorite-smectite | Interstratified | Variable, intermediate between components | Phyllosilicate | Chlorite-swelling [chlorite], Corrensite, Chlorite-vermiculite | Clay | Very low | | unknown | unknown | At very early stages of weathering |

| | | | | | | | | | |
|---|---|---|---|---|---|---|---|---|---|
| Kaolin-smectite | Interstratified | Variable, intermediate between components | Phyllosilicate | Kaolinite-smectite, Halloysite-smectite | Clay | Moderate | | unknown | 30–70 | Moderately drained |
| Palygorskite | Hormite | $Si_8Mg_5O_{20}(OH)_2(OH_2)_4{\cdot}4H_2O$ | Phyllosilicate | (Attapulgite), Sepiolite | Clay | Low, espec. in acid | Fibrous | 140–190 | 3–30 | In dry, usually calcareous regions |
| Imogolite | Short-range order mineral | Si tetrahedra within Al octahedra in tube | Tubular | | Clay | Low | Gel | unknown | pH-variable | Limited, mainly from pumice, also podzols |
| Allophane | Short-range order mineral | Variable, between halloysite & imogolite | Mostly imogolite-like | | Clay | Low | Very small particles | 145–660 | pH-variable | From volcanic ash & in podzols |
| Gibbsite | | $Al(OH)_3$ | Al hydroxide, in sheets | ($\alpha$, or $\gamma$ Alumina trihydrate), Bayerite | Clay | Moderate, except in acid | Hexagonal crystals | unknown | pH-variable | Where Si low; espec. in strongly weathered soils |
| Boehmite | | AlOOH | Al oxyhydroxide | ($\alpha$, or $\gamma$ Alumina monohydrate), Diaspore | Clay | High, except in acid | | unknown, prob. high | pH-variable | Strongly weathered soils; laterite, bauxite |
| Goethite | Iron oxide | $\alpha FeOOH$ | Octahedra in double chains | (Limonite: major component of) | Clay | Moderate | Yellow-brown, antiferro-magnetic | 14–77 | pH-variable | Most common soil Fe oxide |
| Hematite | Iron oxide | $\alpha Fe_2O_3$ | Sheets of edge-shared octahedra | | Clay | Moderate | Bright red, weak or antiferro-magnetic | 35–45 | pH-variable | Soils of warmer climates |
| Lepidocrocite | Iron oxide | $\gamma FeOOH$ | Zigzag sheets of octahedra | Akageneite ($\beta\gamma$-FeOOH) | Clay | Moderate | Orange colour, antiferro-magnetic | unknown, prob. high | pH-variable | Reducto-morphic soils |
| Maghemite | Iron oxide | $\gamma Fe_2O_3$ | Cubic | Titanomaghemite | Clay | Moderate | Ferri-magnetic | unknown, prob. high | pH-variable | Tropical & subtropical soils |
| Ferrihydrite | Iron oxide | $Fe_5HO_8{\cdot}4H_2O$ | Defective hematite-type | Feroxyhite | Clay | Low | Light red | 200–500 | pH-variable | Widespread; where Fe oxidised rapidly |
| Birnessite | Mn oxide | $(Na_{0.7}Ca_{0.3})Mn_7O_{14}{\cdot}2.8H_2O$ | Layered | Todokorite, Hollandite, Lithiophorite, Pyrolusite | Clay | Low | Black | unknown, prob. high | pH-variable | Rare; espec. in 'clean' sites e.g. saprolites |
| Gypsum | | $CaSO_4{\cdot}2H_2O$ | Layered | Bassanite, Anhydrite, Barite ($BaSO_4$), | Sand, silt | Low | Moderate solubility | unknown | unknown | Often in desert soils |
| Halite | | NaCl | Cubic | Epsomite, Thenardite, Mirabolite (sulphates) | Sand, silt | Very low | Confers high osmotic pressure | unknown | unknown | Seasonally dry saline soils |
| Pyrite | | $FeS_2$ | Cubic | Mackinawite, Greigite, Amorphous Fe sulphide | Silt | Very low | Confers high acidity | unknown | unknown | Coastal regions & from some sediments |

Table 3.1. (Cont.)

| Name | Group name | Chemical formula[a] | Structural type | Related phases (or other names) | Usual particle size | Stability | Distinctive physical properties | Specific surface ($m^2g^{-1}$)[b] | **CEC** ($cmol^+ kg^{-1}$) | Soils of main occurrence |
|---|---|---|---|---|---|---|---|---|---|---|
| Jarosite | | $KFe_3(OH)_6(SO_4)_2$ | Cubic | Natrojarosite, Schwertmannite | Silt, clay | Low | Yellow efflorescence | unknown | unknown | Acid sulphate soils |
| Plumbogummite | | $PbAl_3(PO_4)_2(OH)_5 \cdot H_2O$ | Insular, trigonal | Crandallite (Ca), Gorceixite (Ba,Al), Vivianite, Strengite | Silt, clay | High | | unknown | unknown | Rare, rock phosphate breakdown products |
| Anatase | Ti oxide | $TiO_2$ | Tetragonal | | Clay | High | | unknown | unknown | Small amounts, often |

[a] Ideal, or typical formulae; subscripts generally rounded to whole numbers or single figures after decimal point (M = cation).
[b] External specific surface shown; but internal values given by I.
[c] Probably increases with decreasing size.
CEC, Cation exchange capacity.

secondary minerals, their ultimate origin may be confused by the fact that many sedimentary rocks are composed of the products of one or more earlier processes of weathering and perhaps also subsequent diagenesis, leading to the formation of secondary minerals that are thereby recycled as detritus into new soils. The mineral residues of such rocks are strictly secondary only in relation to their original igneous and metamorphic rock sources. Minerals are classified as secondary in Table 3.1 when there is clear evidence for their formation by pedogenic processes and when they do not also occur commonly as primary minerals.

### *Structural type*

Except in some unusual cases, e.g. calcareous soils, in which calcite and/or dolomite are dominant, silicates comprise the predominant mineral components of most soils by weight and volume. Silicates are predominant among primary minerals. Silicates differ in their type according to the number of oxygen atoms that are shared for each silicon atom. Table 3.2 identifies the weakest bond in each of the major types of silicates in soils. The alteration of a mineral begins by the disruption of its weakest bond.

Almost all secondary silicates are phyllosilicates, and the various possible structures for these are given in Fig. 3.1. Secondary phyllosilicates are often known as 'clay minerals'.

### *Related phases or other names*

While some of the phases within this category in Table 3.1 are less common in soils than the relevant main phase, many are rare, or have not even been reported, in soils.

Some names that are given are alternative names for minerals, e.g. chinochlore, endellite, metahalloysite, limonite. Many of these have been discredited for use in the scientific literature. Other names given are more general descriptions, e.g. plagioclases, pedogenic chlorite, hydroxy-interlayered vermiculite (HIV).

### *Common particle sizes in soils*

In the main, primary minerals occur in soils as coarse, i.e. sand- or silt-size particles, where sand-size covers 0.02–2.0 mm equivalent spherical diameter (e.s.d.) and silt-size 0.002–0.02 mm, i.e. 2–20 μm. Occasionally, primary minerals occur as gravels, i.e. >2.0 mm. Table 3.1 records that some primary minerals may also occur as clay-size particles. These include chrysotile, pyrophyllite, talc, (trioctahedral) vermiculite, chlorite and rutile.

Secondary minerals occur predominantly as clay-size particles. Very many secondary minerals occur in particles that are finer than 2 μm with some particles being <0.02 μm.

Table 3.2. *The nature of bonding and weakest bonds in the most common types of silicates in soils*

| Name | Structural type | Formula | Shared O per Si[a] | Weakest bonds | Examples |
|---|---|---|---|---|---|
| Tectosilicates | Framework | $SiO_2$, with Al substitution | 4 | Cations ($K^+$, $Na^+$, $Ca^{2+}$) | Feldspars |
| | | $SiO_2$ | 4 | Si-O bonds | Quartz |
| Phyllosilicates | Sheet | $Si2O_5{}^{2-}$, with Al substitution, joined to Al-, Fe-, Mg-hydroxy octahedra in layers | 3 | Via interlayer cations, usually $K^+$ | Micas |
| Inosilicates | Single chains | $SiO_3{}^{2-}$, with Al substitution | 2.5 | Via divalent, and other, cations | Pyroxenes |
| | Double chains | $Si_4O_{11}{}^{6-}$, with Al substitution | 2 | Via divalent, and other, cations | Amphiboles |
| Nesosilicates | Isolated tetrahedra | $SiO_4{}^{2-}$ | 0 | Via divalent cations | Olivines |

[a]Cyclosilicates (e.g. tourmaline) have two O bonds per Si and sorosilicates (e.g. epidote) have one O bond per Si, but neither type is common in soils (see Table 3.1).

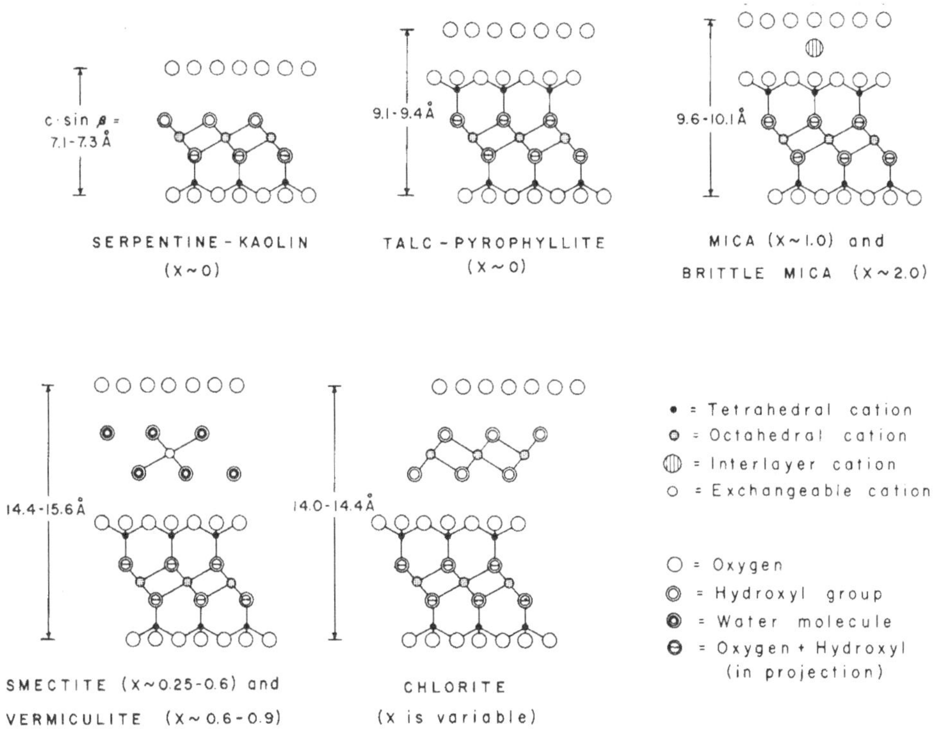


Fig. 3.1. View of structures of major clay mineral groups. From Bailey (1980). Reproduced by permission of the Mineralogical Society, London.

## *Surface areas*

As a first approximation, surface area bears an inverse relationship to the size of particles. Broadly speaking, minerals that occur mainly as sand- or silt-size particles have a negligible surface area. Hence almost all of the surface area of a soil is contributed by its clay-size fraction and especially by the finest clay-size particles. Surface area plays a role in controlling sorption reactions in soils although surface charge usually also plays an important role in sorption. The quantitative contribution of the extent of surface is given by measurements of specific surface, or the surface area per unit mass. Various methods have been used for its measurement, and each have different advantages and disadvantages (e.g. Churchman and Burke, 1991). Some, using polar molecules, also incorporate the influence of charge and provide a (practical) measure of both internal and external surfaces. The distinction is particularly important for expandable layer silicates, especially smectites. Internal specific surfaces for smectites have been calculated from their structures to be 600–800 $m^2g^{-1}$ (Reid-Soukup and Ulery, 2002).

### *Cation exchange capacities*

Almost all primary minerals in soils are uncharged. Secondary minerals may carry either a permanent negative charge and/or a charge whose magnitude and sign both vary with the pH, and usually also the ionic strength of the surrounding solution. Permanent charge, which is always negative in naturally occurring minerals, arises out of the substitution of either aluminium or silicon in an aluminosilicate by a (cat)ion with a lower positive charge than either aluminium (with a charge of 3) or silicon (with a charge of 4). Commonly, magnesium or ferrous iron replaces Al in the octahedral sheet, while Al replaces Si in the tetrahedral sheet. Permanent charge originates within the layer structure.

Variable charges originate in 'broken bonds', leaving unsatisfied charges, on the surfaces of minerals. These either attract excess hydrogen ions from solutions at lower pHs or release them in response to excess hydroxyl ions at higher pHs. As a result, oxides, hydroxides and oxyhydroxides can display a net positive charge at lower pH values, as can some aluminosilicate minerals with little or no permanent charge, e.g. kaolinite. For each of these minerals, there is a characteristic pH value at which it displays no net charge, i.e. where its positive and negative charges are equal. This is its 'point of zero charge' (pzc). Minerals with sufficiently large permanent charges have no measurable pzc since the pH can never fall sufficiently low in reality for the positive charge from interactions between surfaces and hydrogen ions in solution to overwhelm the negative charge developed in the bulk of the mineral. Minerals with high cation exchange capacities (CECs), such as smectites, vermiculites and illites (Table 3.1), always carry a net negative charge. At the pH of almost all soils, minerals are net negatively charged, although a very few highly weathered soils have been found to display a net positive charge at their field pH (Gillman and Bell, 1976). Hence layer silicates are effective for sequestering contaminant cations, including most heavy metals, from polluted water by exchange. By the same token, however, they are not so effective for the uptake and restraint of anions of environmental concern, e.g. phosphates and arsenates. These may be removed from solutions by oxides, hydroxides and oxyhydroxides under acidic conditions, where these solid phases carry a net positive charge.

### *Distinctive physical properties*

The different primary and secondary minerals occur in different shapes, often reflecting their crystal form. In general, primary minerals are harder than secondary minerals, although some primary minerals, such as micas, talc, pyrophyllite and (trioctahedral) vermiculite, and ice can be quite soft.

All secondary minerals – and some primary minerals – adsorb considerable amounts of water, although these amounts vary with mineral type, in particular its surface area and other influential factors, e.g. the nature of the exchangeable cations.

In the case of smectites, zeolites and also, to a lesser extent, vermiculites, substantial amounts of water are taken up into the crystals of the minerals. Water uptake has a profound effect on the physical properties of smectites especially. These can undergo extensive 'crystalline' swelling, which occurs between layers (Olson *et al.*, 2000). The tendency of highly smectitic soils to expand when wet and then shrink when dry leads to such effects as 'gilgai', i.e. land surfaces with a repetitive pattern of small mounds and depressions, landslides and 'soil creep', and also to crack in not only the soils but also in buildings and roads constructed upon them. According to Reid-Soukup and Ulery (2002), 'Expansive soils are responsible for more structural damage than any other natural disaster, including earthquakes and floods'.

The extent to which expansion can occur is dependent very largely on the nature of the exchangeable cations associated with the mineral surface. In soils containing smectites, the extreme of expansive behaviour is that exhibited by the soil minerals which have substantial exchangeable sodium. Unrestrained, there may be no limit to their possible expansion. In contrast, those which have Ca and/or Mg but little or no Na as their exchangeable cations show only limited expansion. Sodium need not be dominant numerically to control this behaviour; even a little Na can have an effect. Its distinctive effect is considered to arise from the predominantly ionic nature of the bonds it forms with the clay surface, whereas Ca and Mg form predominantly polar covalent bonds (Rengasamy and Sumner, 1998). A net negative charge remains on the clay layers that are associated with $Na^+$, so that they tend to move apart from one another, thereby attracting water into the interlayer region, and hence the clays swell. By contrast, the polar covalent bonds formed between the divalent ions and clay surfaces leave effectively no net charge on the layers and there is no net repulsive interaction between them to cause swelling. At the extreme of water uptake, the expansion process leads to a change from swelling of solid material to dispersion of separated particles in a suspension.

### 3.1.2 Processes and occurrence

*Stability of minerals*

Generally, the relative stabilities for minerals given in Table 3.1 relate to their resistance to the usual driving force for weathering, i.e. water, most commonly acidified through the dissolution of atmospheric $CO_2$ and/or organic acids. The chemical reaction involved is of a similar form to that for the hydrolysis of albite by water containing $CO_2$:

$$\underset{\text{albite}}{2NaAlSi_3O_8} + 2CO_2 + 11H_2O = \underset{\text{kaolinite}}{Al_2Si_2O_5(OH)_4} + 2Na^+ + 2HCO_{3-} + \underset{\text{silicic acid}}{4H_4SiO_4} \qquad (3.1)$$

Some minerals will also be particularly unstable in specific environments, e.g. calcite, dolomite and palygorskite in acidic environments, halite in water, pyrite in oxidizing environments, and iron oxides in reducing environments. In different environments, these same minerals may display much greater stabilities. Nonetheless, a stability series for the major types of primary minerals which was drawn up by Goldich in 1938 (Eq. 3.2) remains generally useful as an indicator of the order in which the minerals from rocks become degraded by increasingly aggressive conditions or the passage of time in weathering processes.

$$\longrightarrow$$

Olivines Mg-pyroxenes Mg-Ca pyroxenes Amphiboles
Biotite K-feldspar Muscovite Quartz (3.2)
Plagioclases: Calcic- Calc-alkaline Alkali-calcic Alkaline
*Increasing stability against alteration*

$$\longrightarrow$$

This series presents the minerals in the order in which they crystallized out of a magma as it cooled. This is because the higher the temperature at which a mineral crystallized out of the magma, the further it would be out of equilibrium with conditions at the Earth surface. As stated by Kittrick (1967): 'Fundamentally a mineral is a package for its elements. It will persist in nature only so long as it is the most stable package for those elements in its environment.' Among others, Kittrick pioneered the application to soils of thermodynamic concepts to provide a quantitative basis for the understanding of stabilities of minerals, both individually and in suites with other minerals. The aim of this approach was to enable prediction of the course of the formation of secondary minerals from their primary precursors and of their possible changes to other secondary phases.

Changes almost invariably take place in aqueous solutions, and the thermodynamic approach involves the definition of the 'stability fields' for the various minerals that are found together in the soil or soils of interest. Each mineral's stability field is defined by the solution conditions under which it is the least soluble, hence most stable, of the various minerals competing for the group of elements available in the surrounding solution. The boundaries of the stability fields of a pair of interacting minerals can be calculated from their respective Gibbs free energy of formation, $\Delta G_f^o$, which is assumed to be a unique value at standard temperature and pressure for each mineral. Most commonly in studies of this kind related to soils, $\Delta G_f^o$ values have been obtained experimentally from studies of the dissolution of minerals, usually in mildly acidic aqueous solutions, to give soluble species only.

The validity of the thermodynamic approach, as outlined, has been questioned on a number of grounds. For one, $\Delta G_f^o$ values for particular minerals have been found to vary widely. This has been attributed to one or more factors. These include the impurities often found in minerals and an inherent variability in their composition. They also include problems from relying upon mineral dissolution for obtaining thermochemical data. Among these are the difficulty of reaching and establishing equilibrium in dissolution experiments and the likelihood that many minerals dissolve incongruently, with at least some components remaining in solid phases. Both types of problems are most serious with more complex minerals, particularly secondary phyllosilicates. Thermochemical data for simpler minerals such as quartz, haematite, gibbsite and also some primary minerals have been generally considered to be reliable. Among other minerals, kaolinite has been shown to dissolve congruently, with all components going into solution, but many other common clay minerals have exhibited incongruent dissolution.

The foundation of the thermodynamic approach upon solution concentrations implies that these provide the only important driving force for alterations to minerals. This is to ignore the important role that oxidation and reduction can play in bringing about mineralogical changes in soils. Many primary minerals, and especially those most susceptible to weathering, comprise Fe in its ferrous state (Table 3.1). Hence the thermodynamic approach presents a greatly simplified model of mineralogical changes on weathering.

A more fundamental difficulty is encountered when the thermodynamic approach is applied to real soils because it is based upon the assumption that equilibrium is obtained in reactions between minerals and solutions in soils in the field when these processes are almost certainly irreversible. This objection may conceivably be met if the overall processes are subdivided into a series of partial equilibria, when stability diagrams can be used to describe or predict each of these states of partial equilibrium. However, the complexity of the soil system when changing climatic and biological effects (e.g. Zabowski and Ugolini, 1992) are taken into account means that thermodynamic descriptions of mineral–solution interactions in soils can probably only be macroscopic and qualitative (Churchman, 2000).

These fundamental problems are evident even before practical problems like the difficulties of extracting a soil solution *in situ* have been solved satisfactorily. It may be that thermodynamics can only indicate whether reactions are possible. Consideration of kinetics is required before it can be determined whether they are likely to occur during the time period of interest. Given all of the many difficulties with the thermodynamic approach, its promise of the possibility of the prediction of the course of mineral changes in soils is abandoned in favour of an analysis – in the next section – of the various possible paths of development of minerals from their parent rocks through different stages of soil formation.

### *Alteration, formation, transformation in soils of main occurrence*

Much of our knowledge on mineral changes during the formation and development of soils has come from the study of soils within so-called monosequences, in which the influence of one particular soil-forming factor in Jenny's relationship is varied while the other main factors are held constant.

However, the Jenny equation is too simplistic for explaining mineral development, which requires consideration of the whole profile. As long as there are weatherable primary minerals, or secondary minerals that may be transformed further within any part of the soil profile, including at its base, and provided soil-forming factors are active, then the suite of minerals within the soil is capable of undergoing further changes. It is the influence of water, and particularly its throughflow, that determines above all whether mineral development continues within a soil profile. This is consistent with the central role that water plays in chemical weathering (Eq. 3.1). Furthermore, the hydrological environment of the soil, as determined by both rainfall and drainage, determines the redox environment, which may constitute an important factor in weathering at depth rather than at soil surfaces.

The most common processes of mineral development within soils are physical comminution, salt accumulation, transformation, neoformation, decomposition and inheritance.

*Physical comminution* The absence of free water is a common feature for soil formation in both perpetually cold climates and also in arid regions. The lack of water ensures that little or no alteration of primary minerals has taken place, so that minerals in the finer fractions of soils are present predominantly as comminuted fragments of the primary minerals. Illite (or clay-size mica) and chlorite are common.

*Salt accumulation* Soluble salts, including carbonates, halides and sulphates, are widespread in soils formed in both perpetually cold climates and arid regions. They form by dissolution and then precipitation upon evaporation. They can also appear seasonally, namely during dry summers, in soils with climates of the Mediterranean type within landscapes containing accumulations of the salts within the soil profile.

*Transformation* Acid leaching is common under cool wet conditions, particularly in a forest vegetation regime, when organic matter tends to accumulate and the soil profile becomes acidified. Acid leaching of micaceous parent materials, e.g. granites, many schists and argillaceous sedimentary rocks, leads to the transformation, within the solid phase, of micas to give vermiculite, also sometimes beidellitic smectite, or else an interstratified mineral composed of layers of mica together with those of either vermiculite or beidellitic smectite. Acid leaching occurring in cool, often highland environments gives rise to regularly interstratified phases typified by an X-ray diffraction (XRD) peak for a high spacing that is the sum of those for the two component layer types, e.g. 24 Å, expandable to 27–28 Å on solvation with glycerol, indicating the interstratification of a 10 Å mica layer with a smectite layer that displays spacings of 14 and 17–18 Å before and after glycerol solvation respectively. The removal of $K^+$ from one interlayer brings about structural adjustments within the layers that make it more difficult to remove $K^+$ from the adjacent interlayer than from the more distant interlayers. Since the driving force for the alteration of micas is probably weaker in cold climates than in warmer climates, their transformation occurs slowly enough to proceed step-wise through regular interstratifications. By contrast, interstratified phases formed in warmer climates are most often random in type. Chlorites, which are often less abundant in parent materials than micas, are nonetheless even more susceptible to breakdown under acid leaching and may also give rise to regularly interstratified phases among other products.

Typical mineral transformations occurring as a result of acid leaching processes are:

---

Mica → Mica-Vermiculite → Vermiculite (3.3a)

↓ (3.3b)

Mica-Beidellite → Beidellite (3.3c)

*(3.3a) and (3.3c) involve progressive loss of* $K^+$; *(3.3b) involves acidification*

Chlorite → IHM → Chlorite-swelling Chlorite → Chlorite-Vermiculite (3.4)

*IHM = interlayered hydrous mica*

*(3.4) involves loss of a cation($Mg^{2+}$ or $Fe^{3+}$ from) interlayer hydroxide*

---

Acid leaching typifies the process of podzolization. True podzols comprise a lower B horizon containing an elevated concentration of organic matter together with Al and Fe compounds, often as the short-range order minerals, allophane or imogolite, and ferrihydrite. Therefore podzolization leads to a profile displaying a considerable extent of mineralogical differentiation with depth. In the upper levels of a podzolized soil, Al, and often also Fe, is mobilized relative to Si by acid leaching, and eluviation takes place. Nonetheless, eluvation from upper horizons by acid hydrolysis can bring about mineralogical changes even without the formation of a mature podzol with a definite horizon of accumulation of humic materials and of Al and Fe.

The acid hydrolysis process typical of podzolization often results in acidification of upper horizons to yield pH values that are below the critical value at which aluminium becomes soluble, i.e. ~4.5. However, where pH rises to 4.5 after the process has already effected transformation to give expandable layers, aluminium may precipitate out, as hydroxyl cations. These are particularly strongly attracted to the interlayer regions of expandable aluminosilicate clays (vermiculite and smectite) wherein they bring about a 'reversion' of the opening of the interlayers that occurs in transformations. The result is a 14 Å phase that is neither expandable nor can it be contracted by saturation with $K^+$ and drying: a dioctahedral chlorite, 'pedogenic chlorite', or hydroxy-interlayered vermiculite (HIV), which may also more strictly be described as hydroxy-interlayered smectite (HIS), dependent upon whether the prior transformation had proceeded by Eq. (3.3a) or (3.3b).

*Neoformation* In contrast to transformation within the solid phase, neoformation involves the natural synthesis of a solid mineral phase from the constituents of solutions formed by the dissolution of primary minerals or other secondary minerals. The neoformed minerals differ in structural type or, at least, layer composition, from their antecedent primary or secondary minerals. The formation of minerals with short-range order in the lower horizons of podzols is an example of a neoformation process.

Many studies have ascribed the formation of particular secondary minerals to specific primary mineral precursors and vice versa. Mainly using electron microscopy, hence after sample preparation including drying, many have concluded that feldspars often give rise to kaolin minerals (kaolinite and/or halloysite) and gibbsite on weathering. Following studies of associations between primary and secondary minerals in a deep weathering profile on granite that was aimed to ascribe secondary products to their primary minerals of origin, Eswaran and Bin (1978) concluded that 'irrespective of the primary weatherable mineral, the type of secondary mineral found is a function of the microenvironment'. Where neoformation is involved, all primary minerals undergoing alteration may contribute constituents to all secondary phases that are feasible as their products based on their chemical composition. Nonetheless, there are some common sequences of changes in secondary minerals in soils.

Accordingly, in deep weathering profiles, the following sequence is commonly observed, although halloysite and boehmite are not always present:

$$\text{Halloysite} \rightarrow \text{Kaolinite} \rightarrow \text{Gibbsite} \rightarrow \text{Boehmite} \quad (3.5)$$

Halloysite was only found to form in the deeper parts of such profiles. This is because it is diagnostic of halloysites that they contain, or have contained, water in their interlayers. Hence all halloysites have formed in wet environments. Halloysites lose their interlayer water irreversibly to give a mineral with the same composition as kaolinite, but the dehydration of halloysite alone is insufficient to produce kaolinite. The structures of the two types of kaolin minerals differ from each other and therefore one may change to the other only via dissolution and recrystallization, i.e. neoformation. However, electron microscopy has shown some clear evidence of halloysite tubes forming on kaolinite plates (Robertson and Eggleton, 1991; Singh and Gilkes, 1992), suggesting that kaolinite has transformed to halloysite in the solid phase. This particular change contradicts (a) the usual trend within profiles, (b) thermodynamic data showing that halloysites are generally less stable than kaolinites, and also (c) the generally accepted recognition that halloysites and kaolinites have different structures. One possible interpretation of these apparent transformations is that neoformation of halloysite has taken place in a microenvironment that differs from that under which the underlying kaolinite was formed, leading to the later halloysite being precipitated on a preformed kaolinite substrate. Clearly, at least, these observations show that kaolinites and halloysites can form under very similar conditions. Kaolinite is 'the most ubiquitous phyllosilicate in soils' (White and Dixon, 2002) and, although some kaolinites are inherited, the formation of this type of mineral in soils is usually favoured by warm humid climates and/or strong leaching.

Kaolinites formed by weathering are often more disordered than inherited kaolinites (Garcia-Talegon *et al.*, 1994).

Typically the changes shown in Eq. (3.5) occur under heavy rainfall at high temperatures, namely in a tropical environment, to give profiles characterized as lateritic, although halloysite may not always be present, particularly if seasonal drying of profiles extends into the saprolite zone. Laterites are particularly notable for the reddish colours in their surface and near-surface zones. These are due to the formation of iron oxides.

Leaching leads to desilication of both primary minerals (e.g. Eq. 3.1) and of preformed secondary minerals, and sometimes also their dehydration (Eq. 3.5). In complete contrast, when accumulation of soluble mineral constituents occurs, smectites may be neoformed in the resulting solutions. This occurs when these solutions contain sufficiently high concentrations of the elements, particularly Si and Mg, that are essential for the formation of the smectites, and when they retain these elements during the process of smectite neoformation. While no particular parent rock type is critical for the formation of smectites, poor drainage and an alkaline pH are favourable for their formation (Reid-Soukup and Ulery, 2002). Hence they are commonly found in poorly drained hollows in landscapes from basic and/or calcareous rocks. They have also been found in salt-affected soils, both saline and sodic (Reid-Soukup and Ulery, 2002). In these, evaporation and reduced permeability, respectively, have concentrated Si and basic cations sufficiently well to effect the neoformation of smectites in appropriate zones of the salt-affected soils. The poorly drained, and therefore reducing, conditions under which smectites are formed are antagonistic to the drying, oxidative conditions for the formation of iron oxides and therefore (a) smectites and iron oxides are usually not found together in soils, and hence (b) smectitic soils are not coloured red; rather they are often black, from dispersed colloidal organic matter.

In dry areas, calcareous parent materials can give rise to palygorskite by neoformation. The same type of materials led to the formation of smectites when mean annual precipitation exceeded $\sim$300 mm (Paquet and Millot, 1972).

Illite is often regarded as either a fine-grained form of primary micas, or else a dioctahedral mica that has less K and more associated water than either of the primary minerals but which shows no tendency towards expansion (CMS, 1984; Weaver, 1989). However, in some, generally arid or semi-arid, situations, illites and related phases appear to have originated by neoformation. A lacustrine origin has given rise to an illite with a uniformly fine particle size (Norrish and Pickering, 1983) and neoformation is also suggested by concentrations of illite in surface horizons when transformations in soils generally lead to lower concentrations in surfaces than in subsoil horizons (Norrish and Pickering, 1983). Some vermiculites may also have formed in soils. Smectites can undergo alteration when their environments change. An alteration of smectites towards kaolin minerals appears to occur towards the surface of profiles of some soils containing smectites in their deeper zones. The alteration from smectite to kaolins within these profiles is gradual and has taken place via the formation of interstratified phases comprising kaolin and smectite layers, namely kaolin–smectite interstratified minerals. The main changes are:

---

Smectite → Smectite-kaolin → Kaolin-smectite → Kaolin (3.6)

*where smectite-kaolin and kaolin-smectite do not form discrete phases but indicate that proportions of smectite and kaolin change within interstratified phases*

---

These changes involve desilication, and silica, as opal C-T, has been identified as a by-product of the alteration (Watanabe *et al.*, 1992). The mechanism of the change is uncertain, although a proposed transformation involving an interlayering of the smectitic layers has had support from studies of both geological and synthetic alterations. Its credibility is enhanced by work showing that interlayers in 2:1 expandable aluminosilicates could contain Si as well as Al and therefore could serve as precursors of kaolins (Wada *et al.*, 1991). Interstratifications of kaolins with smectites also often occur within toposequences across rolling undulating landscapes. In many of these, black smectitic soils occur in hollows where element accumulation prevailed in soil formation while red kaolinitic soils occur on humps, where element leaching was the dominant formation process. Soils that are intermediate between these extremes have sometimes been found to contain kaolin–smectite interstratified phases. It has been determined in some cases that the kaolin layers in the interstratified phases comprise halloysite. Kaolin-smectites may be quite common in soils formed on basalts worldwide. Minerals identified positively as halloysite-smectites have been found invariably in soils from volcanic ash (Delvaux and Herbillon, 1995). They appear to form by a process by ‘co-neogenesis’, or ‘syngenesis’, with their component halloysite and smectite layers being neoformed together.

Minerals with short-range order, especially allophane, and sometimes, imogolite, are formed rapidly in the weathering of volcanic ash. Halloysite also forms as a more-ordered secondary product from the weathering of these materials. Up until about the 1970s, it was commonly thought that allophane altered to halloysite with time. However, it has since been shown that allophane can be long-lived and also that halloysite can form directly from volcanic ash. Furthermore, halloysite was found when there was a thick overburden of weathered ash and halloysite formed from ash instead of allophane when the drainage was restricted (Churchman, 2000). These observations all pointed to the formation of halloysite being favoured over that of allophane when silica concentrations were relatively high in the solutions for neoformation (Parfitt *et al.*, 1983). For Parfitt *et al.* (1983), it is the rate of throughput of water, $R_w$, above all other considerations, including the composition of the parent ash, that governs the silica concentration [Si], and hence the principal secondary mineral formed. The minerals that can form range from halloysite when rate of throughput is low, through to gibbsite, when it is very rapid, so that [Si] is extremely low. Furthermore, the weathering of volcanic ash in basins with restricted drainage, i.e. extremely low $R_w$, provides the archetypal conditions for the formation of smectites as 'bentonites' (Reid-Soukup and Ulery, 2002). The particular $R_w$, hence [Si] values, that control the formation of allophane depend upon the nature, i.e. composition, of the allophane phase that forms. There are two main types of allophane by composition, namely Al-rich, with Al:Si $\sim$2, and Si-rich, with Al:Si $\sim$1, although these represent the end points of a range of compositions. The rarer Si-rich allophane (Al:Si $\sim$1) probably forms when $R_w$ and [Si] values are similar to those defining halloysite formation, while the more common Al-rich allophane (Al:Si $\sim$2) forms when $R_w$ and [Si] values are intermediate between those for halloysite and gibbsite. The controls that $R_w$ and [Si] exert upon the minerals formed from volcanic ash can be summarized as follows:

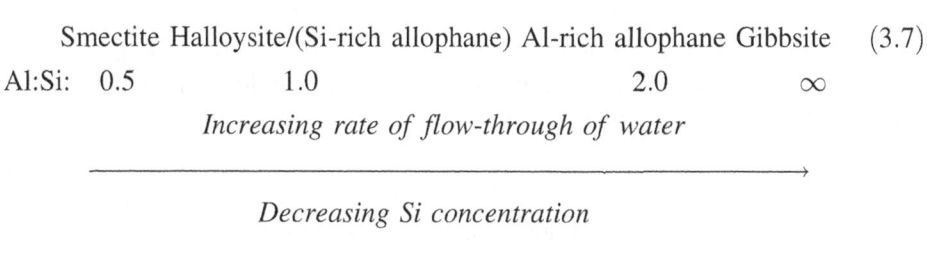

| | Smectite | Halloysite/(Si-rich allophane) | Al-rich allophane | Gibbsite | |
|---|---|---|---|---|---|
| Al:Si: | 0.5 | 1.0 | 2.0 | $\infty$ | (3.7) |

*Increasing rate of flow-through of water*

$\longrightarrow$

*Decreasing Si concentration*

Al-humus complexes can form from volcanic ash in place of allophane when organic matter contents are high, due to the strong affinity of humus for Al. Probably for a similar reason, Al-interlayers were formed within 2:1 (Si:Al) aluminosilicates in place of allophane (or imogolite) in weathered tephra after 10 years following the 1980 eruption of Mt St Helens (Dahlgren *et al.*, 1997a).

Also formed at depth in podzols, Al-rich allophane and imogolite both have a similar, distinctive structure comprising a tubular arrangement of an octahedral gibbsite sheet outside a modified tetrahedral silica sheet that includes hydroxyl groups on the inside surfaces of the tubes. Their dissimilar crystal structures preclude the direct transformation of allophane to halloysite. Although halloysite could arise by neoformation from solutions produced by the dissolution of preformed allophane, the evidence for the environmental conditions governing the formation of either allophane or halloysite from volcanic ash that is represented by Eq. (3.7) is much stronger than that presented in favour of a necessary alteration of allophane to halloysite with time.

*Decomposition* While secondary mineral formation often occurs through an acidification of primary minerals (Eq. 3.1), their breakdown may also occur through acidification associated with alternating oxidation and reduction that occurs when successive (e.g. seasonal) wetting and drying takes place within profiles. In this process, known as ferrolysis (Brinkman, 1970), alternating redox conditions lead to the release of ferrous iron from within aluminosilicate layers, and this is oxidized to ferric iron, as oxides, and hydrogen ions.

Iron oxides occur whenever drying and hence oxidation has occurred in soil formation or development. Most originate ultimately from the breakdown and oxidation of the ferrous iron contained in many primary minerals (Table 3.1). They almost always occur as minor components of soils in terms of their percentage composition, but their content varies considerably between soils. Goethite is the most widespread Fe oxide, forming in all types of weathering regimes. Haematite occurs, usually in association with goethite, in soils in warmer climates. Ferrihydrite forms in soils where rapid oxidation has occurred, particularly where there is substantial organic matter, e.g. in the lower B horizons of podzols, but it is often metastable relative to the other Fe oxide phases. Lepidocrocite occurs where soils are seasonally wet and hence reduction occurs, while maghaemite, also relatively rare, owes its origin to prior heating of other Fe oxides, probably in fires, and/or to the natural oxidation of primary magnetite.

The occurrence of manganese oxides in soils is also indicative that oxidation has occurred, with Mn(II) from the breakdown of primary minerals usually being oxidized to Mn(IV).

While gibbsite may result from neoformation (Eq. 3.5 and 3.7), it may further dehydrate, e.g. in duricrusts at the surfaces of lateritic profile, giving rise to oxides and oxydydroxides of Al such as boehmite and corundum, with the latter possibly formed in bush fires.

Titanium oxides occur in soils, commonly as the result of the oxidative degradation of ilmenite, to give anatase or rutile, often via pseudorutile as an intermediate phase.

Acidification may also decompose soil minerals. Sulphide minerals, especially pyrite, oxidize on aeration to give sulphuric acid, which reacts with soil minerals to give mostly jarosite, natrojarosite and/or gypsum. Anthropogenic sources of acidification, including the release of acid-mine drainage and the oxidation of mangrove swamps by drying, can degrade soil minerals through these types of reactions.

*Inheritance* Many minerals, including important secondary minerals, are inherited from other locations and/or previous cycles of weathering and other processes of alteration of primary or other secondary mineral precursors. This is an important source for kaolinite in soils and part of the reason why this particular mineral is so widespread globally (White and Dixon, 2002). Minerals can also arrive in soils in dust carried both locally and globally (Simonson, 1995).

An understanding of mineral genesis in soils is complicated not simply because some minerals are inherited from earlier stages of weathering, but also because the soils are polygenetic. This may arise because of the addition of new parent materials at different stages during soil development, e.g. successive volcanic tephra, successive loess layers or deposits of materials on top of palaeosols formed on different parent materials (e.g. Jaynes *et al.*, 1989). Interpretations of mineral genesis within profiles need to take cognizance of such discontinuities in this and other soil-forming factors. Jaynes *et al.* (1989) considered that discontinuities between soils resulting from the deposition of new parent materials on palaeosols led to an intensification of weathering that resulted in the formation of a 1:1–2:1 interstratified phase when a discrete 2:1 mineral dominated the fine fractions of the palaeosol, following a similar sequence to that in Eq. (3.6). Besides discontinuities in mineral development, polygenesis from successive inputs of different parent materials often results in minerals towards the base of profiles showing a more advanced stage of weathering than those near the surface when genesis on a single parent material tends rather to give the reverse trend in weathering stages.

### *Associations in soils*

Secondary minerals occur in soils only rarely, if at all, as discrete clay-size particles. It is necessary to apply physical energy and/or chemical disaggregating agents to obtain dispersed clay-size material from soils (Churchman and Tate, 1986). Otherwise, soils would be completely unstable against erosion by water and wind. In soils, individual crystals of secondary minerals, which have high surface areas and are electrically charged, and hence can attract other entities, are found in association with one other, with other inorganic materials, e.g. oxides, and also with organic matter. As summarized by Rengasamy and Sumner (1998) 'soil clay systems ... are complex heterogeneous intergrowths of different clay structures intimately associated with organic (matter) and biopolymers'. Electron microscopy shows that clay minerals often surround and enclose organic matter thereby providing protection against its degradation. Associations of secondary minerals and organic matter form the basis of a stable soil structure.

### 3.2 Future prospects

Our knowledge of soil minerals has been greatly influenced by the analytical techniques that have been employed. X-ray diffraction, which was first employed for soil minerals in about 1930, has been the predominant technique. Generally, powder samples extracted from soils, often following various chemical pre-treatments, and after size fractionation and drying, have been examined using this and other techniques. In particular, organic matter and oxides associated with minerals and also their Al-interlayers may be removed by pretreatments for analyses. The possibility that the materials examined are artefacts of these treatments has to be considered. Further, with $< 2$ μm size fractions commonly being examined, the results often give an average composition of many much smaller particles. This enables generalizations about the bulk behaviour of minerals. The advent of high resolution instruments, especially high resolution transmission electron microscopes (TEM), provides the capacity to examine minerals within undisturbed samples of soils and to discover their natural associations. Future work will enhance our understanding of minerals and their developments at the nanoscale.

# 4

# Soil phases: the organic solid phase

*Claire Chenu*

Organic matter is quantitatively a minor fraction of soil. It typically comprises 0.1 to 10% of the soil mass, and up to 40% in the case of Histosols. However, soil organic matter is quantitatively very important at the scale of the planet. The global soil carbon pool of 2500 Pg ($1 Pg = 10^{15} g$) includes about 1550 Pg of soil organic carbon (SOC) and 950 Pg of soil inorganic carbon (SIC). The soil C pool is 3.3 times the size of the atmospheric pool (760 Pg) and 4.5 times the size of the biotic pool (560 Pg) (Lal, 2004). Hence, the soil C reservoir could be used to control the composition of the atmosphere.

Despite its low relative concentrations in soil, organic matter has major qualitative importance because of its functions. Soil organic matter (SOM) contributes to soil properties and functions. The major, and long-recognized role of SOM is its contribution to soil chemical fertility. SOM is a reserve of nutrients such as N, P, S, which are its constitutive elements and are released with mineralization. Furthermore, SOM is charged and has a high cation exchange capacity (60 to 400 cmol(+) $kg^{-1}$); it thereby retains nutrient cations such as $K^{+}$, $Mg^{++}$, $Ca^{++}$, $Fe^{+++}$ on its negative charges. Thus SOM ensures most of plant nutrition in natural ecosystems as well as in organic farming and low input cultivation systems. This role of SOM is less important in the case of intensive cultivation systems relying on mineral fertilizers. In all soils SOM is the basis of most soil biological activity, being the source of carbon and energy of heterotrophs, from micro-organisms to macrofauna, through food webs. SOM has a major influence on soil physical quality by increasing the retention of water, aggregating mineral particles and thus contributing to a favourable soil structure and preventing soil erosion. This is an increasingly relevant topic, as the low organic matter status of many cultivated soils around the world leads to serious

*Soils: Basic Concepts and Future Challenges*, ed. Giacomo Certini and Riccardo Scalenghe.
Published by Cambridge University Press. 

problems of physical degradation and erosion. Soil organic matter also affects the quality of other resources, such as water, air and crops. Soil organic matter has major environmental roles because it retains organic pollutants, heavy metals and radionuclides, thanks to its high chemical reactivity. As a consequence, fewer pollutants tend to be transferred to water and to plants. SOM mineralization releases $CO_2$, methane and N-oxides, which affect the composition of the atmosphere. Controlling soil organic matter contents in soils through land use and agricultural practices may help in reducing the greenhouse effect and have a crucial impact on the Earth's climate. Soil organic matter is thus a major component of soil quality and it has a central role in the functioning of terrestrial ecosystems.

Although soil organic matter has been very early associated with soil fertility and studied since the nineteenth century, there has been a renewed interest in it during the last 20 years. The crucial need to optimize agroecosystems, maintain soil quality and protect natural resources leads researchers to focus on soil organic matter. New methods or combination of methods have been developed, such as stable isotope tracing, nuclear magnetic resonance spectroscopy (NMR), pyrolysis, electron microscopy, Fourier transform infrared spectroscopy (FTIR), and these methods have enhanced the knowledge of SOM.

Soil organic matter has the following key characteristics:

(1) it is complex, being a mixture of a multitude of different components;
(2) it is closely associated with soil minerals;
(3) it turns over constantly, being renewed by inputs on the one hand and by losses through mineralization on the other;
(4) it is very sensitive to soil management.

This chapter aims at presenting these characteristics, without covering them comprehensively.

## 4.1 Soil organic matter complex composition

Soil organic matter composition depends on the composition of the inputs as well as on the degree and pathways of transformation of these compounds. Given the variety of the inputs (different plant or plant tissues, micro-organisms, fauna), and the different stages of decomposition attained by them, soil organic matter is an extremely complex mixture. The complexity of SOM has historically led to attempts to subdivide it into several fractions by various approaches, each approach providing a picture of SOM. Examples of the types and abundance of fractions separated from SOM by different approaches are presented in Fig. 4.1.

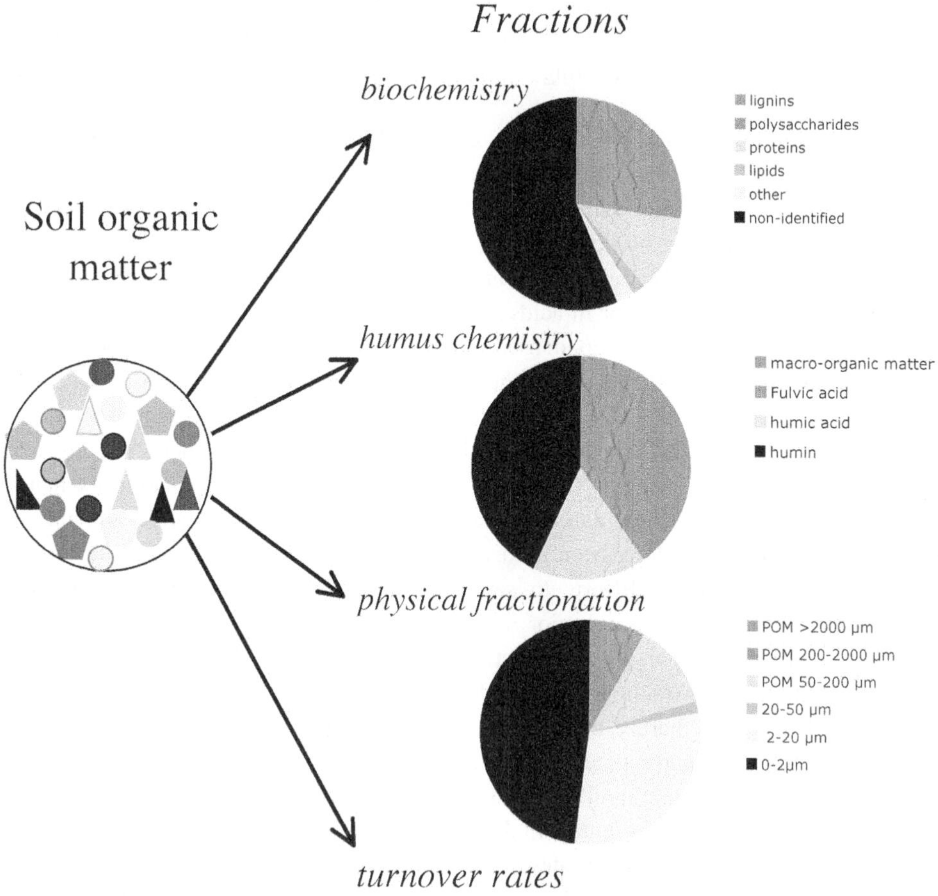


Fig. 4.1. Soil organic matter complexity can be resolved by different approaches of fractionation. The fractions and their relative amount (in % C) are represented here for the A horizon of a Cambisol from a temperate area (various sources).

### *4.1.1 Different approaches to characterize SOM components*

Soil organic matter has early been described as a combination of easily recognizable plant material undergoing decomposition and humus, a dark substance associated with minerals and whose components were visually unidentifiable. Most fractionation approaches have intended to resolve the complexity of humus.

(1) Biochemical fractionation. Classical biochemical fractions, such as lignin, carbohydrates and proteins, can be isolated from soil using biochemical methods. However, various studies have shown that a large part of SOM is not identifiable by these techniques. Two hypotheses may explain why: (a) there are soil organic molecules which have a composition that is original, i.e. different from that of plant or

microorganisms and (b) the association of SOM with minerals prevents its extraction and further identification as biochemical entities. A compilation of various sources is reported in Fig. 4.1, showing the relative abundance of biochemical fractions in SOM.

(2) Humus chemistry. Humus has been chemically fractionated as early as in the eighteenth century (Feller, 1997). The classical method to fractionate humus is first to disperse humic colloids with a dilute alkali, e.g. sodium hydroxide, and then to fractionate the extract by acidifying it, e.g. with HCl. Fulvic acids (FA) remain in solution whereas humic acids (HA) are precipitated. Humin is a complex fraction, neither soluble in alkali nor in acids, which is a mixture of identifiable biochemical components and highly condensed molecules. All fractions can be further fractionated, e.g. on the basis of the molecular weight of their constituents. An example of the abundance of humic fractions is reported in Fig. 4.1. These operationally defined chemical fractions have distinct chemical properties. Compared with HA, FA have a larger O content, smaller C content, larger cation exchange capacity and lower molecular weight (Stevenson, 1994). The relative amounts of humic fractions have long been used as criteria for soil and humus classification.

(3) Physical fractionation. Efforts to separate SOM fractions corresponding to true moieties in soil, without major chemical alteration due to extractants, have successfully relied upon physical fractionation methods since the 1980s (Christensen, 1996). Particle size fractionation and density fractionation after adequate dispersion of the soil release purely mineral fractions (as coarse or dense fractions), purely organic fractions (particulate organic matter (POM), macro-organic matter or light fraction), and organomineral fractions (silt and clay-size fractions). An example of such fractionation is presented in Fig. 4.1. Different sizes of particulate organic matter have distinct turnover rates in soils: e.g. Balesdent (1996) found that in some cultivated temperate soils POM $> 2$ mm has a mean residence time (MRT) $= 1$ year, POM 0.2–2 mm MRT $= 3$ years and POM 0.05–0.2 mm MRT$=$ 10 years. The clay-sized fraction is more heterogeneous in terms of mean residence time as it contains stable carbon with MRT $>$ centuries, as well as young C and N that turn over fast (Balesdent, 1996).

It can be noted that all fractionation approaches end with a fraction that is complex, often undefined and represents about half of the soil C: the unidentified residue after applying biochemical methods, the humin with humus chemistry and the clay-size fraction with particle size fractionation. The multifaceted composition of soil organic matter is responsible for the variety of the fractionation approaches. Other methods have been used, such as fractionating SOM with respect to its location in the soil structure: inside or outside aggregates of different sizes and stabilities. These approaches (e.g. Puget *et al.*, 1995; Six *et al.*, 2000) have permitted a novel understanding of the relationships between soil structure and soil organic matter dynamics, which will be described below.

### 4.1.2 Major SOM components

*Particulate organic matter (POM)*

Decomposing plant debris can be separated from the soil matrix by either size fractionation (as macro-organic matter >50 μm or >200 μm), density fractionation (as light fraction, e.g. <1.6 g cm$^{-3}$), or by a combination of size and density fractionation (Christensen, 1996). An example of POM separated after dispersion of soil, sieving, and flotation is given in Fig. 4.2. Particulate organic matter represents typically 10–25% of the soil organic carbon (Gregorich *et al.*, 1995). Relatively fresh plant debris is usually easily separated from the soil matrix, whereas more humified material becomes occluded in soil aggregates (Golchin *et al.*, 1998). Particulate organic matter is a fraction of soil organic matter which is very sensitive to management, and is thus used as an early indicator of changes in SOM status (Gregorich *et al.*, 1995). In terms of functions, POM represents a labile source of C and N usable by micro-organisms and fauna; thus, it is a 'biologically active fraction' and a short-term nutrients reservoir. POM has an important role in soil aggregation, the particles being the nuclei for the formation of soil aggregates (Golchin *et al.*, 1998). POM also exhibits very high retention rates of pesticides (Benoit *et al.*, 2000) and heavy metals (Besnard *et al.*, 2001), which are ascribed either to the physicochemical reactivity of POM or to the absorption of pollutants by micro-organisms attached to it.

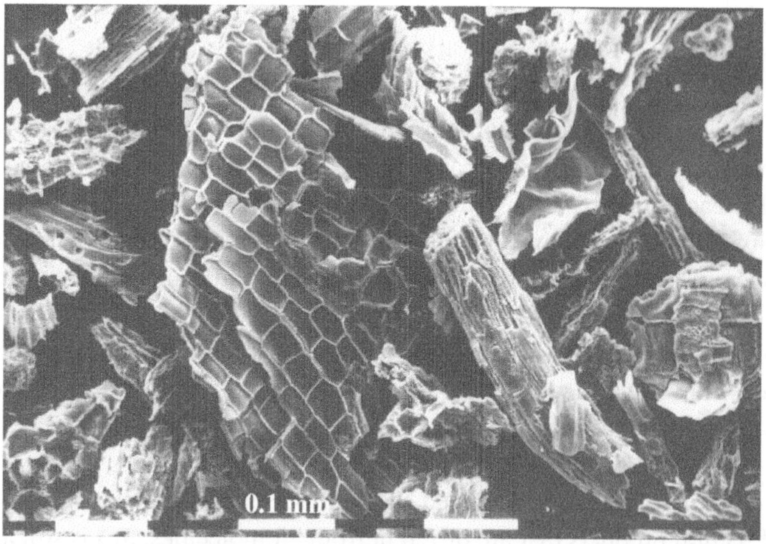


Fig. 4.2. Scanning electron microscopy observation of particulate organic matter separated from soil (Puget and Chenu, unpublished).

### *Biota and biomolecules*

Living organisms are an integral part of soil organic matter. Microbial biomass, which can be measured by several methods, typically accounts for 1–3% and 2–6% of soil organic C and N, respectively. Microbial biomass is both an agent for the transformation and cycling of nutrients and organic matter in soils and a labile pool of organic matter itself. It is thus used as an indicator of N mineralization and as an early indicator for changes in soil organic matter status or soil pollution (Gregorich *et al.*, 1995). Living organisms such as roots, bacteria and fungi exude biomolecules in soils. Among these, enzymes and polysaccharides reveal particular functional importance, although such exudates represent only a few % or a few ‰ of soil organic carbon. Extracellular polysaccharides play a prominent role in soil aggregation by adsorbing to clay particles and bonding them (Chenu, 1995; Baldock, 2001).

### *Soluble organic matter*

A minor fraction of SOM is present as soluble molecules in soils. This fraction, operationally separated as organic matter $< 0.45\,\mu m$, typically accounts for 1–5% of SOC. Soluble OM compounds are either products of the decomposition of plant residues and micro-organisms, or are exudates from micro-organisms and roots. These are organic acids, phenols, carbohydrates, amino acids and proteins, and fulvic acids. Soluble OM is a dynamic and functionally important fraction, as it is available organic C and N for micro-organisms, contributes to mineral weathering and podzolization, has an important reactivity towards heavy metals, organic pollutants and mineral surfaces, and it is susceptible to leaching.

### *Humic substances*

Humic substances comprise 60–80% of C of the humus. They are random amorphous organic macromolecules formed by polyaromatic building blocks bridged to each other by ester, ether and C links and carrying variable proportions of carboxyl, hydroxyl, amine and other hydrophilic groups. The essence of humic substances resides in the combination of extreme molecular diversity and pronounced chemical reactivity. Their diversity is such that writing a molecular structure has no meaning, because it is unlikely that there would exist two humic molecules that are exactly the same (McCarthy, 2001). However, humic substances from different soils and land uses display remarkable uniformity in their gross properties (Preston *et al.*, 1994).

Although it is generally accepted that humic substances constitute a distinct class of natural products that are different from biomolecules, this is still a subject of debate (Burdon, 2001). The macromolecular character of humic substances is also under question. Recent work suggests that humic substances are not large macromolecules but rather supramolecular associations of small molecules derived from the biodegradation of dead biological material and which are bound by weak chemical forces (Piccolo, 2002).

Their fractionation being based on their chemical properties, humic substances are entities with distinct chemical and physicochemical properties. As such, they are used in speciation models when dealing, for example, with the fate of heavy metals in soils. However, humic chemistry has proven unable to separate fractions which are kinetically distinct (Balesdent, 1996), or functionally distinct in terms of their biological and physical properties. The relative abundance of the different humic substances in soils has, thus, little practical value as an indicator of soil organic matter functions.

## 4.2 Organomineral associations

One of the prominent features of soil organic matter is its association with minerals. It is quantitatively very important: about 40–80% of soil carbon is present in the clay-sized fraction of soils and cannot be separated from minerals (Christensen, 1996). Soil organic matter associations are essential because they affect both the properties of the minerals and the dynamics of the organic moieties.

### *4.2.1 Definitions and composition*

The term 'organomineral association' refers currently to different degrees of interaction from loosely bound organic and mineral particles in a soil clod, to tightly bound complexes (e.g. the adsorption of a protein to a smectite clay). Organomineral associations cover a wide range of spatial scales from the decimetre (e.g. a soil ped) to the nanometre (e.g. the protein–clay complex already cited).

Primary organomineral associations are associations of organic matter with individual mineral particles, e.g. a smectite–protein complex or a humic coating on a sand grain. The isolation of primary organomineral associations requires complete soil dispersion. Secondary organomineral associations, namely aggregates, are made by the grouping together of several particles and primary organomineral associations. Secondary organomineral associations are separated after incomplete soil dispersion, for example by wet sieving. However, the distinction between primary and secondary organomineral associations is difficult to establish when dealing with very small spatial scales; in fact, the observation of clay-sized fractions in soil shows that many clay particles are microaggregates. Several examples of primary and secondary organomineral associations are given in Fig. 4.3.

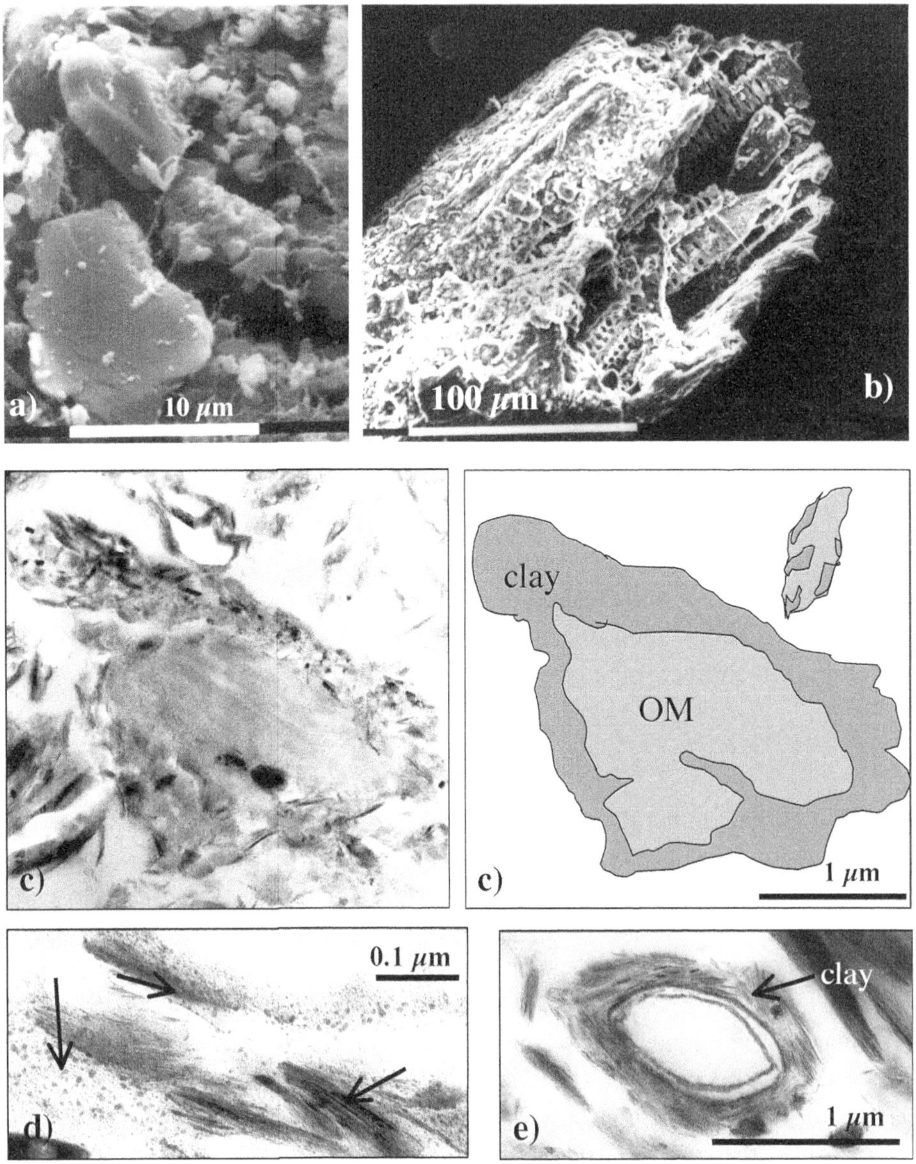


Fig. 4.3. Examples of primary and secondary organomineral associations in soils (Chenu, unpublished). (a) Organic matter adsorbed to silt particles (primary); (b) POM encrusted with minerals (secondary); (c) microaggregate of amorphous organic matter and clay minerals (secondary); (d) polysaccharides adsorbed to clay mineral (stained with black dots of silver) (primary); (e) clay particles adhering to bacterial remnant (secondary).

### *4.2.2 The formation of organomineral associations*

At the molecular scale, physicochemical interactions are the rule. Organic compounds can interact with mineral surfaces through a variety of mechanisms from electrostatic and coordination bonds to weak hydrogen bonds and van der Waals forces. The type of bond between organic and mineral moieties and the surface areas engaged in the contact will determine the cohesion of the organomineral association. These interactions may take place between soluble organic compounds and minerals (adsorption), or between solid organic particles, such as bacteria or POM and minerals (adhesion). Given their large surface area and their charge density, clay minerals are particularly involved in organomineral interactions.

A conceptual model linking soil organic matter dynamics and the formation, stabilization and destabilization of aggregates has been progressively developed through the work of a number of researchers (Golchin, 1994; Puget *et al.*, 1995; Angers *et al.*, 1997; Six *et al.*, 2000). Schematically, fresh organic matter entering soil is responsible for aggregate formation as it locally stimulates microbial activity. Microbial binding agents are exuded and they impregnate and aggregate the surrounding clay particles. Fungi directly enmesh the particles. As a result, the plant fragments (POM) become encrusted in clay particles and new aggregates are formed: POM become the nuclei for aggregation. As decomposition proceeds, microbial activity diminishes and the newly formed aggregate eventually loses its stability. With time the stabilization takes place first at a macroaggregate scale (>200 μm) and then at a smaller one (Angers *et al.*, 1997). External stresses, such as tillage, affect the persistence of aggregates and the protection of SOM from decomposition in microaggregates (Six et al., 2000).

### *4.2.3 Consequences of organomineral associations*

The binding of organic matter to mineral particles directly affects the properties of minerals such as their cation exchange capacity (Thompson *et al.*, 1989) and their wettability (Chenu *et al.*, 2000). Water retention, porosity and interparticle cohesion may also be increased (Chenu and Guérif, 1991; Emerson, 1995).

Furthermore, linkage to minerals decreases the biodegradation rate of organic matter. The adsorption of organic molecules to mineral surfaces tends to reduce their availability to micro-organisms (Scow and Johnson, 1997), a process termed physicochemical or chemical protection. Organic matter in aggregates is protected from decomposition, a process that is named physical protection. The entrapment of organic matter in microaggregates reduces their physical accessibility to micro-organisms or restricts the availability of oxygen to microbial decomposers (Baldock and Skjemstad, 2000).

## 4.3 Soil organic matter dynamics

### *4.3.1 Quantification and conceptual description*

Soil organic matter contents result from the dynamic equilibrium between the inputs and outputs by mineralization and leaching. The turnover of an element in soil organic matter is generally described by its mean residence time (MRT), which is the average time the element resides in a given pool at steady state. Soil organic matter transformations are generally assumed to follow first order kinetics. In first order kinetics the rate of decomposition of organic matter is proportional to its amount. Mean residence time is measured by different approaches, including the modelling of C and N stocks and pools with first order kinetics, $^{14}C$ dating and $^{13}C$ natural abundance. Furthermore, tracer techniques ($^{14}C$, $^{13}C$, $^{15}N$) have been used to measure the decay of organic material added to soil (Coleman and Fry, 1991). The dynamics of C and N have received most interest compared with those of P and S.

Mean residence times of bulk organic carbon in soil obtained by the different methods range from a few decades to thousands of years and vary with climate, soil type and land use (Six and Jastrow, 2002). Soil organic matter being very heterogeneous, its dynamics are generally described using different pools, i.e. different conceptual subsets of SOM, each characterized by a given MRT. For example, the Rothamsted model considers five pools of soil organic carbon (Jenkinson and Rayner, 1977): decomposable plant material (with a half-life of two months), resistant plant material (half-life: 2.3 years), soil biomass (half-life: 1.7 years), physically stabilized organic matter ('slow C', half-life: 50 years), and long-term chemically stabilized organic matter ('inert pool', half-life: about 2000 years). The correspondence between these kinetic pools with measurable SOM fractions or functional pools is still very difficult to establish. One example is given in Fig. 4.4. POM fractions have clearly distinct turnover; total POM can be used as a surrogate to the resistant plant material in the Rothamsted model (Balesdent, 1996). However, pools such as the slow pool and the inert pool, which are crucial to predict long-term changes of C stocks in soil, have not yet any clear correspondence with measurable fractions.

### *4.3.2 Factors affecting soil organic matter dynamics*

Soil organic matter dynamics depend on the one hand on the rates of organic matter input to soil (thus, on primary production) and the biochemical quality of these inputs and, on the other hand, on the activity of microbial decomposers. The latter is influenced by a variety of environmental factors, which will only be cited

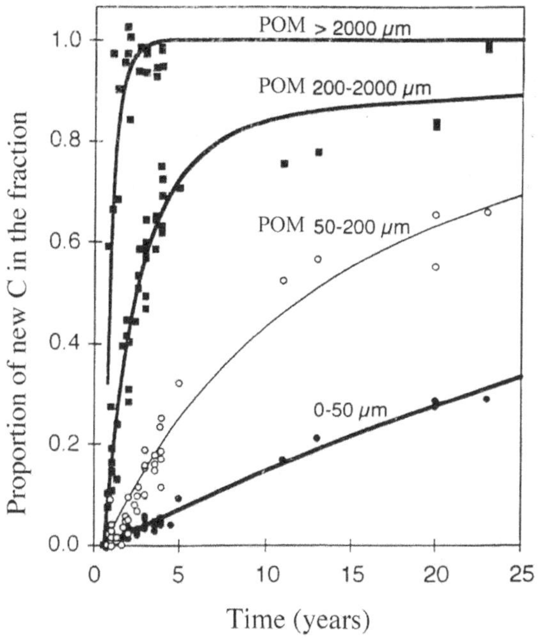


Fig. 4.4. Incorporation of new C4-derived carbon in C of different particle size fractions from cultivated soils from temperate areas. Particulate organic matter (POM) are labile fractions (>2000 μm) and slowly renewed ones (50–200 and 200–2000 μm), whereas the fine-sized <50 μm fraction exhibits mean residence times of 70 years.

here: climate (temperature, water availability), oxygen availability, soil mineralogy, soil pH and soil management (e.g. fertilization, tillage).

Future advances in soil organic matter dynamics concern mainly (a) the coupling of different processes such as C, N and P dynamics, C and N dynamics with water flow (Garnier *et al.*, 2001), C dynamics with soil structure characteristics (Plante and McGill, 2002; Six *et al.*, 2002), and (b) the identification of stable C (Falloon *et al.*, 1998). One emerging hypothesis is that the explanation for the stabilization of OC in soil does not arise from a single process, such as chemical recalcitrance, physicochemical protection through adsorption, chemical protection through interaction with metals or molecules, and physical protection due to soil structure, but rather to a combination of these processes.

# 5

# Soil phases: the liquid phase

*Randy A. Dahlgren*

If one considers soil '*the excited skin of the Earth*', then the soil solution is its blood. Just as analysing blood chemistry can tell you much about the health of a human, analysing soil solution can tell you much about soil quality and the processes occurring in the soil ecosystem. Water plays a vital role as a transporting agent and chemical solvent in soil processes and ecosystem functions. All major pools and fluxes interact through the soil solution (Fig. 5.1) making it dynamic and sensitive to changes occurring in the soil ecosystem. Major fluxes are regulated by a combination of equilibrium (exchange and sorption reactions) and kinetic (mineralization, chemical weathering, and kinetically constrained mineral equilibria and redox reactions) processes that interact with hydrological processes and nutrient cycling by vegetation and soil biota. Thus, examination of soil solutions provides a means of elucidating the fate, behaviour and transport of dissolved and colloidal constituents in the soil environment. Studies utilizing soil solutions have been used to elucidate pedogenic processes, equilibrium and kinetic factors, solute transport, soil fertility, nutrient cycling, and the fate and transport of environmental contaminants (Wolt, 1994).

While studies of extracted soil solutions have been utilized since the 1860s, soil solution chemistry did not emerge as a subdiscipline of soil chemistry until the late 1960s (Wolt, 1994). Early studies emphasized plant nutrient availability and equilibria between solid and liquid phases to predict changes in mineralogy through thermodynamic calculations. More recently, soil solution studies are commonly used to understand *in situ* soil processes, effects of perturbations and rates of change in soil properties. Soil solution dynamics reflect current biogeochemical processes occurring in soil ecosystems. In contrast, interpretation of processes based on solid-phase analyses is confounded because solid-phase properties integrate all soil processes that have occurred to date in the development of the soil. Similarly, in

*Soils: Basic Concepts and Future Challenges*, ed. Giacomo Certini and Riccardo Scalenghe.
Published by Cambridge University Press. 

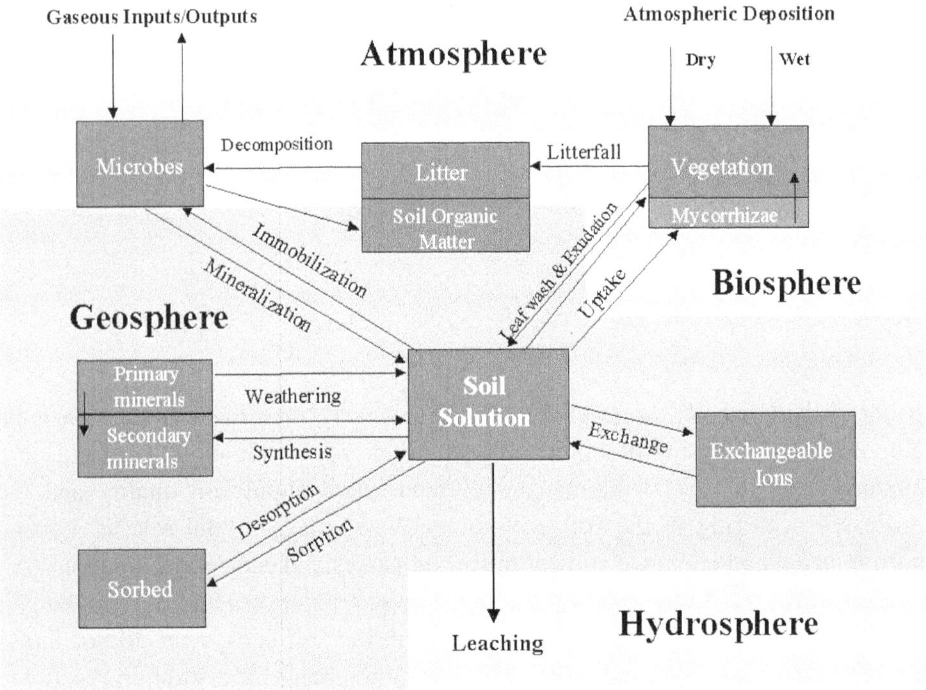


Fig. 5.1. Elemental cycles consist of a series of interrelated processes occurring within and between the atmosphere, hydrosphere, biosphere and geosphere. The soil solution serves as the hub connecting all these interactions.

response to soil ecosystem perturbations, it is very difficult to measure small differences in large elemental pools in the soil solid-phase, especially given the high spatial variability that commonly exists in soils at the landscape scale. As a result, soil solution studies often provide a more powerful and sensitive approach for monitoring, assessing and interpreting soil processes and ecosystem perturbations (Ugolini, 2005). While soil solution studies can be a powerful experimental approach, several challenges exist with this approach, including collection of a representative and chemically unaltered soil solution, spatial heterogeneity of soil properties, dynamically changing hydrologic flowpaths (e.g. matrix versus preferential flowpaths), and difficulties in acquiring soil solutions at low moisture contents.

The primary objective of this chapter is to provide a brief review of theory, methodologies, and applications of soil solution techniques in the environmental sciences, with a particular focus on pedogenesis. Comprehensive reviews dealing with the subject matter of this chapter include Lindsay (1979), Sposito (1981), Wolt (1994), and Snakin *et al.* (2001).

## 5.1 The liquid phase of soils

Soil water is an important component facilitating interactions within and between the atmosphere–biosphere–geosphere ecosystem components (Fig. 5.1). The soil solution is an aqueous solution composed of liquid water and dissolved solids and gases. Dissolved solids include neutral and charged species, and can be either complexed or non-complexed. The soil solution is linked to the gaseous and numerous solid phases via the transport of energy and matter: solid $\leftrightarrows$ liquid $\leftrightarrows$ gaseous. Because changes in the composition of the soil solution depend on interactions with the solid and gaseous phases, one cannot consider soil solution processes or composition in isolation from the other soil components. Water plays many critical functions in the soil ecosystem including roles as a transporting agent, chemical solvent, readily available nutrient pool, source of water for metabolic activities of soil biota and vegetation, and as a factor affecting soil air composition and soil temperature.

The variable amount of water contained within a soil and the energy state of water in soil are important factors affecting availability of water for plants and the flow of water through a soil. Numerous soil properties depend on water content, such as consistency, plasticity, strength, compactibility, penetrability and stickiness. Measurements of soil wetness (water content) commonly report the fractional content of water in terms of either mass or volume ratios. Mass wetness is defined as $\omega = M_w/M_s$, where $\omega$ is mass wetness, $M_w$ is mass of water and $M_s$ is mass of dry soil defined as the mass of soil dried to equilibrium at 105 °C. Volume wetness is defined as $\theta = V_w/V_t$, where $\theta$ is volumetric water content, $V_w$ is volume of water and $V_t$ is total volume of the soil. The mass and volume water contents are related to each other through the soil bulk density ($D_b$): $\theta = \omega \times D_b$ ($\mathrm{Mg\,m^{-3}}$). The degree of saturation is expressed as the water volume present in the soil relative to the total pore volume, $s = V_w/V_p$, where $s$ is the saturation index, $V_w$ is the volume of water and $V_p$ is the volume of pores (total porosity). The relative humidity of soil air is expressed as the ratio of the vapour pressure of the soil air to the saturation vapour pressure at the same temperature, $\mathrm{RH} = C/C_0$, where RH is relative humidity of soil air, $C$ is vapour pressure of soil air and $C_0$ is the saturation vapour pressure. Direct and indirect methods for measuring soil moisture content include gravimetric (sampling and drying), neutron scattering, gamma-ray absorption, and time-domain reflectometry (see Hillel, 1998, for a rigorous review of these techniques).

The total soil-water potential (a measure of the relative energy level of soil water) is affected by matric forces (responsible for adsorption and capillarity), osmotic forces (water attraction to dissolved solids), and the force of gravity.

Total soil-water potential can be expressed as:

$$\Psi_t = \Psi_g + \Psi_m + \Psi_o + \ldots$$

where $\psi_t$ is the total potential, $\psi_g$ is gravitational potential, $\psi_m$ is matric potential, $\psi_o$ is osmotic potential and the ellipsis signifies that additional terms are theoretically possible. In unsaturated conditions, matric and osmotic potentials are the dominant energy components and the thermodynamic potential can be expressed as $\psi_t = \psi_m + \psi_o = RT\ln(RH)/M$, where R, T, RH and M are the universal gas constant, absolute temperature, relative humidity and molar mass of the water, respectively. Soil-water potentials can be measured using techniques such as tensiometers, electrical resistance blocks, thermocouple psychrometers, heat dissipation, centrifugation, and pressure membrane equilibration (Hillel, 1998).

The energy level of water is strongly related to the soil moisture content (Fig. 5.2). As the water content of the soil decreases, the remaining water is held closer to soil surfaces and in smaller pore diameters, resulting in a stronger attraction to soil particles. Because of the high surface area and dominance of smaller pores in clay-rich soils, they hold more water at a given potential than does a loamy or sandy soil. Similarly, at a given soil-moisture content, water is held more strongly in the clay than in a loam or sand. The concentration of

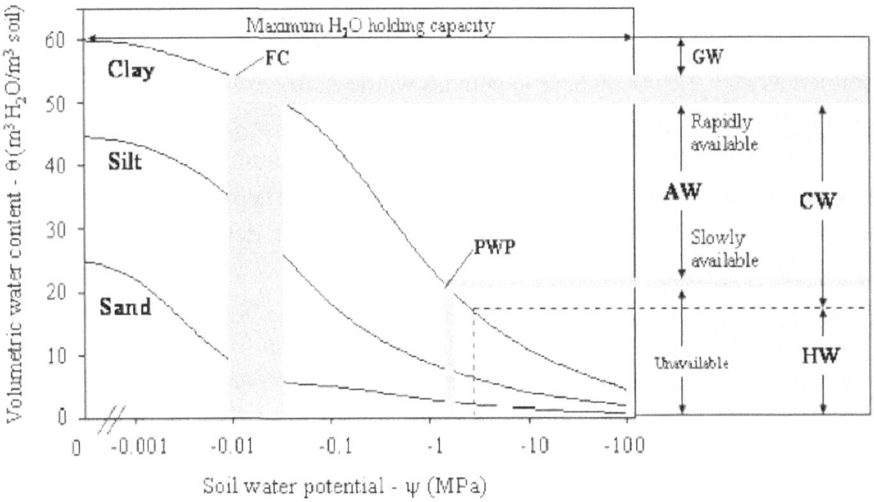


Fig. 5.2. Soil water potential curves for three representative mineral soils of varying texture. The water content–matric potential curve for the clay is used to illustrate the different terms for describing water in soils (GW, gravitational water; CW, capillary water; HW, hygroscopic water; AW, plant-available water; FC, field capacity; PWP, permanent wilting point). The shaded area depicting field capacity and permanent wilting point indicates that these concepts occur over a range of water potential values for contrasting soils.

dissolved constituents also affects the energy level of water due to the attraction of water molecules by solutes, thereby lowering the osmotic potential. The osmotic potential of soil solution can be approximated through its relationship with the electrical conductivity (EC) of soil solution:

$$\Psi_o\ (\mathrm{J\,kg^{-1}}) = -36\ \mathrm{EC}\ (\mathrm{dS\,m^{-1}}).$$

Soil water is often categorized into three fractions based on matric potential: hygroscopic (adsorbed, $\psi_m \leqslant -3$ MPa), capillary ($\psi_m \approx -3$ to $-0.03$ MPa), and gravitational ($\psi_m > -0.03$ MPa) (Fig. 5.2).

While the potential values separating these categories of water are only approximations and vary among soils, these boundaries assist in the qualitative description of different forms of water in soils. The hygroscopic water is so strongly adsorbed by soil surfaces that it is considered non-liquid (or non-solvent water) and can move only in the vapour phase. Gravitational water is loosely held in soil pores and moves due to the force of gravity. This water fraction is temporary in nature (following periods of soil saturation), but very important in transport of solutes and colloids within the soil profile, and from the soil to surface and ground waters. Capillary water is retained in soil pores against the force of gravity and is the most important source of water and nutrients for vegetation and soil biota. Plant-available water holding capacity (water accessible for plant utilization) is often considered to be the difference in water contents between field capacity (between − 0.01 and − 0.03 MPa; water content following free drainage of soil by gravity) and permanent wilting point (− 1.5 MPa; water content at which plants cannot extract water fast enough to maintain their metabolic processes) (Fig. 5.2). With respect to soil texture, plant-available water holding capacity follows the general relationship: loam, silt loam, clay loam > clay > sandy loam > sand.

The primary forms of water movement through soils are saturated flow, unsaturated flow and vapour movement. Saturated hydraulic conductivities range widely across the various textural classes: clay = $10^{-5}$ to $10^{-3}$ cm h$^{-1}$, silt = $10^{-3}$ to $10^{-1}$ cm h$^{-1}$, sand = 1 to 400 cm h$^{-1}$. Water conductivity decreases sharply (4–5 orders of magnitude) as soil water content decreases from saturation ($\psi_m \approx 0$ MPa) to the limits of plant available water ($\psi_m \approx -1.5$ MPa). Hydraulic conductivity decreases because pores become empty and unavailable for water transport as the soil dries. Vapour transport is generally negligible below the upper few centimetres of soil. Low vapour movement within the soil profile results from the low gradient in relative humidity (generally between 98 and 100%) and slow diffusion through the soil matrix.

The type of water flow through a soil can have a tremendous impact on the chemical composition of soil solution. The chemical composition of capillary water can be quite heterogeneous at small spatial scales (within a soil horizon) as a result of non-equilibrium processes, such as diffusion gradients resulting from nutrient uptake or emissions by plants and soil biota. As soils dry, the water transport slows appreciably, allowing kinetic-limited reactions greater time to influence pore water chemistry. Large soil aggregates limit intra-aggregate water penetration resulting in strong intra-aggregate diffusion gradients that greatly alter solution chemistry. Similarly, gas transport into soil aggregates is often limited, resulting in decreasing oxygen concentrations toward the centre of the aggregate. As a result of changing redox conditions within an aggregate a gradient in concentrations of redox-sensitive species may exist. Therefore, isolation of soil solutions from such spatially heterogeneous systems provides only an average condition (bulk soil condition) that might not accurately represent parameters such as bioavailability, elemental activities or solute transport for a given soil horizon, flowpath, or pore.

Due to the short relative residence times of gravitational waters and transport along preferential flowpaths (root channels, biopores), these waters often have considerably different chemical composition from associated capillary waters. Layers of low hydraulic conductivity common in subsoil horizons (e.g. argillic horizon, duripan) may also induce downslope lateral flow through more permeable surface horizons. Shifting hydrologic flowpaths can dramatically alter the chemistry of surface waters at the catchment scale. For example, waters percolating through the vadose zone (the unsaturated soil region above the permanent water table) are often enriched in base cations and bicarbonate, while waters originating as subsurface, lateral flows through the upper soil horizons are often enriched in nutrients and dissolved organic carbon (DOC). Thus, to predict solute transport at the catchment scale, it is not sufficient only to understand soil solution composition at the soil horizon scale – one also has to determine hydrologic flowpaths. Both soil solution chemistry and the hydrologic flowpaths change dynamically during a storm event, making modelling of catchment-scale chemical dynamics from pedon solution chemistry very difficult.

The interaction of dissolved chemicals in soil solution with chemicals associated with the solid phase is a fundamental aspect of soil chemistry research dealing with availability, mobility and distribution of chemicals in soil (Wolt, 1994). For example, soil solution composition proves to be the most directly correlated soil index of bioavailability. However, it is a simplification to state that the soluble nutrient concentrations fully dictate soil fertility because the soil solution only contains a small portion of most plant-available nutrients. The concentration (or activity) of a nutrient in soil solution is a measure of the *intensity* factor. The source of soil solution nutrient replenishment is known as the *quantity* factor and involves all solid-phase forms of the nutrient that are released to solution (primarily exchange and adsorbed forms). The replenishment factor (nutrient buffering capacity) plays a decisive role in the sustainability of plant nutrition.

## 5.2 Methods of soil solution characterization

Investigations examining soil solution chemistry involve two contrasting approaches: isolating soil solutions for chemical analyses, or analysis of *in situ* chemistry using electrochemical techniques (Snakin *et al.*, 2001). Methods for obtaining chemically unaltered soil solutions are fraught with problems. Two of the most serious problems are separating a representative and chemically unaltered soil solution from the solid phase and obtaining sufficient solution volume for analysis at low moisture contents. Laboratory methodologies for obtaining soil solutions may be broadly defined as displacement techniques and include column displacement (pressure or tension with or without a displacing solution), centrifugation (with or without immiscible liquid), and saturation extracts (see comprehensive reviews by Wolt, 1994 and Snakin *et al.*, 2001). The intent of these laboratory methods is to obtain a soil solution that is in quasi-equilibrium with the soil solid-phase. Depending on methodology, these techniques are capable of extracting capillary water held between field moisture content and about $\psi_m = -1.8$ MPa. The recovery of solution is highly dependent on soil texture and can be estimated from soil moisture release curves and the applied displacement pressure/tension.

Comparisons among the chemical composition of solutions obtained by displacement methodologies show a potentially wide range of variability among methods and a strong interaction with soil type (Dahlgren, 1993; Wolt, 1994; Snakin *et al.*, 2001). A comparison of five extraction techniques with three soil types showed that solution pH varied from 0.5 to 3 units (Dahlgren, 1993). Because pH is a master variable regulating the solubility and speciation of many solutes, serious artefacts may result. Immiscible liquids are often used in association with some displacement techniques. These immiscible liquids (non-polar solvents) can solubilize high concentrations of DOC that can subsequently partition into the aqueous phase. DOC, like pH, plays an important role in regulating soil solution composition through its role as an anion, metal chelator and pH buffer. In soils with elevated $pCO_2$ and $HCO_3^-/CO_3^{2-}$ as major anions, simply removing the solid phase from its soil environment drastically changes soil solution chemistry due to degassing of $CO_2$ from samples before soil solution extraction (Dahlgren *et al.*, 1997b). Water extracts of soils have been widely used to represent soil solution composition. The soil:solution ratio (commonly saturated, 1:1 and 1:5) can strongly affect the chemical composition, especially in poorly buffered systems (low quantity/intensity factor ratio). In summary, soil solutions are operationally defined by the method of soil solution displacement and serious artefacts may result from some methodologies. Furthermore, the interaction with soil type makes comparisons among contrasting soils tenuous.

Field methods for sampling soil water are generally grouped under the category of lysimetry (Wolt, 1994). These methodologies are characterized as (a) monolith (collection from undisturbed soil block with a box built around the block), (b) filled-in (collection from disturbed soil block placed in a column or box), (c) tension (using a porous media and vacuum), and (d) zero-tension (collection of gravitational water from trench). The *in situ* solutions collected by these devices represent a mixture of low-tension waters ($\psi_m \approx 0$ to $-0.03$ MPa) moving through the soil matrix and waters moving through preferential flowpaths that bypass appreciable interaction with the soil matrix. Thus, the chemical composition of waters collected by lysimetry will depend on antecedent conditions (e.g. water content and degree of equilibration), hydraulic residence time, and hydrologic flowpaths. As a result, these waters may not represent a quasi-equilibrium with the solid phase, but rather a measure of the chemical transport through the soil.

Nevertheless, lysimetry has several advantages with respect to laboratory displacement methods in determining soil-forming processes, ecosystem perturbations, and leaching of chemical constituents. Because lysimetry methods utilize devices that are emplaced in the soil for a long period of time, it is possible to obtain a long-term temporal record of chemical transport for a given volume of soil, eliminating spatial variability as a factor. In contrast, laboratory methods require destructive sampling of the soil and sampling of different soil masses (susceptible to spatial variability) to acquire a temporal record. Continuous collection of solutions by lysimeter results in a time-weighted or volume-weighted chemical composition that means all temporal events are included in the resulting chemistry. Laboratory methods provide only a snapshot of soil solution chemistry for the instance in time that the sample was collected.

There are several potential limitations to lysimetry that must be addressed to assure meaningful interpretation of soil chemical composition. Chemical interactions with the sampling device (especially tension lysimeters) can cause serious alterations to solution chemistry. While Teflon® samplers impart no appreciable alterations, ceramic materials may leach several constituents while at the same time adsorbing constituents such as DOC, pesticides and trace elements (Wenzel *et al.*, 1997). Leaching of contaminants can be partially overcome by rigorous acid washing prior to installation, but adsorption requires long-term exposure in the soil environment to reach a quasi-equilibrium with dissolved chemicals. Depending on construction, many types of lysimeters are susceptible to $CO_2$ degassing once the solution is isolated from the soil mass. One result of $CO_2$ degassing is an increase in solution pH that may change chemical speciation and solubility (Suarez, 1987).

Soil disturbance associated with installation of lysimeters has been shown to create artefacts such as increased nitrate leaching from disturbance-stimulated mineralization/nitrification, for up to a year after installation. Spatial heterogeneity in soil properties makes it difficult to obtain a representative soil solution sample due to the limited spatial extent of soil contributing to the sampler. A large number of replicate samplers is necessary to ensure that one is capturing the spatial variability at larger scales. Most lysimetric methods preferentially sample gravitational water (low tension waters; $\psi_m \approx 0$ to $-0.03$ MPa) and they may miss the fraction associated with unsaturated flow. Similarly, lysimetry methods are not effective in dry regions where soils remain unsaturated for the majority of the time. While monolith and filled-in lysimetry techniques are capable of measuring the water volume associated with the collected soil solutions, tension and zero-tension lysimetry methods cannot accurately quantify water flow volumes.

Given the limitations to isolating a chemically unaltered soil solution from the solid phase, much research has been directed at using electrochemical techniques for studying soil in its undisturbed natural state (reviewed by Snakin *et al.*, 2001). Commercially available electrodes are able to measure activities of $H^+$, $Na^+$, $K^+$, $NH_4^+$, $Ca^{2+}$, $NO_3^-$ and $Cl^-$, as well as redox status (Eh) in the liquid phase of soils. Coupling these ion selective electrodes with a microprocessor theoretically allows collection of long-term continuous data. While there are several advantages associated with these electrochemical techniques, the technology associated with many applications requires further development for reliable field application. Temperature compensation, changing moisture conditions, electrode specificity and interferences, electrode sensitivity, electrode poisoning, and long-term stable calibration are some issues that require further investigation. The micro-heterogeneity of soils is also an issue with electrochemical techniques as they tend to measure only a small volume of soil. As with lysimetry, a large number of electrodes would be required to adequately characterize the larger soil mass. In addition, there remain some theoretical considerations associated with the notions of separate ion activities and the possibility of distortion induced by the charged suspended particles and the gas phase while carrying out measurements in suspensions of soils ('the suspension effect'; Snakin *et al.*, 2001).

## 5.3 Application of soil solution studies to pedogenesis

### 5.3.1 *Deciphering mechanisms of soil formation*

The collection, analysis and interpretation of soil solution chemistry can be a powerful approach for deciphering pedogenic processes and ecosystem disturbance. As an example, Ugolini and co-workers (Ugolini and Dahlgren, 1987; Ugolini *et al.*, 1988) used lysimetry to examine the soil-forming processes of podzolization and andosolization in young volcanic soils where allophane and imogolite dominated in B horizons. It is often difficult to distinguish between podzolization and andosolization during the earlier stages of soil formation. In particular, the controversy surrounding the origin of allophane/imogolite in Spodosol and Andisol B horizons (translocation of proto-imogolite sols versus *in situ* formation) was addressed.

Soil solution studies clearly demonstrated the existence of two chemical weathering compartments in the soil profile of Spodosols. The upper compartment (O, E, and Bhs horizons) is dominated by organic acids consisting primarily of fulvic acids with low concentrations of low-molecular-weight organic acids (Fig. 5.3). The pH is depressed ($pH < 4.5$) by organic acids which restrict the dissociation of carbonic acid ($pK_a = 6.3$). The low pH coupled with the metal complexing properties of fulvic acids create an intense weathering environment in which iron and aluminium released by weathering are rendered largely unavailable for mineral synthesis due to complexation by organic acids. Dissolved Al and Fe are transported through the upper profile primarily as metal-organo complexes until they become immobilized at the Bhs/Bs horizon boundary by adsorption/precipitation reactions. Removal of organic acids at the Bhs/Bs boundary results in an increase in the pH ($pH > 5$) of the solution entering the lower chemical weathering compartment.

In the lower weathering compartment, consisting of the Bs, BC and C horizons, carbonic acid becomes the dominant proton donor. The carbonic acid weathering regime is conducive to formation of allophane/imogolite because it is characterized by non-Al-complexing inorganic proton donors, low concentrations of complexing organics, and pH values greater than 5, which favour Al hydrolysis and polymerization. Soil solutions are generally unsaturated with respect to common soil minerals in the organic acid weathering compartment (E and Bhs horizon; Fig. 5.4).

These data indicate that imogolite is highly undersaturated in the organic acid weathering compartment. Thus, proto-imogolite sols would not be expected to form in these horizons, arguing against the possibility that they could participate in transport of Al to the lower soil horizons. In contrast, soil solutions approached or exceeded mineral saturated indices for these minerals in the carbonic acid

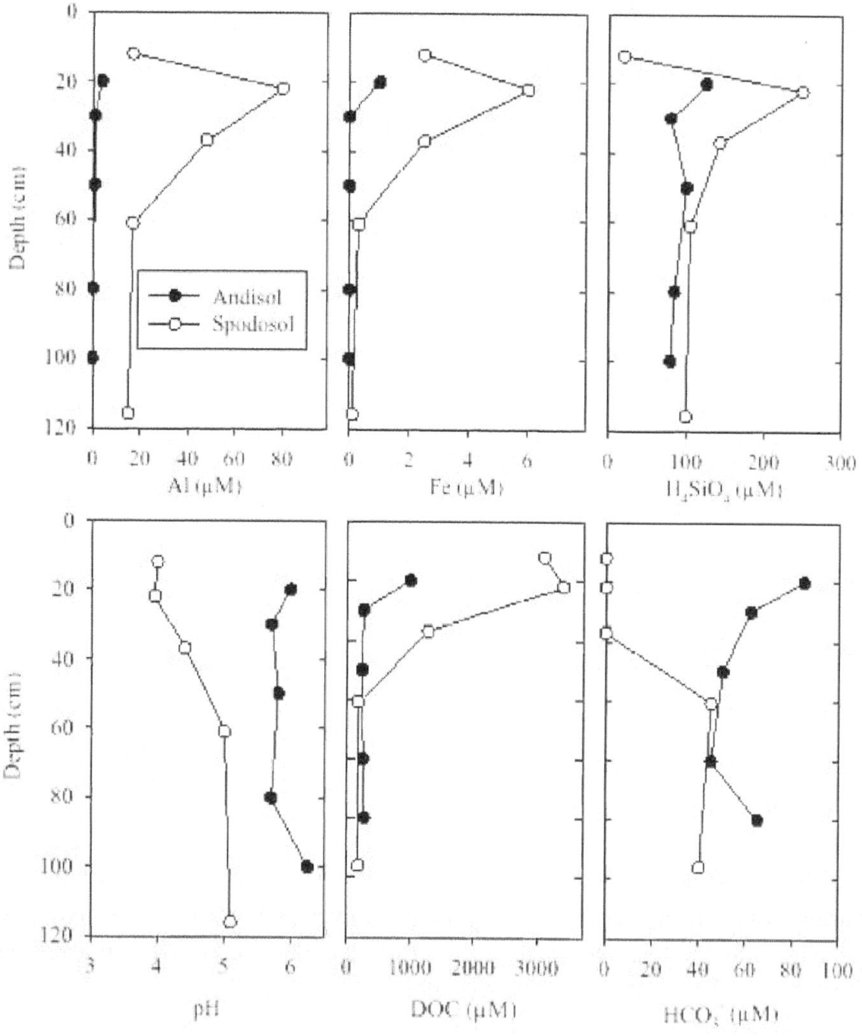


Fig. 5.3. Soil solution signatures for major constituents active in podzolization and andosolization. Data represent solutions exiting the A1, A2, A3, Bw1 and Bw2 horizons of an Andisol, and Oa, E, Bhs, Bs1 and C horizons of a Spodosol, respectively.

weathering compartment. This suggests that conditions for imogolite formation were favourable within the carbonic acid weathering compartment. Thus, it was concluded that imogolite in the lower B horizons formed via *in situ* weathering rather than by translocation of proto-imogolite sols from the upper soil horizons. Further soil solution investigations suggested a simultaneous equilibrium between dissolved $Al^{3+}$ from hydroxy-Al interlayers ⇆ imogolite ⇆ Al-humus complexes ⇆ exchangeable $Al^{3+}$ (Dahlgren *et al.*, 2004). Because of the rapid

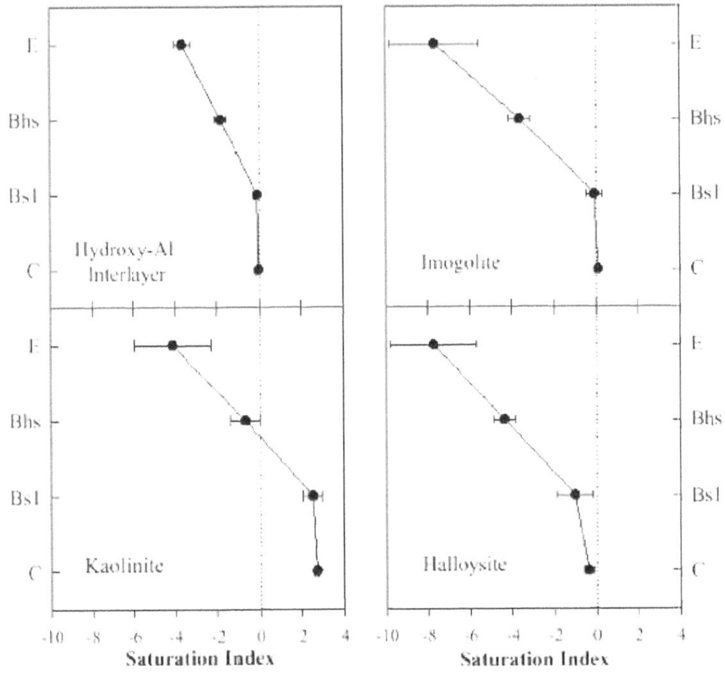


Fig. 5.4. Saturation indices of selected soil minerals calculated for solution exiting the soil horizons of a Spodosol. Positive, negative and zero values suggest that a solution is oversaturated, undersaturated or in equilibrium, respectively, with the mineral phase of interest.

release/retention associated with exchangeable $Al^{3+}$ and Al-humus complexes, apparent equilibrium occurs quickly in the lower B horizons. The hydroxy-Al interlayers are believed to provide the ultimate control of Al activities while imogolite regulates $H_4SiO_4$ activities through its equilibrium with hydroxy-Al interlayers.

In contrast to podzolization, andosolization is characterized by an accumulation of Fe, Al and DOC in A horizons with little translocation of these components into B horizons (Fig. 5.3). The lack of appreciable organic acids results in higher pH values (pH > 5) throughout the entire soil profile. Again, origin of allophane/imogolite in B horizons is the result of *in situ* weathering rather than translocation from the upper soil horizons.

### *5.3.2 Identifying sources of soil acidity and nutrient leaching*

The origin of small (<10 ha) patches of extremely acidic soils (pH < 4.5) in the Klamath Mountains of northern California remained a mystery for many years (Dahlgren, 1994). These acidic regions were devoid of coniferous vegetation,

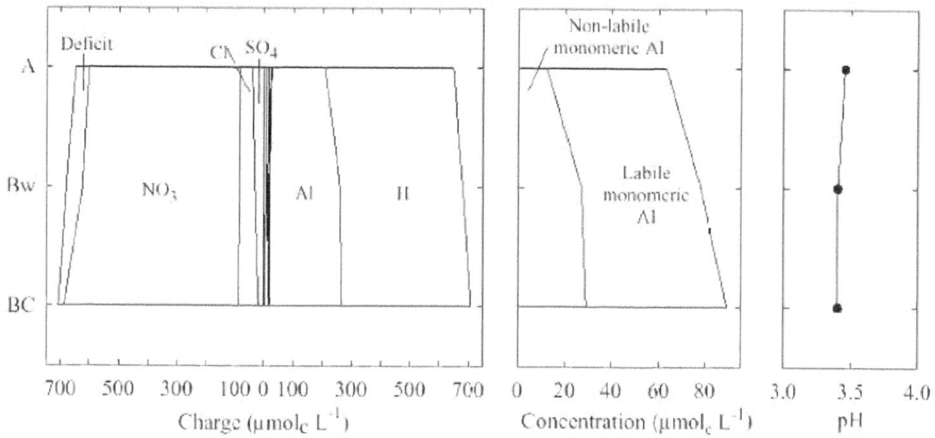


Fig. 5.5. Soil solution chemistry collected by centrifugation from strongly acidic soils devoid of coniferous vegetation. The width of each compartment is in proportion to the ion's charge contribution. Non-labile aluminium is primarily organic complexes while labile aluminium is primarily $Al^{3+}$ and its inorganic complexes with $F^-$ and $SO_4{}^{2-}$.

although surrounded by healthy coniferous forest. The source of the extreme acidity was revealed by analysis of soil solutions collected by centrifugation. Soil solutions showed that anions were dominated by $NO_3{}^-$ (>87%) and cations by $H^+$ and $Al^{3+}$ (Fig. 5.5). Extracted soil solutions had a pH ranging between 3.4 and 3.5. This soil solution signature indicates that soil acidity was largely derived from nitric acid and that organic acids (as estimated from the anion charge deficit) were only a minor contributor. The consequent acidity mobilizes potentially toxic levels of $Al^{3+}$ and causes intense leaching of nutrient cations (e.g. $Ca^{2+}$, $Mg^{2+}$, $K^+$). Having identified nitric acid as the source of the extreme soil acidity, the solid phase was examined to determine the origin of the nitrogen. The source of nitrogen was attributed to the mica schist parent material that contains 2700 mg N $kg^{-1}$ rock, believed to be in the form of $NH_4$-substituted mica. Upon nitrification of $NH_4{}^+$, nitric acid was produced resulting in soil acidification, solubilization of $Al^{3+}$ and base cation leaching with $NO_3{}^-$.

### *5.3.3 Ecosystem perturbations*

Soil solution studies have been used to follow the response of soil ecosystems to perturbations, such as volcanic ash deposition, forest clear-cutting and acidic deposition. Here we provide an example of volcanic ash deposition from the 1980 eruption of Mt St Helens on a subalpine forest ecosystem in the Cascade Range of western Washington, USA. Soil solution composition provided an instantaneous

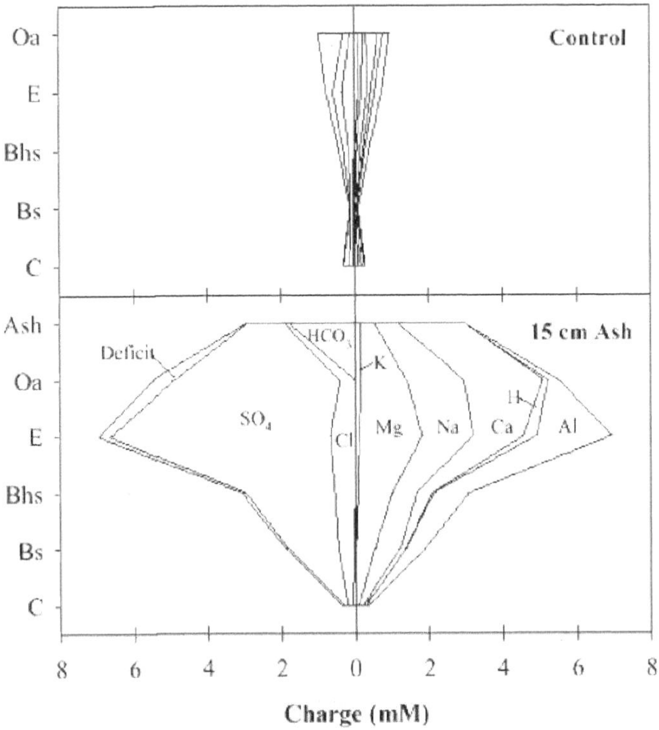


Fig. 5.6. Soil solution charge balance for soils receiving 15 cm of Mt St Helens ash and controls receiving no ash. Soil solutions represent the first month of leaching after addition of fresh ash to the surface of a Spodosol. The width of each compartment is in proportion to the ion's charge contribution. The anion deficit was assumed to be the contribution of dissociated organic ligands.

record of the response of soil processes to the added volcanic ash (Dahlgren and Ugolini, 1989b).

Soluble salts, composed of basic cations ($Ca^{2+} \approx Na^{+} > Mg^{2+} > K^{+}$) and strong acid anions ($SO_4^{2-} > Cl^{-}$), were initially leached from the ash layer (Fig. 5.6). Base cations in the leachate displaced $H^{+}$ and $Al^{3+}$ from the exchange sites in the upper soil horizons of the buried soil, resulting in extremely acidic (pH < 4) solutions entering the B horizons. Protons were neutralized in the B horizons by protonation of variable charge colloids, sulphate sorption and Al dissolution. The high anion sorption by non-crystalline materials in the B horizons resulted in virtually no losses of sulphate from the C horizon. The enhanced $H^{+}$ and $Al^{3+}$ concentrations returned to ambient levels about six months after volcanic ash addition.

Following leaching of the neutral salts from the ash, the majority of the base cations released from the ash were retained on cation exchange sites in the buried organic horizons as the $HCO_3$ anion was lost to protonation ($HCO_3^- + H^+ = H_2CO_3$) (Fig. 5.7). Thus, release of nutrients from intermittent air-fall tephra additions provides an important source of slowly available nutrients that helps to sustain forest productivity in these strongly acidic soils.

### *5.3.4 Weathering rates*

The initial stages of chemical weathering in air-fall tephra deposits were examined by quantifying elemental fluxes from the tephra layer over a four-year period. The primary proton donor in the fresh tephra layer was carbonic acid. Solutions leached from the tephra layer indicated incongruent dissolution resulting in formation of a cation-depleted, silica-rich leached layer on the glass and mineral surfaces. Products of incongruent dissolution reactions consist of both dissolved species and an altered solid phase. Due to the near neutral pH values and low concentrations of complexing organic ligands, aluminium and iron were relatively insoluble and accumulated in the tephra layer rather than being leached. Field weathering rates, calculated as yearly averages, ranged between $10^{-18}$ and $10^{-17}$ mol cm$^{-2}$ s$^{-1}$ for sodium, calcium and silicon. While these rates are 1–3 orders of magnitude less than those determined for glass and plagioclase minerals in laboratory dissolution experiments, they are remarkably

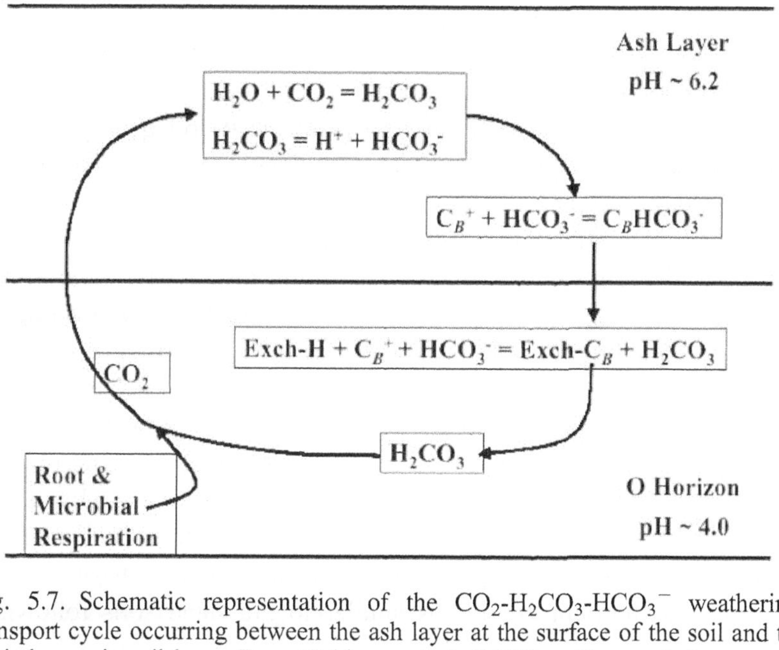


Fig. 5.7. Schematic representation of the $CO_2$-$H_2CO_3$-$HCO_3^-$ weathering/transport cycle occurring between the ash layer at the surface of the soil and the buried organic soil layer. From Dahlgren *et al.* (1999), with permission.

high given that soil temperatures were near 0 °C for nearly six months of the year when a snowpack persists.

The majority of the carbonic acid in the tephra layer originates from upward transport of $CO_2$ from the organic-rich soil horizons (~15 cm of O horizon) of the buried soil. Elevated concentrations of $CO_2$ beneath the tephra layer originate from biological respiration (e.g. roots and micro-organisms) and from protonation of $HCO_3^-$ leaching from the overlying tephra layer. Cations released by weathering in the tephra layer (pH $\approx$ 6–7) leach downward with bicarbonate to the acidic organic horizons (pH $\approx$ 4) where $H_2CO_3$ re-equilibrates with the high $pCO_2$ (Fig. 5.7). At pH 4, $H_2CO_3$ decomposes to $CO_2$ and $H_2O$ and the gaseous $CO_2$ diffuses upwards to take part in another cycle of weathering and transport. This example of weathering in air-fall tephra deposits demonstrates a unique weathering pathway in which the buried organic-rich soil pumps protons upward to the tephra layer, which acts as an alkaline trap for $CO_2$. Thus, the overall process of tephra weathering appears to be controlled to a large extent by solute/gas transport of the carbonate system ($CO_2$-$H_2CO_3$-$HCO_3^-$).

### 5.3.5 Vegetation effects

The influence of oak trees on soil quality and fertility was examined in oak woodlands of northern California using soil solution chemistry. In California oak woodlands, scattered trees create a mosaic of open grasslands and oak/understorey plant communities. We compared soil solution composition beneath the oak canopy with adjacent grassland soils not influenced by the oak canopy, and changes in soil solution composition in the year following removal of oak trees. Oak trees were shown to create islands of enhanced soil quality and fertility beneath their canopies (Dahlgren *et al.*, 1997c). Concentrations of plant nutrients, such as potassium, nitrate and phosphate, were much higher in soil solutions beneath the oak canopy compared with grassland soils (Fig. 5.8).

In contrast, the non-essential nutrient, sodium, displayed similar concentrations for oak canopy and grassland soils indicating that nutrient cycling by oak trees was an important factor enhancing soluble nutrient concentrations beneath the oak canopy. Increased cycling of base cations by oak trees resulted in higher base saturation and pH (0.5 to 1 unit) in soils beneath the oak canopy. The ability of the oaks to create islands of enhanced soil quality and fertility results primarily from additions of organic matter and nutrient cycling. Oak tree removal resulted in an immediate shift in soluble nutrient concentrations towards that of the grassland soils (Fig. 5.8). These data indicate that islands of soil fertility quickly revert to nutrient conditions similar to grassland soils following tree removal.

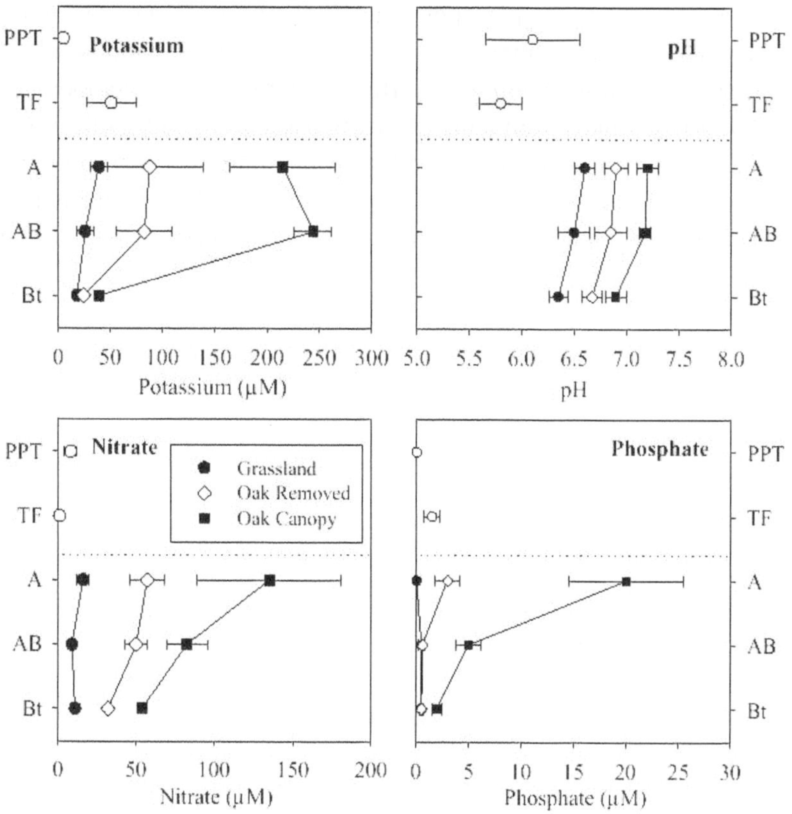


Fig. 5.8. Concentrations (means ± standard deviation) for selected constituents from precipitation (PPT), canopy throughfall (TF) and soil solutions from A, AB and Bt horizons in a California oak woodland. Soil solutions were collected from soils beneath the oak canopy, from adjacent grasslands, and from sites where oaks were removed in the year before collection.

## 5.4 Conclusions

Water plays many critical functions in the soil ecosystem including roles as a transporting agent, chemical solvent, readily available nutrient pool, and as a source of water for metabolic activities of soil biota and vegetation. The soil solution is an aqueous solution composed of liquid water and dissolved solids and gases. Because changes in the composition of the soil solution depend on interactions with the solid and gaseous phases, the soil solution is dynamic and sensitive to changes occurring in the soil ecosystem.

The examples provided in this chapter demonstrate a few applications of soil solution studies to pedogenesis. While care must be taken in interpretation of soil solution chemistry within the context of methodological limitations, this approach does provide a potentially powerful tool for distinguishing current soil processes and soil ecosystem response to disturbance. Future technologies in electrochemical techniques may further enhance our understanding of soil processes by providing a technique that offers a more sensitive temporal analysis while at the same time minimizing artefacts associated with collection of soil solutions. Long-term monitoring of soil solution chemistry may provide a sensitive method for detecting ecosystem response to gradual shifts in environmental forcing factors, such as acidic deposition or global climate change.

# 6

# Soil phases: the gaseous phase

*Andrey V. Smagin*

Soil acts as a global source, sink and reservoir of gaseous substances contributing to the control of the composition of the atmosphere and affecting the climate conditions of the planet. Despite the importance of the soil gas phase, the study of the processes of production, consumption, and transport of gases in soils still suffers from many uncertainties, particularly methods of measurement. Most frequently, soil gases are monitored as fluxes, or net gas flows at the surface, from which the soil capacity to adsorb or release some gaseous substances is assessed. Surface flux measurements ignore the processes that operate in the soil, and thus many questions remain regarding the mechanisms controlling the fluxes. Another shortcoming is that often the gaseous phase of soil is studied separately from the liquid and solid phases, resulting in serious errors in quantitative evaluation of the soil's capacity to produce, absorb, release and accumulate gaseous substances. The problem concerning the mechanisms and forms of gas transport in such a complicated porous medium also remains open and this restrains the modelling of the gaseous phase dynamics, its vertical and lateral distribution in different types of soils. The spatial and temporal irregularities in gas dynamics require changes to the standard approach of the field studies being carried out only in warm vegetated seasons. Some of these problems will be discussed in this chapter as related to quantitative analyses of the gaseous phase composition and its state in the soil.

## 6.1 Gaseous components of soil

Soil atmosphere is a mixture of gases and vapours, filling water-free pore space and interacting with soil liquid and solid phases. The composition of the soil

*Soils: Basic Concepts and Future Challenges*, ed. Giacomo Certini and Riccardo Scalenghe.
Published by Cambridge University Press. 

gaseous phase is very dynamic and diverse, containing both natural components and pollutants. Inorganic gases ($N_2$, $O_2$, $CO_2$, etc.), vapours ($H_2O$, $NH_4$, etc.), and volatile organic components (carbohydrates, organic acids, alcohols, oils, pesticides, etc.) are commonly present. Some of these components are relatively well known and studied, but a large fraction of the gaseous compounds remains poorly investigated and their role and behaviour in the soil is unclear. In particular, trace gases and vapours, which have concentrations in the soil air of 10–100 ppm (parts per million), are particularly poorly studied. However, modern techniques such as gas chromatography and infrared spectroscopy allow solutions to this problem (Smagin, 2003). For example, high-performance gas chromatography on capillary columns enabled the identification of more than 50 natural volatile organic compounds, such as butane, propane, pentane, octane, their chlorine and methyl derivatives, cyclohexane, methanol, ethanol, higher alcohols, isoprene, benzene, etc. (Fukui and Doskey, 1996). These gaseous components are first of all the products of soil metabolism, but some of them supposedly carry out the functions of regulation and signals (information transfer) in the soil as a bio-inert system. In this case a study of the natural volatile organic compounds will be one of the more important tasks in future for soil science, since it enables understanding of complicated mechanisms of the soil and ecosystem organization and management.

Analysis of the soil atmosphere ($X$) expresses the amount of the gas of interest ($V_g$) as a ratio to the sampled volume ($V_t$): $X\% = 100V_g/V_t$, (0.01% = 100 ppm). However, because the gaseous phase volume strongly depends on the atmospheric pressure ($P$) and temperature ($T$), it is more accurate to report the mass concentration of the gas:

$$C_g\,(\mathrm{g\ m^{-3}}) = X\%\ PM/(100RT) \tag{6.1}$$

where $P$ is the atmospheric pressure (Pa), $M$ is the molar mass (g m$^{-3}$), $R$ is the universal gas constant (8.31 J mol$^{-1}$ K$^{-1}$), $T$ is the absolute temperature (K).

For example 1% $CO_2$ under normal atmospheric pressure (101.3 kPa) and 20 °C (293 K), according to Eq. (6.1), gives:

$$C_g = 1 \cdot 101.3 \cdot 1000 \cdot 44/(100 \cdot 8.31 \cdot 293) = 18.3\ \mathrm{g\ m^{-3}}.$$

For vapour, which has a critical content or saturated concentration ($C_g^0$), it is convenient to use a relative concentration ($C_g/C_g^0$). For example, the relative humidity of water (RH $= C_g/C_g^0$) theoretically varies from 0 to 1.

A variety of methods have been developed to sample the gaseous phase of soil. The simplest way is direct probe sampling using thin tubes or needles placed at different depths of the soil and samples collected by vacuum pump or syringe. More complicated techniques require the application of stationary sample devices (chambers). The membrane diffusion technique is based on the capacity of polymeric membranes to pass air and dissolved gaseous components into an isolated chamber. Alternative techniques convenient for wetlands have superseded probe sampling. In these techniques, a small chamber is filled with an inert gas or atmospheric air and is then placed in wet soil until the concentrations of gases within the encapsulated chamber space (air bell) and outer soil air reach equilibrium. Then the gases are collected from the chamber, and the chamber is filled by a new aliquot of inert gas.

Biophysical and biochemical processes of gas production, consumption and mass-transfer in the soil determine the composition of the soil gaseous phase. Due to intensive biological activity, the composition of air in soil differs strongly from that of the atmosphere. Dry atmospheric air consists of 78.1% $N_2$, 20.9% $O_2$, 0.95% Ar, and 0.07% other gases, including 0.036% $CO_2$. The variation in $N_2$ and Ar contents in soil air is relatively small (77–82% for the sum of these gases), however the concentrations of $CO_2$ and $O_2$ can vary from 0.03% to 20% for $CO_2$ and 0.05% to 20.5% for $O_2$. Typical gas profiles for different soils and wet porous media are shown in Fig. 6.1. The content of C-gases ($CO_2$ + $CH_4$) regularly increases with depth and reaches maximum values (4–6%, up to 12%) at depths of 30–80 cm in hydromorphic soils (Smagin *et al.*, 2000). In contrast, $O_2$ shows maximum concentrations in topsoil layers since the atmospheric air is the source of this gas. The concentration of $O_2$ gradually decreases with depth to levels of 15–18% in well-drained soils and 3–6% in wetlands. The relative humidity (RH) or water vapour concentration in the soil is mostly near 1, and only in dry arid soils do RH values reach 0.7–0.5 or less.

## 6.2 Sources, sinks and transport of gases in the soil

Consumption and generation of gases are controlled by the biological activity of soil as well as by abiotic physicochemical processes. Among abiotic processes dissolution and sorption strongly influence the state of gaseous components in the soil. The simplest equilibrated thermodynamic approach allows estimation of the distribution of gas between liquid, soil and gaseous phases by the following models:

$$C_\mathrm{l} = a_\mathrm{s} C_\mathrm{g} \tag{6.2a}$$

$$C_\mathrm{s} = K_\mathrm{h} C_\mathrm{g} \tag{6.2b}$$

where: $C_{\mathrm{g,l,s}}$ ($\mathrm{g\,m^{-3}}$) are the concentrations of the gases in the corresponding phases, $a_\mathrm{s}$ the specific solubility, $K_\mathrm{h}$ the Henry's constant of gases sorption, which usually varies from 1 to 5 in mineral dry soils and reaches 10–60 in organic

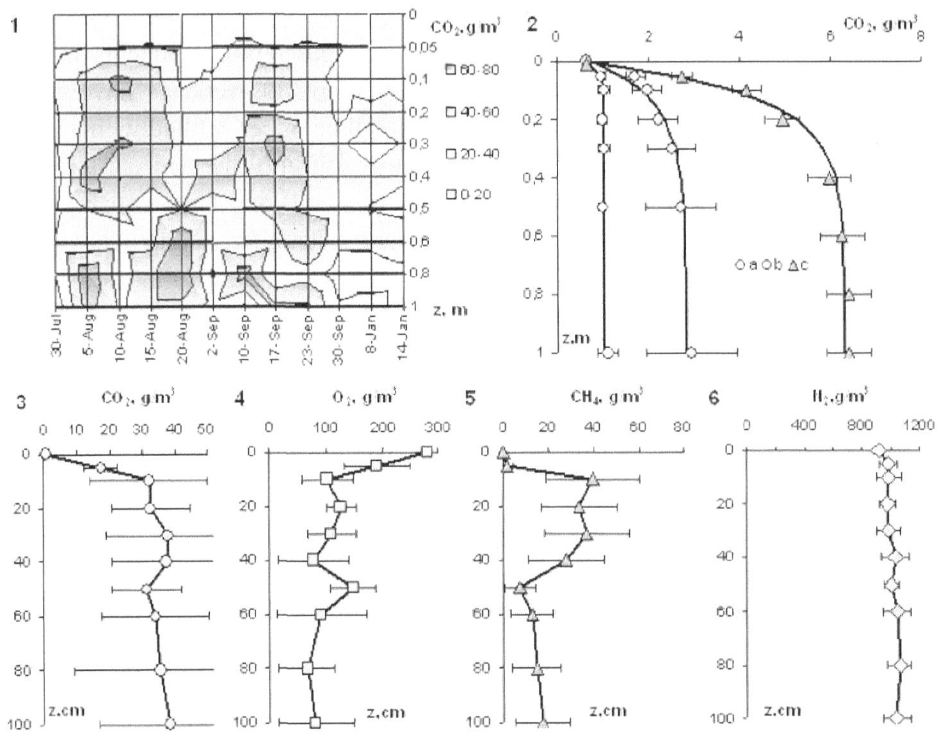

Fig. 6.1. Measurement of gases distribution in different soils: (1) $CO_2$ dynamics in soddy-alluvial soil (Moscow region, in 2000); (2) average profiles of $CO_2$ in sandy forest soddy-podzolic soil of the Moscow region (a – winter 1990, b – summer 1989), and in loamy arable soddy-podzolic soil (c – summer 1997); (3–6) average profiles of various gases in Bakcharsky bog (West Siberia, in 2000).

samples such as humus and peat. Some of the gases are dissolved in the water in appreciable amounts ($SO_2$, $NH_3$, $H_2S$, $Cl_2$, $CO_2$), but most of them have a small solubility ($O_2$, $N_2$, $CH_4$, CO, NO, $H_2$). The solubility of many gases in deionised water is well known and easy to obtain from specialized handbooks as a function of temperature. Solubility of gases strongly depends on the composition of the soil liquid phase. For $CO_2$ the dependence of solubility on pH of the water is:

$$a_s = a_t(1 + K_1/10^{pH} + K_1K_2/10^{2\,pH}) \tag{6.3}$$

where $a_t$ is the tabulated solubility (in deionised water), $K_1$ and $K_2$ are the tabulated constants of dissociation of carbonic acid at first ($HCO_3^-$) and second ($CO_3^{2-}$) stages in deionised water. According to this equation, the specific solubility of $CO_2$ in acid soils (pH = 5) is the same as the tabulated value, but for alkaline soils (pH = 8–9) $a_s$ exceeds $a_t$ many times. For example, the solubility of $CO_2$ in deionised water (pH = 6.5 at 20 °C) is 0.88, while it is 38 for pH = 8 $\{a_s = 0.88(1 + 0.42 \cdot 10^{-6}/10^8 + 0.42 \cdot 10^{-6} \cdot 0.40 \cdot 10^{-10}/10^{16}) = 37.99\}$. Therefore, at pH = 8 and under equilibrated conditions the $CO_2$ concentration in the liquid phase is approximately 38 times higher than in the gaseous phase.

The application of the equations (6.2) and (6.3) is possible if the time of establishment of the interphases equilibrium is relatively small as compared with the dynamics of gases in soil. Otherwise, more complicated models of gases dissolution and sorption should be applied. An analytical solution of these models takes the form:

$$C(t) = C_e + (C_0 - C_e)\exp(-kt) \tag{6.4}$$

where: $C(t)$ is a variable from time concentration of the gases at liquid or solid phases, $C_e$ its equilibrated value, which is given by Eq. (6.2), $k$ a kinetic constant. Experimental evaluation shows that $k$ in separated liquid phase as well as in wet soils varies from 0.15 to 0.35 $h^{-1}$. In dry porous medium the value $k$ ranges from 2–5 $h^{-1}$ (mineral soils) to 10 $h^{-1}$ (peat). For example, it may be necessary to estimate how much $CO_2$ will be dissolved into the fresh rainwater (pH = 6.5, $C_0 = 0.03\% = 0.54$ g $m^{-3}$), which has penetrated the soil during 2 hours, if the concentration of $CO_2$ in the soil air is $C_g = 20$ g $m^{-3}$, temperature is 15 °C, and kinetic constant is $k = 0.2$ $h^{-1}$. From equations (6.2) and (6.3) it is easy to calculate the specific solubility ($a_s$) and after that to obtain the equilibrated concentration of $CO_2$ in the soil solution: $C_e = a_s C_g = 44.2$ g $m^{-3}$. Then the current concentration of $CO_2$ in the soil solution can be evaluated by Eq. (6.4) as: $C(t) = 44.2 + (0.54 - 44.2)\exp(-0.2*2) = 14.9$ g $m^{-3}$. This is only a third of the value at the equilibrium (44.2 g $m^{-3}$) obtained by Eq. (6.2).

The complete description of gas interactions in wet porous medium is based on a combination of the kinetic model of dissolution of gases with sorption on solid particles following mass-transfer through the water films (McCoy and Rolston, 1992).

### 6.2.1 Biological consumption and generation of gaseous components

Biological consumption and generation of gaseous components of soil generally predominate over physical (sorption, condensation and dissolution) and physicochemical (chemisorption and reactions in the bulk) mechanisms; in fact, soil sterilization results in an abrupt decrease in gas exchange by the medium. However, neglecting abiotic processes results in serious errors in the interpretation of experimental data about the soil capacity to adsorb, accumulate and emit gaseous substances.

In the case of the laboratory experiment, the specific rate of gas emission ($U$) is calculated from the net change of concentration ($\Delta C_g$) in the vessel airspace:

$$U = \Delta C_g / (\Delta t \cdot m_s) \tag{6.5}$$

where $m_s$ is the weight of dry soil. The gas produced from the sample during the incubation is partly dissolved in the soil moisture and adsorbed by the solid phase. Therefore, the real emission rate ($U^{real}$) is higher than the gas increment measured. On the contrary, in field experiments, the $CO_2$ production may be overestimated, because a part of the gas adsorbed and dissolved in the soil passes into the airspace of the vessel. The discrepancies with the real emission rate can be considerable in both cases (Table 6.1). In the laboratory, the correction is made experimentally upon degassing (desorption) of the sample by rapid heating by microwave or by evacuation.

The biological consumption and production of gases are strongly controlled by such thermodynamic factors as temperature ($T$) and water content ($W$) of the soil. It is intuitive that the highest exchanges ($U_{max}$) occur at the optimal temperature and moisture for biological activity (usually $T_m \approx 25$–30 °C and $W_m \approx 0.6$–0.8 $W_s$, where $W_s$ is the water content at saturation). For organic matter decomposition and $CO_2$ emission the following equations are suggested:

$$U_{(T,W)} = f_{(T)} \cdot f_{(W)} \cdot U_{max},$$

$$f_{(T)} = Q_{10}^{\,0.1(T-Tm)}, \quad f_{(W)} = (W/W_m)^a\{(1-W)/(1-W_m)\}^b \tag{6.6}$$

where $W_m$ = a/(a+b) is the moisture at maximum biological activity ($U_{max}$), a, b are empirical constants, $Q_{10} \approx 2$ is the *temperature coefficient*, definable as the change in the rate of a process as a result of increasing the temperature by 10 °C.

The models of biogenic production and consumption of the gases must thus include biophysical and biochemical kinetic mechanisms, like the simple linear kinetic models or the non-linear Michaelis–Menten's dependency of microbial growth and rate of fermentative reactions with the substrate concentrations (Cho *et al.*, 1997; Kruse *et al.*, 1996; Rudolf *et al.*, 1996). Another approach is through the application of kinetic models of organic matter transformation in the soil in order to forecast emission of gases. For example, the kinetic constant of carbon decomposition in peat compost is $k = 0.5$ y$^{-1}$ and the initial store of organic carbon is $C_0 = 5$ kg m$^{-2}$. Here, using a well-known linear model for organic matter decay in the soil ($C(t) = C_0 \exp(-kt)$), one could estimate annual emission of $CO_2$ from the soil as $(C_0 - C(t))44/12 = (5 - 5 \cdot \exp(-0.5 \cdot 1))44/12 = 7.2$ kg $CO_2$ m$^{-2}$ y$^{-1}$ (44 and 12 are the molar masses of $CO_2$ and carbon, respectively). Note that the value obtained exceeds the contribution of automobile exhaust to total $CO_2$ emission in urban areas (that is $10^5$ tonnes per $10^5$ ha, or

Table 6.1. *Emission of $CO_2$ measured (U) and real ($U^{real}$) in four soils of the Moscow region, Russia (in mg $kg^{-1}$ $h^{-1}$)*

| Layer (cm) | $U$ | $U^{real}$ |
|---|---|---|
| Soddy-podzolic forest soil (laboratory experiment) | | |
| 0–10 | 1.30 ± 0.28 | 2.21 ± 0.43 |
| 10–20 | 0.57 ± 0.17 | 0.86 ± 0.25 |
| 20–30 | 0.31 ± 0.05 | 0.49 ± 0.08 |
| 30–40 | 0.27 ± 0.05 | 0.35 ± 0.05 |
| 40–50 | 0.23 ± 0.09 | 0.31 ± 0.11 |
| Soddy-podzolic arable soil (field experiment) | | |
| 0–10 | 3.12 ± 0.59 | 2.35 ± 0.55 |
| 10–20 | 1.33 ± 0.41 | 0.61 ± 0.21 |
| 20–40 | 1.27 ± 0.22 | 0.52 ± 0.16 |
| 40–60 | 0.41 ± 0.07 | 0.20 ± 0.03 |
| 60–100 | 0.20 ± 0.04 | 0.15 ± 0.04 |
| Alluvial-bog silty-peat-gley forest soil (field experiment) | | |
| 0–10 | 32.33 ± 8.93 | 8.30 ± 2.39 |
| 10–20 | 30.04 ± 9.12 | 3.95 ± 1.20 |
| 20–40 | 24.21 ± 6.57 | 3.79 ± 1.25 |
| 40–60 | 14.13 ± 4.28 | 3.57 ± 1.02 |
| 60–100 | 9.51 ± 2.87 | 1.32 ± 0.41 |
| Eutrophic humus-peat arable soil (laboratory experiment) | | |
| Surface | 4.8 ± 0.37 | 10.6 ± 0.49 |
| 0–5 | 1.06 ± 0.15 | 4.86 ± 1.15 |
| 5–10 | 2.26 ± 0.15 | 7.00 ± 0.25 |
| 10–20 | 1.31 ± 0.06 | 3.68 ± 0.35 |
| 20–30 | 1.46 ± 0.37 | 4.13 ± 0.68 |
| 40–50 | 0.56 ± 0.38 | 2.46 ± 0.73 |
| 60–70 | 0.17 ± 0.05 | 0.63 ± 0.35 |

*Source:* Smagin (2000).

0.1 kg $m^{-2}$ in Moscow) by approximately 70 times! More complicated non-linear models of organic matter dynamics and gases production in bio-inert systems with vibrations and trigger regimes provide more plausible data (Smagin, 1994, 1999).

### *6.2.2 Transport of gases and vapours in the soil*

The problem of mass transfer of gases in soil remains open. Conventional concepts assign the major role of gas transport to molecular mechanisms, in particular isothermal diffusion, a mass transfer proportional to a concentration gradient ($dC/dz$):

$$q = -D dC/dz \tag{6.7}$$

where $q$ is the gas flux and $D$ the diffusion coefficient. This coefficient or its dimensionless analogue – relative gas diffusivity ($D/D_0$) – depends on air-filled porosity ($\varepsilon_a$). Linear, polynomials, or exponential models are used for approximating this dependence (Campbell, 1985). The simplest Penman equation for macroporous media has the form $D/D_0 = 0.66\varepsilon_a$. The diffusion coefficient of gaseous components in the atmosphere ($D_0$) is on the order of $10^{-5}$ m$^2$ s$^{-1}$ (1.77, 1.39, and $2.13 \cdot 10^{-5}$ m$^2$ s$^{-1}$ for $O_2$, $CO_2$, and $H_2O$ under standard conditions $T = 273$ K, $P = 101.3$ kPa) and may be corrected with temperature and air pressure: $D_0 = D_0^{st}(T/273)^n(101.3/P)$, $n = 2$ for $O_2$, $H_2O$, and 1.75 for $CO_2$ (Campbell, 1985). In water the gas diffusion strongly decreases ($D_l = 2 \cdot 10^{-9}$ m$^2$ s$^{-1}$ for $O_2$ and $CO_2$) and water-saturated soil layers are considered to show gas mass transfer near zero (Campbell, 1985). However, experiments show that in some water-saturated soils ($\varepsilon_a \rightarrow 0$), the diffusion coefficient can be 3–10 times higher than in the pure water probably because of the surface diffusion.

Temperature gradients in the soil ($dT/dz$) lead to a combined thermodiffusion molecular mass transfer:

$$q = -D \cdot dC/dz - D \cdot k_T \cdot (dT/dz)/T \tag{6.8}$$

where $k_T$ is a thermodiffusion constant proportional to gas concentration. Theoretically, the contribution of the temperature gradient is lower than that of the concentration gradient, and can be neglected for gas mass transfer in soils, except for gases that have low gradients ($dC/dz \rightarrow 0$) near the soil surface ($N_2$, $O_2$) and for water vapour, where the thermodiffusion can be 3–4 times higher than the simple diffusion flux if the temperature and concentration gradients are co-directional. Other reasons for enhanced thermal diffusion can be the macroscopic movement due to the convective fluxes of soil air in the gravity field (natural convection) and the thermal circulation of air (thermal sliding) in capillary-porous media, which are not well understood in soil science (Smagin, 2000).

Convective macroscopic mass transfer occurs both in the soil air and moisture, with viscous hydrodynamic media capable of moving in the gravity field (natural convection) or under the effect of external pressure differentials (induced convection). Natural convection in the gaseous phase can involve both the downward movement (gravitation flow) of denser air (cold or enriched in heavy gases) and the upward flows of rarefied air (heated or enriched in light components). The reasons for the difference in density of the soil air and the atmosphere – solar radiation (heating) and biogenic activity (composition changes) – are almost constantly present in the soil. Natural convection can be an essential factor of mass transfer, because its rate is comparable to diffusion and its duration is unlimited. Theoretical analysis shows that for evaporation from surfaces with a characteristic length of about 1 m, the convective currents can exceed diffusion more than 100 times. The same theoretical result is obtained for $CO_2$ gravitation flow in macropores. Another well-known example of natural convection is the upward transport of gases by bubbles to the surface of wetlands and paddy soils. Its contribution to the total mass transfer through the flooding layer is estimated at 10–30%. The biophysical mechanism of bubbles formation under unsaturated conditions is determined by the microbial kinetics of gases production, which lead to local saturation of the solution near the colony of micro-organisms where bubbles are formed. However, in a porous medium such as a wet soil bubbles can only travel short distances because of restriction by the solid phase.

Induced convection in the gaseous phase (transfer under the effect of a pneumatic pressure gradient) is apparently less significant, because its causes (wind gusts, variations in the groundwater table, and movement of rainfall front) are generally episodic. Convective flux of a gas (vapour) with a concentration $C_g$ in the liquid phase can be described by the following equation:

$$q = -(K/\eta)C_g\mathrm{d}P_g/\mathrm{d}z \qquad (6.9)$$

where $K$ is the air permeability of the soil, $\eta$ the dynamic viscosity of the air, and $\mathrm{d}P_g/\mathrm{d}z$ the pneumatic pressure gradient. $K$ depends on the diffusion coefficient and on air-filled porosity or water content in the soil. According to Kozeni–Karman's theory, the second dependence has the form $K = m \cdot \varepsilon_a{}^n$, where $\varepsilon_a$ is the air-filled porosity and $m$ and $n$ are empirical constants ($n = 0.5 - 2$ for macro-porous media like sands and peat, $n = 2 - 10$ for loamy and clay soils) (Lauren, 1997). In solidly built substrates like soil screens, which cover urban wastes in landfills, air permeability should be low over a large interval of porosity. This guarantees low methane emissions from the landfill until the $\varepsilon_a$ value is less than 30–40%. In natural soils, even minor pressure gradients (1 Pa $m^{-1}$) at the usual levels of air permeability ($K = 10^{-10}$ $m^2$) and air viscosity ($\eta = 10^{-5}$ Pa·s) can induce fluxes of $10^{-5}$ m $s^{-1}$, which are comparable to the diffusion rate of gases in the atmosphere. However, the evaluation of these fluxes in the soil is hampered by the lack of sensors sensitive to minor variations in air pressure.

The physical mechanisms determine a further form of mass transfer of gases: preferential (local) transport. If diffusion was the predominant movement mechanism, gases should be uniformly distributed in the soil. However, when the soil is waterlogged and, hence, the diffusion rate decreases by several orders of magnitude, the movement of gases becomes discontinuous and represents a preferential transfer by the broadest channels in the structure of a porous body, as well as inside and along the surface of plant stems and roots (the so-called 'vascular' transport). In well-drained soils, zones of preferential gas transfer can also form because of irregular moistening, compaction, and the presence of air-bearing channels.

### 6.3 Agroecological evaluation of the soil air

The composition and state of the soil air strongly influence fertility and productivity of the soil. Air and water in the soil should be balanced and a lack of aeration, as in the case of high water content, suppresses the growth of most terrestrial plants and aerobic microflora. In particular, oxygen in the soil is utilized in respiration to provide the energy for roots to grow and to uptake nutrients. Bacteria, fungi and other soil micro-organisms oxidize organic matter aerobically and therefore consume a large amount of oxygen. In order to evaluate aeration of the soil and its gas regime, it is important to know $O_2$ and $CO_2$ concentrations or store in the soil, air-filled porosity ($\varepsilon_a$), water saturation degree ($W/W_s$), relative gas diffusivity ($D/D_0$), air entry potential of soil water ($P_e$) and some other characteristics (Smagin, 2003). For many crops, symptoms of oxygen starvation and root intoxication manifest themselves when the concentration of $O_2$ is below 15–17% and $CO_2$ content in the soil air is higher than 3–4% (Campbell, 1985). In wetlands the content of $O_2$ in the gaseous phase can drop to 3–6% or less; however, a certain amount of $O_2$ is in the dissolved state. For example, in a 10 cm surface peat layer with porosity 90%, air filled porosity 5% and $O_2$ volume 6%, the store of gaseous $O_2$ content would be approximately 400 mg m$^{-2}$; in these conditions the amount of dissolved oxygen in gas equilibrium with the atmosphere reaches 1200 mg m$^{-2}$, that is three times more than that in the soil air. The concentration of gases in the soil is not a general criterion for poor aeration because, the soil being an open system, the demands of roots and micro-organisms can be satisfied even under low $O_2$ content if the velocity of its transport from the atmosphere is equal to the consumption rate. Traditionally, the diffusion of gases in soil is regarded as a general mechanism of aeration. Hence the relative gas diffusivity (ratio between the gas diffusion coefficient in the soil, $D$, and that in the atmosphere, $D_0$) is a widespread index of aeration in the soil. For many crops, $D/D_0$ values lower than 0.06 indicate the first symptoms of $O_2$ starvation in soils. Crop damage and growth limitation as well as anaerobic conditions in the soil arise if $D/D_0$ decreases to 0.02 (Letey, 1985; Smagin, 2003). The range of $D/D_0$ from 0.02 to 0.06 usually corresponds to an air-filled porosity of 6–10%. Taking into account the relationship between air-filled porosity ($\varepsilon_a$), water content ($W$), soil bulk density ($\rho_b$), particle density ($\rho_s$) and water density ($\rho_l$):

$$\varepsilon_a = 1 - \rho_b/\rho_s - W\rho_b/\rho_l, \tag{6.10}$$

it is easy to calculate, for a wide interval of soil bulk density ($1.0 < \rho_b < 1.6$ Mg $m^{-3}$), that values of water saturation degree ($W/W_s$) higher than 0.85–0.90 will be critical for the plantation because the soil is poorly aerated (Smagin, 2003). The water regime analysis of several urban loamy soils in Moscow shows that poor aeration lasts for 10–25% of the field season from May to October, especially for soils with high bulk density (e.g. paths, sports-grounds, lawns). These soils should therefore be managed to provide an adequate amount of $O_2$ for plant and microbial respiration by tillage, drainage, structure optimization and other measures. If the water regime of the soil is controlled by tensiometers, the lack of aeration can be diagnosed simply based on the air entry water potential (potential at which the largest water-filled pores drain). According to Campbell (1985) this value strongly depends on particle-size distribution and structure of the soil and varies from −0.6 to −9.0 J $kg^{-1}$. Aeration of the soil and renewal of soil air control the biota respiration and trace-gases emission. These processes as well as the global gas function of soils will be regarded in the next section.

## 6.4 Gases emissions and global ecological functions of the soil

Emission ($Q$) is the gas flux from the soil surface to the atmosphere, which can be determined as the mass of gaseous substance ($m$) that passes through the cross-section area of the soil surface ($S$) per unit of time ($t$): $Q = m \cdot S^{-1} \cdot t^{-1}$. Methods used for its measurement can be divided into two main groups: chamber methods and micrometeorological methods.

Chambers can be either closed or open (flow-through). They are installed at the soil surface for a short time ($\Delta t = 10$–$20$ min) to allow gas accumulation inside the chamber. After determination of the gas concentration increment ($\Delta C_g$) within the closed chamber or in the stream of air (carried gas) that is drawn through the open chamber, the gas emission can be calculated as:

$$Q = \Delta C_g V/(S\Delta t) = \Delta C_g h/\Delta t \text{ (closed chamber)} \tag{6.11a}$$

$$Q = f\Delta C_g/S \text{ (open chamber)} \tag{6.11b}$$

where $V$, $S$, $h$ are parameters of the chamber (volume, area, height) and $f$ is the air mass flow (mol $s^{-1}$). For longer time intervals the gas losses by diffusion from the closed chamber should be taken into account, otherwise the results can be underestimated significantly. Another reason why $Q$ can be underestimated using the closed chamber method is the use of simple chemical traps, such as soda lime, rather than instrumental $CO_2$-analysers. Use of chemical traps has been shown to have a tendency to underestimate soil-surface $CO_2$ fluxes by 10–100% (Norman *et al.*, 1997).

Micrometeorological methods allow integral evaluation of the fluxes of gases at landscape scales (areas of the order of 10 000 $m^2$), but they require expensive highly sensitive analytical equipment. The concentrations of gases ($C_1$, $C_2$ ...) emitted from the soil are measured in the atmosphere at different heights ($z_1$, $z_2$ ...) and their fluxes from soil are calculated using a simple model of turbulence gas transfer, $Q = D_T \Delta C_g / \Delta z$, where $D_T$ is a turbulent diffusion coefficient which requires knowledge of meteorological data, such as wind speeds, temperature and water vapour gradients, for evaluation (Campbell, 1985). One of the most popular simple micrometeorological methods, the so-called eddy correlation, allows the calculation of gas emission from the experimental data considering the simultaneous fluctuations of the wind speed ($U'$) and the gas concentration ($C_g'$) at a distance around 1–2 m from the soil surface and using the following formula: $Q = \rho_a U' C_g'$, where $\rho_a$ is the air density (Sitaula *et al.*, 1995). A comparison of the different methods for measuring fluxes of $CO_2$ from soil shows that a good correspondence between the methods is not always observed (Norman *et al.*, 1997). Among the reasons for such discrepancies are the scale effects and spatial irregularity of the gaseous emission.

Soil respiration ($CO_2$ emission and $O_2$ consumption) has been studied in detail, but fluxes of trace gases and organic vapours remain poorly investigated. Carbon dioxide is emitted from the soil due to plant and microflora respiration. Usually, in natural well-aerated soils micro-organisms uptake $O_2$ and produce $CO_2$ at double the rate of plant roots. Both autotrophic respiration by roots and heterotrophic respiration by soil organisms are strongly correlated to annual soil surface $CO_2$ flux across a wide range of forest ecosystems (Bond-Lamberty *et al.*, 2004). However, the proportion changes during the field season and depends on the availability of labile organic matter in the soil, especially that derived from fresh plant residue and manure. Moreover it is practically impossible to separate respiration of roots from the respiration of rhizospheric micro-organisms that are utilizing root metabolites. The evaluation of carbon dioxide emission or soil respiration usually varies from 100 to 1000 (2000) mg $CO_2$ $m^{-2}$ $h^{-1}$. There is often a diurnal maximum in the afternoon and a minimum in the morning (4–6 a.m.) due to the dynamics of soil temperature, biological activity and evapotranspiration. However, the presence or absence of roots and a fluctuating soil water content impacts the respiration rate more strongly than any diurnal changes. Seasonal dynamics of soil respiration depend on moisture and temperature regimes of soil. In winter, respiration is often suppressed because of low soil temperatures (Certini *et al.*, 2003), which can even imply low $O_2$ penetration through a frost-bound surface of the soil. However, in some soils the organic matter decay and $CO_2$ emission over the cold period from November to March can reach 30% of the annual values (Smagin, 1994). In the summer, low water content may inhibit root and microbial respiration. The highest respiration rates (1–2 g $CO_2$ $m^{-2}$ $h^{-1}$) have been observed usually when fresh organic matter (litterfall, root exudates, manure, etc.) is incorporated into moist and warm soil.

Let us turn to the examination of a global role for soils in regulating the atmospheric air composition. Fig. 6.2 shows the results of the statistical processing of information concerning the global aspects of the carbon cycle that have appeared in recent decades. The total amount of organic carbon held in the soil, 1480 Pg C, means that soil is of third importance after the lithosphere and oceans. However, the soil carbon reserve is higher than that in plant phytomass by 2–3 times. The mean residence time of the organic matter in soil is long, ranging from several years for the litter layer and the labile humus to several hundred or thousand years for the most stable humus fractions (Scharpenseel, 1993). A loss of stability in the biocenosis-soil system under anthropogenic pressures can result in a catastrophically rapid mineralization of humus accumulated over centuries, as confirmed by multiple observations of the dynamics of organic matter in arable soils or reclaimed peatlands. On a global scale, the loss of organic carbon from soils during the last 130 years is estimated at 40 Pg C (Houghton *et al.*, 1992).

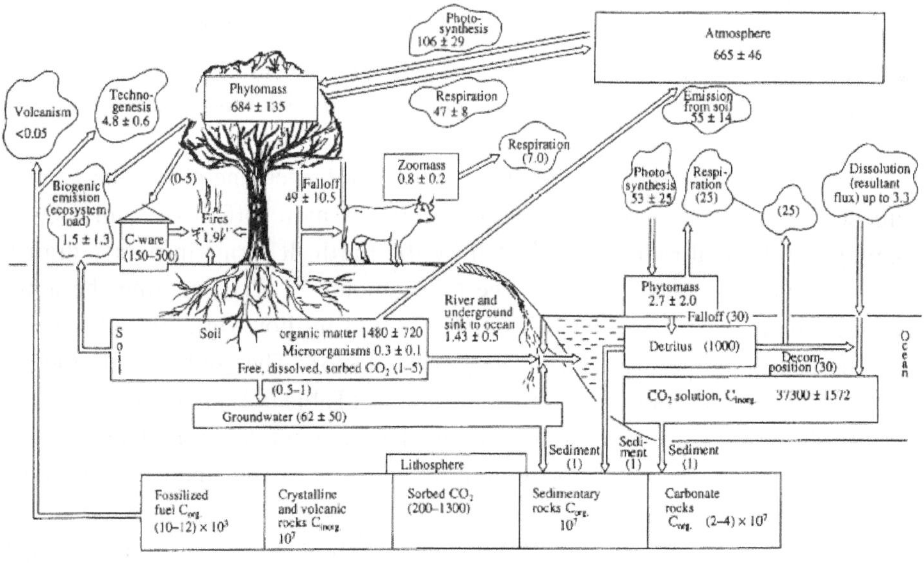


Fig. 6.2. The global carbon cycle. The scheme presents the main carbon reservoirs (the carbon content is expressed in Pg C, 1 Pg = $10^{15}$ g = 1 billion Mg) as well as the fluxes between them (in Pg C $yr^{-1}$). The conventional signs in cloud-shaped contours mark the organic carbon fluxes in the gaseous form ($CO_2$). The data in parentheses refer to poorly studied processes and reservoirs; they were obtained either from a single literature source or by rough calculations.

Table 6.2. *Main sources and sinks of atmospheric methane.*

| Sources/sinks | Mean flux (Tg $CH_4$ $y^{-1}$) |
|---|---|
| *Natural sources* | |
| Wetlands | $115 \pm 60$ |
| Termitaries | $20 \pm 10$ |
| Ocean and fresh water | $20 \pm 10$ |
| *Anthropogenic sources* | |
| Industry, transport | $110 \pm 50$ |
| Rice growing | $60 \pm 40$ |
| Ruminant fermentation | $80 \pm 20$ |
| Animal and human waste | $50 \pm 10$ |
| Refuse dumps | $30 \pm 20$ |
| Biomass and agricultural waste combustion | $45 \pm 15$ |
| *Sinks* | |
| Atmospheric oxidation | $470 \pm 50$ |
| Absorption by soils | $30 \pm 15$ |
| Increase in the atmosphere | $30 \pm 5$ |

*Source:* Smagin, (2000).

It has contributed to more than 25% of the increase in the concentration of C in the atmosphere over the same time period (Smagin, 2000). Currently, the global annual emission of $CO_2$ from soils is evaluated at $55 \pm 14$ Pg C, which is near 30% of total global emission, and exceeds 10 times the technogenic contribution.

Among the gaseous microcomponents of the soil, methane exerts the greatest influence on the greenhouse effect, being much more effective than $CO_2$ in trapping the thermal radiation reflected by the Earth. It forms in poorly drained hydromorphic soils and wetlands. The total amount of $CH_4$ entering the atmosphere annually is 515–560 Tg (1 Tg = $10^{12}$ g = 1 million Mg), 70% of which is from biogenic sources (Bridges and Batjes, 1996). The global emissions of methane by hydromorphic soils and paddy fields (Table 6.2) are estimated at 115 and 60 Tg $CH_4$, respectively, which together amount to more than 30% of the total influx of this gas into the atmosphere and exceed industrial emission (105–115 Tg) by 1.5 times (Houghton *et al.*, 1992).

Methane emission from soils varies significantly in space and time. It ranges from 0.02 to 200 mg $m^{-2}$ $d^{-1}$, although peak emissions up to 1000 mg $m^{-2}$ $d^{-1}$ can occur when there is an excess of organic substrate under optimal temperature conditions (near 30 °C).

In paddy soils and wetlands $CH_4$ emission can vary from 1 to 50 mg $m^{-2}$ $h^{-1}$ ($3 \pm 1$ mg $m^{-2}$ $h^{-1}$ is the median value from log normal distribution of 221 data from different publications), while after heavy rains in periodically moistened soils of the temperate zone it ranges from 0.8 to 26.7 mg $m^{-2}$ $h^{-1}$ (Murase and Kimura, 1996; Toop and Pattey, 1997; Smagin, 2003). The biogenic oxidation of $CH_4$ to $CO_2$ in the soil is relatively weak (from 3.5 to 10% of the global methane sink). However, stable isotope carbon analyses ($^{13}C/^{12}C$) have shown that in some Siberian wetlands from 30% to 80% $CH_4$ can be transformed to $CO_2$ during its movement from methanogenetic horizons to the soil surface because of methanotrophic activity and the large amount of oxygen (60–100 g $m^{-3}$) in locally unsaturated soil conditions.

The potential absorption of carbon monoxide by soils can vary from 2–20 to 100 mg $m^{-2}$ $h^{-1}$, and the global estimate of its consumption by the pedosphere is no less than 450 Tg $y^{-1}$ versus a total influx from natural and anthropogenic sources of about 600 Tg $y^{-1}$. Therefore, soils are considered to be one of the most effective regulators of the content of this dangerous pollutant in the atmosphere (Ingersoll *et al.*, 1974; Seiler, 1974).

Gaseous nitrogen compounds ($N_2$, NO, $NO_2$, $N_2O$, $NH_3$) are formed and transformed in the soil due to nitrogen fixation, ammonification, nitrification and denitrification. On the whole, the rates of these processes are estimated at 0.1–10 mg $m^{-2}$ $h^{-1}$, which are hundreds of times lower than the usual rate of soil respiration. Among the gaseous products of the nitrogen cycle, nitrogen monoxide ($N_2O$) constitutes the highest environmental hazard. The global $N_2O$ emission was estimated 10–20 Tg N $y^{-1}$, 50–60% of which has a pedogenic origin (70–80% considering also the fertilizers) (Smagin, 2000). Croplands are responsible for more than 80% of anthropogenic $N_2O$. Field measurements give the values of specific $N_2O$ emissions from soils in the range 0.003–3(10) mg $m^{-2}$ $h^{-1}$ (Smagin, 2000). It is especially important to consider the seasonal dynamics of gas emission from soils because a major share of nitrogen (up to 30% and more) is frequently lost during the short time periods after heavy rainfalls, application of fresh fertilizers, or soil thawing in spring (Bandibas *et al.*, 1994; Nyborg *et al.*, 1997). The problem of determining the role of soil in the turnover of $N_2O$ and other gaseous nitrogen compounds is still far from being solved and estimates are highly variable (Smith, 1981; Bowden, 1986). The simple calculation of gaseous N-losses following annual mineralization of 77 Tg N fertilizer results in the release 50–60 Tg $y^{-1}$ of gaseous nitrogen compounds into the atmosphere.

Two main gaseous compounds traditionally represent the sulphur cycle: biogenic hydrogen sulphide ($H_2S$) and anthropogenic sulphur dioxide ($SO_2$). The consumption of $SO_2$ by soils is characterized by rates of 20–60 mg $m^{-2}$ $h^{-1}$, or up to 80% of the global anthropogenic emission of this pollutant which is 50 to 55 Tg $y^{-1}$ (Smith, 1981; Smagin, 2000). Similar estimation is given for $H_2S$ emission, which in wet soils can reach significant values (up to 60–100 mg $m^{-2}$ $h^{-1}$).

There are hypotheses about the important role of organic sulphur gases (dimethyl sulphide, $CH_3$ $SCH_3$, carboxyl sulphide, COS, carbon bisulphide, $CS_2$) and sulphur hexafluoride, $SF_6$, although their fluxes are estimated by much lower values than for $SO_2$ and $H_2S$ (Smith, 1981; Smagin, 2000).

The emission of pesticides and other fumigants from the soil is estimated to range from 10 to 100 μg $m^{-2}$ $h^{-1}$, which generally represents 40–60% (up to 90% for methyl bromide) of the added amount (Smagin, 2000). Several mathematical models have been put forward for the reliable prediction of the behaviour of volatile pollutants in the soil and their release into the environment (Boesten and Van der Linden, 1991; McCoy and Rolston, 1992; Amali *et al.*, 1996; Petersen *et al.*, 1996).

# 7

# Soil phases: the living phase

*Oliver Dilly*
*Eva-Maria Pfeiffer*
*Ulrich Irmler*

This chapter provides some concepts of the importance of living components that fall in the size range between 0.2 μm and several millimetres on soil ecosystem functioning. The focus is given to soil micro-organisms due to their key role in elemental cycling. Soil fauna and plant inputs via roots and exudates are considered to include major biotic components in soil functioning.

The rate of soil formation with the initial colonization of bare rocks by macro-, meso- and micro-organisms is affected by mineralogical attributes such as the mineralogical composition, the type of cementing components and the porosity and permeability of the parent material. In addition, the variation of environmental conditions such as temperature, humidity and air pollution affect the dissolution of rocks. Calcareous and especially mica- or clay-cemented sandstones favour biocorrosive, chemo-organotrophic bacteria in comparison with siliceous sandstones (Warscheid *et al.*, 1991). Coarse-grained sandstones generally promote, due to their high permeability for water, a temporary microbial colonization, while fine-grained sandstones favour the long-term establishment of micro-organisms because they preserve humidity. If a sandstone contains appreciable amounts of feldspars and clay minerals like illite and chlorite, optimal conditions for the growth and biocorrosive activity of chemo-organotrophic bacteria emerge. This is explained by both the increased amount of extractable nutrients and the enlargement of the inner surface area. Bacteria and other micro-organisms may be abundant and preserved in altered rock fragments (Agnelli *et al.*, 2001).

Soil fauna plays a minor role in the colonization of stones and clastic sediments. However, animals become abundant when pore- or channel-containing unconsolidated sediments such as dunes and litter are transformed to habitats. For example, the density of microarthropods in dunes varies between 175 000 and

*Soils: Basic Concepts and Future Challenges*, ed. Giacomo Certini and Riccardo Scalenghe.
Published by Cambridge University Press. 

1 400 000 individuals $m^{-2}$, which is several times higher than what is observed in many types of soils (André *et al.*, 1994).

Autotrophic plants colonize soil, use the available resources (both abiotic and biotic, e.g. mycorrhiza), and represent a major external input. As for animals, plant roots are stressed for mechanical impedance by stones and dense clastic sediments (Marschner, 1995).

On a holistic level, the soil represents a pool of genetic information related to the presence, abundance and physiology of its organisms (Table 7.1)

## 7.1 Physiological capabilities of soil organisms

Organisms living in soil express the abundance and activities dependent on environmental factors and human impacts. In general, there is an inverse relationship between the size of organism and its relative abundance and total biomass (Table 7.2). Since the biomass-specific activity increases with decreasing size of the organism, the relative energetic requirement and contribution for compound transformation rate is increased with high abundance of small organisms.

Fungi are regularly assumed to represent the highest biotic component in soil but the combined activity of bacteria may be greater. As a general rule, the total activity of soil organisms is controlled by the efficiency in using the essential elements and nutritional compounds. The use efficiency is controlled by the combination of environmental stressors and substrate accessibility (Dilly, 2005).

Nutrient availability generally limits soil biotic activity. Thus, the growth of soil organisms is usually high close to litter components and roots (Dilly *et al.*, 2001) and after the addition of external readily available nutrients (Dilly, 2005). Throughout the growing season, a major part of the catabolic and anabolic C requirements of soil organisms is ensured by the root turnover and exudates. The faunal physiological capabilities are generally much lower than those of microorganisms. This is reflected in both biomass (Fig. 7.1) and biomass-specific activity (Table 7.2). For two adjacent ecosystems in the Bornhöved Lake District in Germany (54° 06′N, 10° 14′E), the faunal biomass was 0.6 and 6.6% in the agricultural and forest soils, respectively. The agricultural soil had a lower faunal but a higher total biotic proportion in the organic C. In this study, the organic C content was 12 and 24 g $kg^{-1}$ in the A horizon of the agricultural and forest soils, respectively (Dilly and Munch, 1995).

Species make-up of the faunal communities tends to show a sequence of succession in response to changing composition of organic matter. In line with the succession on fresh litter and increasing N accessibility, the species diversity of soil animals increases. Studies in forest soils demonstrated that N released by

Table 7.1. *Genome size of several groups and species*

| Taxonomic rank | Latin name | Common name | Haploid chromosome | Nucleotide base pairs ($\times 10^6$) | Genes ($\times 10^3$) |
|---|---|---|---|---|---|
| *Prokaryotae* | | | | | |
| Archae[a] | 12[a] | archael micro-organisms | – | 1.6–3.0 | 1.5–2.7 |
| Bacteria | 40[a] | bacterial micro-organisms | – | 0.6–7.0 | 0.5–6.6 |
| Bacteria | E. *coli*[b] | no common name | – | 4.6 | 4.3 |
| *Eukaryotae* | | | | | |
| Yeast | *Saccharomyces cerevisiae*[b] | baker's or budding yeast | 16 | 12 | 6 |
| Nematoda | *Caenorhabditis elegans*[b] | nematode | 5/6 | 97 | 19 |
| Insect | *Drosophila melanogaster*[b] | fruit fly | 4 | 180 | 13.6 |
| Annual plant / Angiosperm | *Arabidopsis thaliana*[b] | arabidopsis | 5 | 125 | 25.5 |
| Annual Angiosperm | *Oryza sativa*[b] | rice | 12 | 400 | ? |
| Annual Angiosperm | *Zea mays*[b] | maize | 10 | 2400–3200 | ? |
| Forest tree / Angiosperm | *Eucalyptus*[c] | eucalypts | 11 | 340–580 | ? |
| Forest tree / Gymnosperm | *Pinus*[c] | pines | 12 | 20 000–30 000 | ? |
| Mammals / Rodent | *Mus musculus*[b] | mouse | 20 | 3500 | 21–30 |
| Mammals / Primate | *Homo sapiens*[b] | human | 23 | 3400 | 26–31 |

[a]*Number of species with completely sequenced genomes.*
[b]*Species with completely or almost completely sequenced genome.*
[c]*Data are based on several species.*

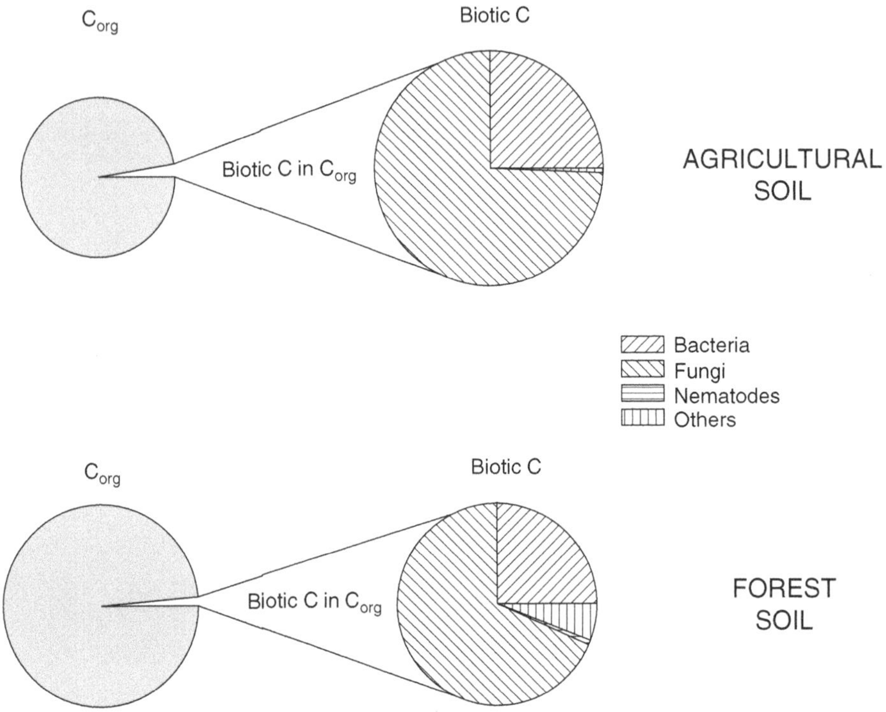


Fig. 7.1. Organic C in total soil mass ($C_{org}$), biotic C in $C_{org}$ ($C_{biotic}\ C_{org}^{-1}$) and composition of biotic C in the A horizon of the agricultural and forest soils of the Bornhöved Lake district. After Irmler *et al.* (1997); Dilly (2005).

soil animals increased during the first year and reached a steady state in the second year (Irmler, 2000). Animals either adjust feeding habits or move towards fresh litter to fulfil their nutritional requirements (Anderson, 1975). Consequently, three groups of the soil fauna can be differentiated: (1) permanent colonizers switching between food components, (2) early colonizers preying on fungi, and (3) late colonizers feeding preferentially on humus particles. The presence and abundance of organisms during colonization of fresh litter is frequently correlated to the C/N ratio and species susceptible to high N availability occur later when litter has a low C/N ratio (Irmler, 2000).

Macrofauna and mesofauna redistribute soil components and are considered as 'engineers' in soil. The movement within the humus horizon (epigeic), in the topsoil (endogeic) or between several soil horizons (anoecic) is characteristic for the different species and is related to soil properties. The bioturbation between soil horizons seems to be particularly important in calcareous habitats. Scheu (1987) determined a faeces production of 3 kg dry soil $m^{-2}\ y^{-1}$ in a calcareous beechwood soil that is equivalent to the reallocation of a 4.3 mm soil layer per year.

Table 7.2. *Biomass and specific activities of organisms abundant in soil*

| | Biomass (kg C $ha^{-1}$) | Specific activity (μl $O_2$ $g^{-1}$ C $h^{-1}$) |
|---|---|---|
| Macrofauna | 50 (equivalent to 1 cow, 10 sheep) | 5[a] |
| Mesofauna | 5 (equivalent to 1 sheep) | 0.6[b] |
| Microfauna | 3 (equivalent to <1 sheep) | 1.4[c] |
| Fungi | 375 (equivalent 75 sheep) | 50[d] |
| Bacteria | 125 (equivalent 25 sheep) | 500[d] |

[a]At optimized conditions and assuming that C content is 50% of organic matter.
[b]For *Collembola*: using the formula of Ryszkowski (1975): μl $O_2 \cdot h^{-1} = 0.357 \cdot 1^{0.813}$.
[c]After Foissner (1987).
[d]Assuming that soil microbial biomass contains 25% bacteria and 75% fungi.

## 7.2 The role of organisms for soil functions

Organisms are relevant for a range of soil functions and their activity is particularly addressed in programmes of the European Commission (2002) dealing with soil erosion, soil contamination and loss in soil organic matter.

The stability of soil structure is affected by the interaction between the soil mineral component, humus and products derived from the activity of the soil biota. Biotic activity promotes the stability of the aggregates by the production of polymeric substances, such as carbohydrates. Autotrophic cyanobacteria colonize the soil surface and thus reduce soil erosion and increase water infiltration. Both soil erosion and water infiltration are particularly relevant for C-poor soils in regions receiving intense rainfall. Under humid conditions, a high biotic biomass and activity can be considered to favour soil stabilization. The biotic interactions with humus components such as leaf litter, fruits, roots and wood lead to the formation of surface organic horizons, which stabilizes the soil surface.

The soil fauna modulates the soil structure by digging holes or burrows, mixing soil components and depositing faeces in the soil or on the soil surface (Table 7.3). In turn, the characteristics of soil organisms such as individual biomass, eyes, pigmentation and extremities are affected by soil conditions and structures. Soil animals and especially large mammals and earthworms are mainly responsible for the bioturbation and the development of mineral-humus complexes. The activity of large animals is visible in C-rich and seasonally dry soils such as a Chernozem. Large pores of 2.5 to 11 mm rely on earthworm holes and small pores between 0.003 to 0.06 mm remain in excrements.

Soil fauna can have positive effects on both soil porosity and microbial activities. Epigeic species such as *Lumbricus rubellus* enrich bacteria and actinomycetes

Table 7.3. *Characteristics of the midden of* Lumbricus terrestris *and the adjacent litter*

| | Midden | Litter | *t*-value |
|---|---|---|---|
| Carbon content (mg $g^{-1}$ soil) | 113 | 78 | ***49.8 |
| Nitrogen content (mg $g^{-1}$ soil) | 6.8 | 5.8 | *8.2 |
| C/N ratio | 16.6 | 13.6 | ***26.8 |
| Microbial biomass (mg $C_{mic}$ $g^{-1}$ soil) | 2.3 | 1.4 | ***53.0 |
| Basal respiration (μg $CO_2$-C $g^{-1}$ soil $h^{-1}$) | 10.5 | 4.9 | ***47.0 |
| Specific respiration (mg $O_2$ $g^{-1}$ $C_{mic}$) $h^{-1}$) | 4.5 | 3.3 | *9.2 |
| Gamasina (individuals $m^{-2}$ soil surface) | 1859 | 1070 | *8.5 |
| Uropodina (individuals $m^{-2}$ soil surface) | 1160 | 100 | *5.4 |
| Collembola (individuals $m^{-2}$ soil surface) | 52481 | 32174 | *5.9 |
| Oribatida (individuals $m^{-2}$ soil surface) | 6344 | 6953 | 0.3 |
| Nematoda (individuals $g^{-1}$ soil) | 26.4 | 20.3 | 3.2 |
| Nematoda (μg $g^{-1}$ soil) | 5.9 | 2.0 | *8.3 |

*Source:* Maraun *et al.*, (1999).

during the passage through the gut and endogeic species such as *Aporrectodea caliginosa* reduce the abundance of these two groups (Kristufek *et al.*, 1992). Thus, microbial communities in casts can be enriched or depleted by passage through the gut. The faeces of isopods and myriapods may support the development of a high and diverse microbial biomass (Zimmer and Topp, 1998). The high amount of coryneformes bacteria present in faeces pellets may differ from the litter microbial community that entered the gut (Byzov *et al.*, 1996).

The grazing effects of the protozoa on micro-organisms depend on aggregate structures and the soil water content since protozoa cannot move between aggregates under dry conditions. Protozoa are not able to prey on bacteria located in inaccessible pores or hydrophobic areas of soil aggregates (Hattori, 1992).

## 7.3 Aerobic and anaerobic metabolisms in soil

Fresh plant litter underlies initial biochemical decay and the mechanical breakdown by the soil fauna. Under aerobic conditions, heterotrophic and saprophytic organisms of the soil microflora and microfauna ensure the almost complete decomposition. A small fraction of the total carbon input can be stabilized against further decay through physical protection, adsorption to clay minerals and transformation to humic substances. Aerobic processes consume $O_2$ as the terminal electron acceptor and $CO_2$ evolves to both soil and atmosphere (Fig. 7.2). During the mineralization of soil organic matter, nitrogen is released as $NH_4^+$ and then nitrified to $NO_3^-$, whereas sulphur and phosphorus are oxidized to $SO_4^{2-}$ and $H_2PO_4^-$ or $HPO_4^{2-}$, respectively. Concurrently, nutrients such as K, Ca and Mg are released and leached from the soil or bonded to the clay fraction and humus.

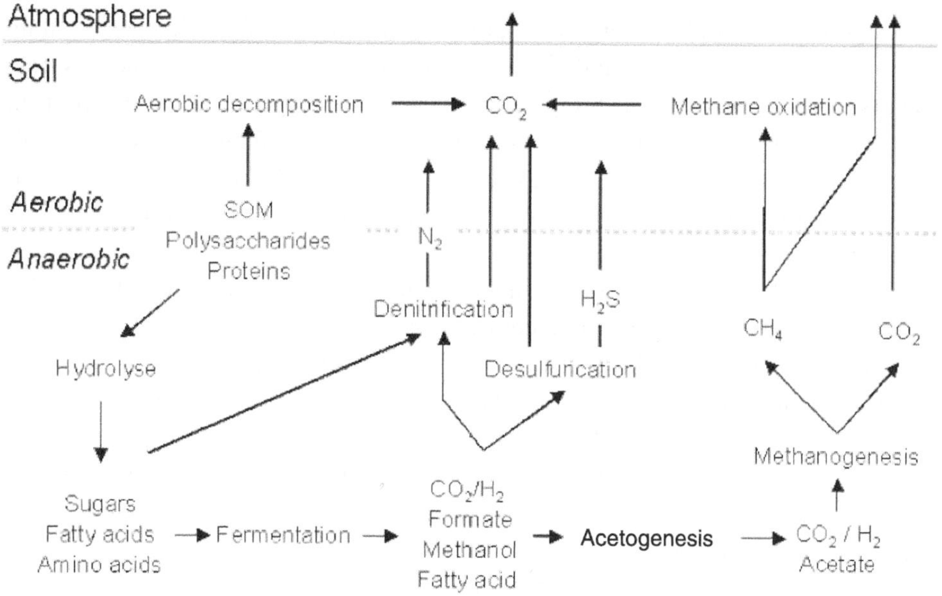


Fig. 7.2. Aerobic and anaerobic C cycling in soil.

The rate of soil organic matter decomposition is affected by the water and oxygen availability, temperature, proton concentration and nutrient supply. The C/N ratio is used as an index for substrate quality and the rate of organic matter decomposition. Values $>25$ and $\leq 10$ refer to low and high activities, respectively. But fresh organic material supports higher activity than old organic matter at an identical C/N ratio due to the presence of more available compounds (Dilly *et al.*, 2003).

The strictly anaerobic Archaea release $CH_4$ (Fig. 7.2). Their simple metabolism can be separated into two main pathways: (1) the reduction of $CO_2$ to $CH_4$ using $H_2$ as energy source and (2) the fermentation of acetate to $CH_4$ and $CO_2$. For the first pathway, organic carbon is not required (Deppenmeier *et al.*, 1996). Archaea are active under anaerobic conditions in hydromorphic soils, peat lands, marshes, swamps, marine and freshwater sediments, flooded rice paddies, geothermal habitats and in the arctic tundra. They transformed in freshwater marshlands 1.1% to 2.4% of the soil organic C to $CH_4$ (Wagner *et al.*, 1999).

Methane is oxidized by proteobacteria in aerobic soil horizons, in oxic microaggregates and in the rhizosphere (Hanson and Hanson, 1996). Methane represents both carbon and energy source, leading to the liberation of $CO_2$. Methanotrophic bacteria survive under unfavourable conditions as spores. The difference between methane production and oxidation determines the net methane emission rates from soil. Up to 90% of the produced methane is reported to be removed by methane oxidation in soils (Khalil *et al.*, 1998).

Large reservoirs of soil organic matter are stored in cool and wet ecosystems like peatlands, marshlands and Arctic tundra. The build-up of organic C stocks occurs when factors such as temperature, water and oxygen availability limit the complete biological oxidation of residues. The accumulation of soil organic matter can also occur at sites favouring plant growth but retarding residue decomposition, such as under climatic conditions with wet springs and dry autumns.

The degradation of soil organic matter can be promoted by natural bioturbations and anthropogenic disturbances such as ploughing. The improved oxygen supply stimulates microbial metabolism. Conversion of grassland and forests to arable systems favours humus degradation, with a rapid humus decrease occurring particularly during the first years. Also, natural changes in environmental conditions such as higher temperature and optimal water availability may stimulate the humus degradation, releasing carbon dioxide and methane and also nitrogenous compounds.

## 7.4 The living phase indicates soil quality

Soil quality was defined as the capacity of a soil to function within ecosystem boundaries to sustain biological productivity, to maintain environmental quality, and to promote plant, animal and human health (Doran and Parkin, 1996). For the evaluation of the soil quality for future generations and for monitoring the soil status, a set of indicators referring to essential soil functions were selected and long-term observation plots in natural, agricultural, forest and urban ecosystems were installed.

Microbial measures are sensitive to changes in environmental conditions and are used as soil quality indicators. In contrast, soil chemical and physical properties such as soil organic C pool, pH value or water holding capacity are used to evaluate the status of soil quality over the long term.

The microbiological indicators most frequently used refer to biomass, diversity and activity, and their interactions with plants (Bloem *et al.*, 2005). A range of microbial estimates was applied to soils under precision (conventional) and organic farming practices. In a study at Scheyern in southern Germany, the organic farming systems stimulated soil microbial biomass and activity in soil. The low serration of the star plot of six characteristics of the soil biota indicated that these characteristics were positively linked (Fig. 7.3). A higher degree of serration was observed for an earlier study of a soil under monoculture in the Bornhöved Lake District in northern Germany (Dilly and Blume, 1998). At Scheyern, the organic farming practices can be considered as more important in terms of nutrient turnover than the precision (conventional) system.

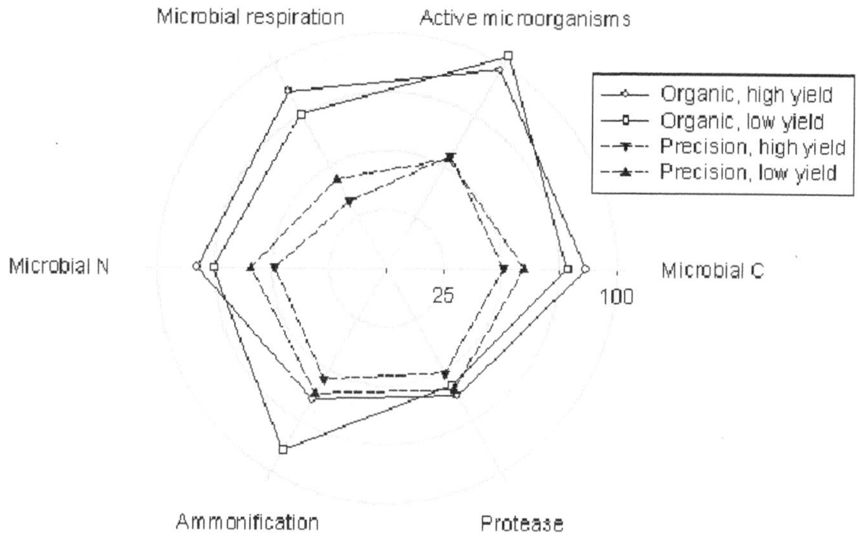


Fig. 7.3. Star plot of microbial indicators in agricultural soils under organic and precision (conventional) farming systems in Scheyern, southern Germany.

Mobile soil organisms are also used for helping to evaluate human and environmental impacts on soil quality. As with the microbial indicators, species composition and relative abundance are addressed. Soil biomass of both the fauna and the microbiota may vary seasonally. Temporal variations may occur particularly in litter horizons with high variability in temperature, water availability and organic matter composition. In contrast, changes in biomass are frequently low in deeper soil horizons due to balanced environmental conditions and limited nutrient availability. Fig. 7.4 shows the typical seasonal pattern for the soil fauna in a sandy soil in a temperate climate. The relevance of the soil water availability for the occurrence of indicative species is shown in Fig. 7.5.

In forests, the soil fauna interacts with humus characteristics (Schaefer and Schauermann, 1990). The contribution of mesofauna and macrofauna to the entire fauna depends on soil horizons, and the vertical distribution of the two groups affects the decomposition process. Soil fauna has the highest functional importance in the 'mull' (a type of humus consisting of a crumbly intimate mixture of organic and mineral materials, formed mainly by earthworms) and lower importance towards the 'moder' (a product of advanced but incomplete humification, despite good aeration) and 'raw humus' (consisting predominantly of well-preserved, though often fragmented, plant remains with few faecal pellets).

Key species and communities were used as indicators for environmental conditions. However, this approach is problematic (Vegter *et al.*, 1988) since species and communities vary according to humus type (Fig. 7.5) and

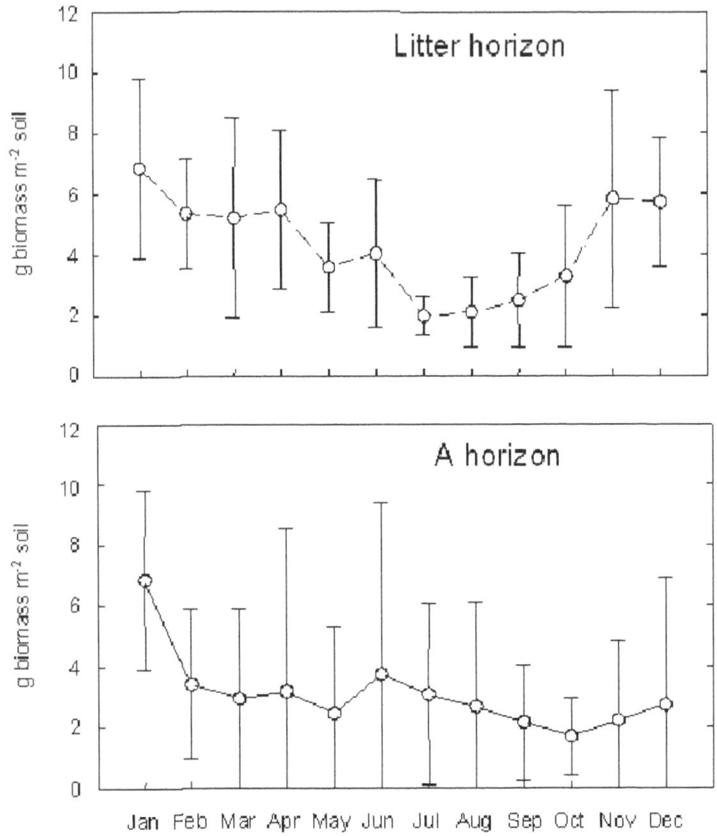


Fig. 7.4. Seasonal dynamics of faunal biomass in the litter and A (0 to 4 cm) horizons of an Arenosol under beech in a temperate climate. Bars indicate standard deviations.

geographical location (Wauthy *et al.*, 1989). Beside the humus type, components of the fauna community such as microarthropods are controlled by chemical properties of the soil. In natural systems, some animals showed maximal abundance at high proton concentration and are thus referred to in the arthropod acidity index (van Straalen and Verhoef, 1997). In agricultural systems, liming and fertilization stimulate some faunal groups that contribute to the amelioration of soil physicochemical properties.

## 7.5 Modification of biotic communities during soil degradation

The degradation of soil can be defined as the regressive development associated with the loss of buffering capacity and resilience (resilience means return to good soil condition). During soil development, several stages related to the local

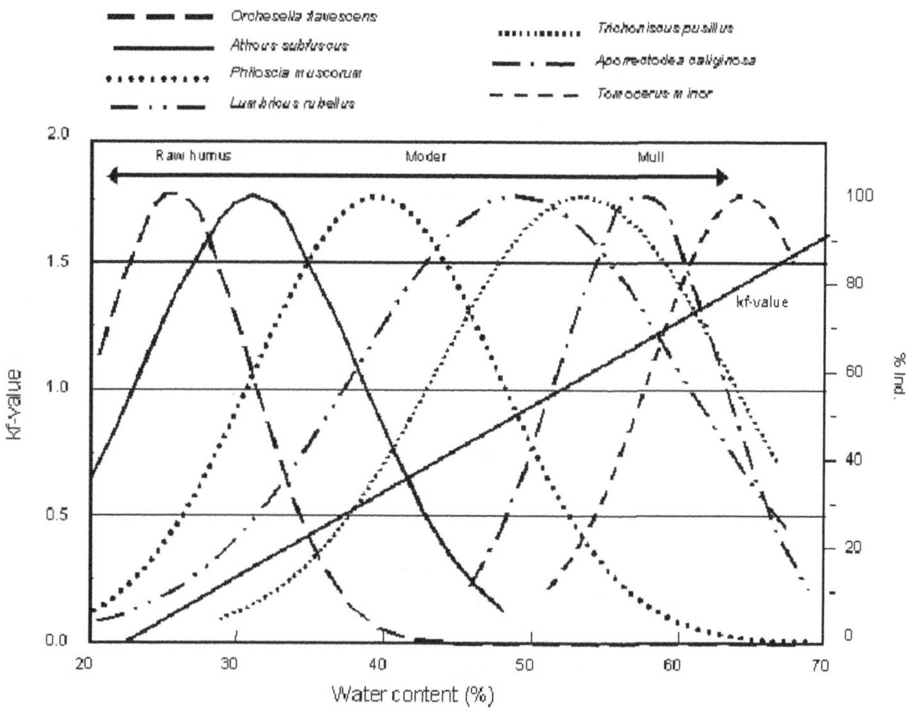


Fig. 7.5. Relationships between the decomposition rate of soil organic matter attributed to faunal activity (kf-value) and the soil water content or species abundance (% ind.). From Irmler *et al.* (1997).

conditions are passed through and are associated with the replacement of the primary vegetation by secondary vegetation. At the same time, soil macrobiota and microbiota are modified and this modification impacts on the humus stocks and composition. Human activity perturbs the native soil structure, e.g. by tillage. Tillage systems lead to the loosening and homogenization of soil and thus hamper autochthonous stability. Soil structure varies according to biotic and abiotic soil constituents. Consequently, degraded soils may not be rapidly restorable, particularly when nutrients and key organisms have been lost.

Soil degradation is generally related to the loss of soil organic matter and living organisms and also shifts of the biotic fraction.

Nitrogen transformation rates are frequently increased by favouring bacteria in the microbial community at the expense of the fungal community and organisms from the higher levels of the food web, e.g. mesofaunal and macrofaunal species. The effect of exogenous N input is reflected in the fauna, with an increased number of bacteria-feeding nematodes (Hyvönen and Huhta, 1989; Mamilov and Dilly, 2002). Ammonium deposition has led to acidification of soils during recent decades. A range of macrofaunal groups are largely dependent on calcium supply and earthworms, woodlice and millipedes tend to disappear from such soils. Fertilizers may favour both soil biocoenosis and plant growth but may also lead to an imbalance in nutritional conditions.

Natural soil resources were consolidated over long periods and can be considered as highly organized with an adequate archive of biodiversity. Soil development may occur rapidly during some decades on waste deposits in urban regions and on marshlands that have been reclaimed. Over short periods, high biotic activity can develop but the buffering capacity of soil in terms of nutrient export may be limited due to the thinness of the horizons (Dilly *et al.*, 2005).

The most important strategy for soil protection is to avoid the destruction of specific horizons and horizon functions. In Germany, human activity destroys currently approximately 1 $km^2$ per day (German Advisor Council on Global Change, 2003). For beneficial remediation of sites which have suffered considerable human impact, e.g. in post mining regions, materials from upper soil horizons should be replaced. Deep sediments with pyrite carried to the soil surface may liberate acids by the activity of pyrite-oxidizing micro-organisms. Therefore, such sediments should be reburied to avoid acidification. The pool of nutrients and active microbial communities control the speed of recovery of the vegetation and the progress towards effective land use.

For soil amelioration, earthworms have occasionally been added. Their beneficial effects on water infiltration may even help to restore water exchange on compacted soils that are frequently waterlogged.

Where soil destruction is unavoidable, the topsoil should be exploited for nutrients and diversity of organisms. Deeper soil horizons frequently contain few biotic resources and thus are of less significance in terms of biodiversity and nutrient conservation.

# 8

# The State Factor theory of soil formation

*Ronald Amundson*

Pedology is the branch of the natural sciences that deals with the biogeochemical processes that form and distribute soils across the globe. The word 'pedology' was coined by the German scientist F. A. Fallou in 1862 (Tandarich and Sprecher, 1994; Oxford Dictionary, 1966), but the point in time that it became a true science rests upon when theories of the soil and its formation were developed and adopted by the early pedological community (Kuhn, 1962). An exhaustive historical analysis of the development of pedology has not been conducted, though we know that numerous 'scientific' advances were made during the nineteenth century by scholars in several countries. For example, Eugene Hilgard arguably was the premier soil scientist of the nineteenth century. His breadth of knowledge, scientific techniques (chemical and physical), and his mastery of multiple disciplines made him a truly formidable presence on the international geological and agricultural science scenes (Jenny, 1961; Amundson and Yaalon, 1995; Amundson, 2005a). Yet, as his biographer Hans Jenny (1961) concluded, Hilgard did not distil his vast knowledge and concepts into neat conceptual models or theories. Although one can easily see that Hilgard clearly understood the important controls on soil formation, an elucidation of these into a concise intellectual package was ultimately accomplished by others. First, Vasily Dokuchaev and his Russian colleagues identified and discussed what we now know commonly today as the 'factors of soil formation' (Dokuchaev, 1883). Second, the American scientist Hans Jenny rigorously defined or redefined the meaning of the factors, and more importantly, added the new concept of the *soil system*, which when combined with the factors provided a powerful conceptual framework through which nature could be probed and understood.

*Soils: Basic Concepts and Future Challenges*, ed. Giacomo Certini and Riccardo Scalenghe.
Published by Cambridge University Press. 

The Russian concept of soils, and the factors that form them, had an enormous organizing effect on Russian pedology and eventually (after a long delay) impacted American soil science when C.F. Marbut introduced an English translation of one of the Russian texts (Helms, 2002). In particular, Marbut chose to unveil a new scheme of soil classification, inspired by Russian pedological concepts, during the post-conference field trip of the First International Congress of Soil Science, which was held in Washington DC, in the summer of 1927. A participant on this field trip – a six-week-long rail excursion across North America – was a young, bright, and (by his own depiction) somewhat brash scientist by the name of Hans Jenny. Jenny had recently received his doctoral degree in soil physical chemistry from the Swiss Federal Institute of Technology under the direction of one of the world's premier chemists, Georg Wiegner, and was in the United States on a Rockefeller Fellowship to work in the laboratory of the future Nobel Prize laureate Selman Waksman, at Rutgers University. Although trained in colloidal chemistry (a field in which he remained an active and important participant for decades; Jenny, 1989; Amundson, 2005b), as a student Jenny had taken the initiative to conduct weekend field trips, activities which led to his collaboration with the young plant ecologist Braun-Blanquet, and to ideas that later proved fruitful in Jenny's growing pedological research directions.

Great scientific advances likely occur in many ways and along many avenues, though some important and notable breakthroughs occur as so-called 'Eureka' events (the scholar Archimedes, while immersing himself at the town baths, suddenly came upon the idea of specific gravity, and rushed home shouting 'Eureka!, Eureka,!' – 'I have found it!, I have found it!'; Singer, 1959). While travelling across the continent as part of that great field excursion in 1927, Jenny later recalled the intellectual transformation that he underwent (Fig. 8.1):

> To me, the tour was a thrill and opened a new world. I was impressed when we went from Washington south and saw the southern red soils and, a few weeks later, the black soils of Canada. I searched for a connection between the two . . . The red soils of the South and the black soils of Canada were showcases of the climatic theory of soil formation. No doubt, our European textbooks by Ramann and Glinka were right in principle . . . The rolling plains, I fancied, must harbor the secret of mathematical soil functions. At times I could hardly sleep thinking about it.
>
> (Jenny, 1989).

What Jenny recognized in his own 'Eureka' moment was the underpinnings of what he would later call the 'Factors of Soil Formation'.

The goal of this chapter is to briefly outline Jenny's ideas of both the soil and the state factors that control its formation and distribution. One of the most important tasks is to clearly define terms and concepts, many of which have a

Fig. 8.1. A sketch of Kansas made by Hans Jenny during the 1927 Transcontinental Excursion.

vernacular usage that is inadequate for our purposes. As Jenny wrote in the first sentence of the formal introduction of his new (or depending on opinion, revised) theory (Jenny, 1941): ‘As a science grows, its underlying concepts change, although the words remain the same.’

## 8.1 The soil system

Soil is the subject of study in pedology, and while the science of pedology itself has a definition that commands some general agreement, there is no precise definition for soil, nor is there likely to ever be one. The fundamental reason for this paradox, as Jenny took pains to emphasize, is that soil is part of a continuum of materials at the Earth surface. At the soil's base, the exact line of demarcation between ‘soil’ and ‘non-soil’ will forever elude general agreement, for the chemical and physical changes induced by pedogenesis disappear gradually, commonly over great vertical distances. Similarly, an identical problem confronts one attempting to delineate the boundary between one soil ‘type’ and another. Soil properties, such as horizonation, commonly change gradually and continuously in a horizontal direction, lending credence to the view that the Earth contains an infinite variety of soils (Jenny, 1941). This dilemma, a science with a ‘poorly’ defined object of study, is not peculiar to pedology. As Charles Darwin (1985) recognized in the *Origin of Species*, an exact and unambiguous definition of a ‘species’ eludes the evolutionary biologist, yet the development of that field of science has progressed unimpeded.

The scientific path out of this disconcerting situation is to divide the soil continuum, albeit arbitrarily, into *systems* that suit the need of the scientist. This intellectual step is one of Jenny's great conceptual contributions to pedological thought. The ecologist M. J. Canny (1981) made the interesting statement that 'A universe comes into being when a space is severed', a statement with a profound connection to systems and the very nature of soil. Systems are human constructs that confine our focus and allow us to develop quantitative tools to evaluate a portion of the soil continuum. These systems are necessarily open to their surroundings – they are part of our imagination after all – and through them pass energy and matter. Jenny, drawing on his training in the physical sciences, noted that the properties of an open soil system (dependent variables) will vary in response to a certain set of factors (independent variables):

$$\underbrace{\textit{soil system}}_{\text{dependent variables}} = f\underbrace{\begin{pmatrix} \textit{initial state of system,} \\ \textit{surrounding environment,} \\ \textit{elapsed time} \end{pmatrix}}_{\text{independent variables}}$$

This formulation of soils in relation to their surroundings forms the foundation of his book, which has been called 'a brilliant synthesis of field studies with the abstract formalism of physical chemistry' (Sposito *et al.*, 1992), and which establishes the study of soil on the same conceptual footing as better established physical sciences. In line with these comments, Jenny strongly emphasized that the model is 'formalistic' – e.g. a theory whose relationships between dependent and independent variables are derived from observational measurements of soils. In other words, the model provides information as to *how* the soil system varies with external factors, but not necessarily *why*. Jenny emphasized that the model is not mechanistic, and independent studies or models involving process and mechanism are ultimately needed (in addition to state factors) to arrive at the reasons for the observed relationships. Thus, one will note, on close reading of the book, that virtually no mention of mechanisms or processes are introduced, even though Jenny vigorously approached these questions in other areas of his research. To emphasize the formalistic basis of his model, he wrote, 'We endeavor to determine how soil properties vary with soil-forming factors. We shall exhibit but little curiosity regarding molecular mechanism of soil formation and thus avoid lengthy excursions into colloid chemistry, microbiology, etc. Such treatments will be reserved for a later occasion.'

Jenny recognized that the independent variables on the right side of the equation encompassed important soil-forming factors first identified by Russian scientists years before. Jenny strove to rigorously redefine these factors and their relationship to the soil system, and after initial discussions, arrived at his famous equation:

$$s = f\,(cl, o, r, p, t, \ldots)$$

'which we shall designate as the fundamental equation of soil forming factors' (Jenny, 1941), where *cl* is external climate, *o* is biotic potential, *r* is topography, *p* is parent material, *t* is time, and the elipses are additional factors. These variables are the *state factors*, in that they *define the state of the soil system* (Fig. 8.2). Below is a more detailed definition of the variables in Jenny's model, and how they differ from common usage or assumptions.

Jenny's 'fundamental equation of soil forming factors' ultimately provides a definition of soil that, while not precise in identifying its boundaries (which ultimately is our choice), is internally consistent with our theoretical framework:

> 'Soil is those portions of the Earth's crust whose properties vary with soil forming factors'.
>
> (Jenny, 1941).

A full understanding of the depth and breadth of this equation, and the concepts behind it, requires repeated readings of Chapter 1 (and all chapters) of Jenny's book *Factors of Soil Formation. A System of Quantitative Pedology*. Jenny frequently remarked to colleagues that, 'it [the equation] looks simple, but it is not'. Numerous misinterpretations of Jenny's model pervade the literature, one of the most common being the meaning of 'independent variables'. Independent variables are (a) independent of the system being studied and (b) in many parts of the Earth the state factors may vary independently of each other (though of course,

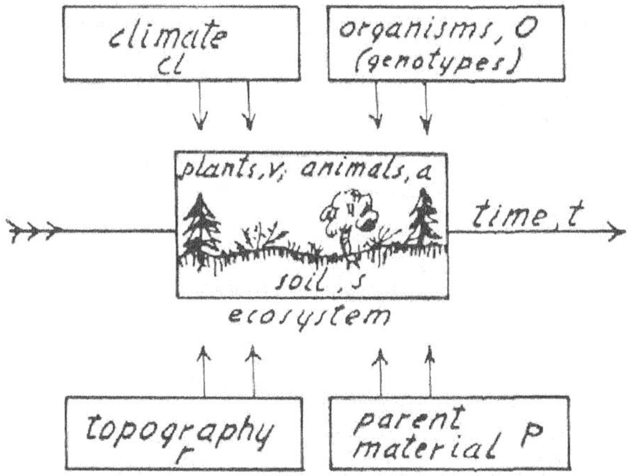


Fig. 8.2. A schematic representation of the soil system and ecosystem changes as a function of the values of state factors. Illustration by Hans Jenny in Amundson and Jenny (1991).

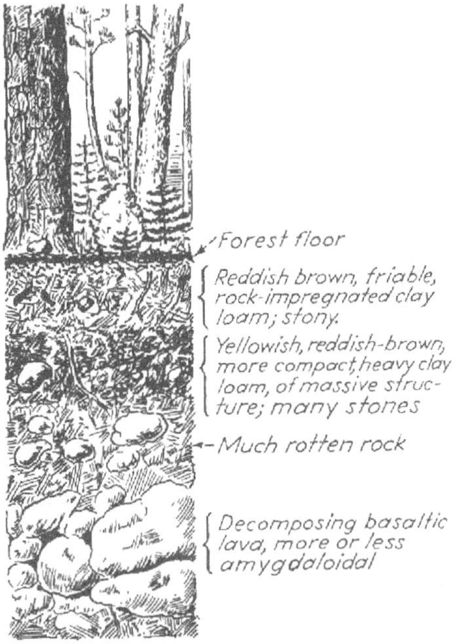


Fig. 8.3. Depiction of Jenny's 'larger system', or ecosystem. From Jenny (1941), Fig. 8.

not always). As a result, through judicious site (system) selection, the influence of a single factor can be observed and quantified in nature. These observational studies in nature are called *sequences* (e.g. climosequence, chronosequence, etc.), and serve to allow scientists to unravel the individual effects of a given state factor on soil systems.

In concluding this section, I will point out one additional attribute of Jenny's model. Jenny clearly recognized that the 'soil system' can be enlarged to encompass soil, plants, animals, etc., a system Jenny first called the 'larger system' in his book (Fig. 8.3). This 'larger system' is, in retrospect, entirely analogous to the concept of 'ecosystem' (Tansley, 1935), and arguably is much better defined than Tansley's description.

## 8.2 State factors

The terms used to represent the state factors have been used by pedologists for more than a century, and their meanings can vary considerably depending on the author, time, and place. Here, the factors are defined in the sense of Jenny, and some challenging issues are identified regarding their role on soils and ecosystems.

The state factor *climate* (*cl*) comprises the meteorological conditions which surround the soil system (temperature, rainfall, humidity, etc.), and is consistent with the concept of regional climate. The climate *within* a soil system or ecosystem is a dependent variable or system property, and is consistent with the concept of microclimate. This internal climate is of course dependent on the surrounding climatic conditions, but it can also be greatly affected by other state factors (e.g. topography, landform age, etc.).

Many pedological studies are established along sequences of rainfall and/or temperature, and climosequences are therefore one of the most common forms of state factor analyses. Studies of the relation of soil C and N to climate are currently of great interest in relation to global change (e.g. Post *et al.*, 1982, 1985). Climosequences have served to provide some of the strongest evidence on the potential responses of soil C to global warming (Trumbore *et al.*, 1996). In general, 'sequence' studies of all types have important roles in assessing the response of soils and ecosystems to disturbances of many kinds.

The *biotic factor* (*o*) is defined by Jenny as the potential biota of the system. The biotic factor therefore constitutes the microbial, plant and animal gene flux that enters the system from the surroundings. The vegetation (*v*) or animals (*a*) that actually survive or reproduce in the system are dependent ecosystem properties, and may not directly reflect the potential biota due to the effect of the remaining state factors. The biotic factor is one of the most conceptually challenging of the state factors, and Jenny (1958) later expanded a discussion of this factor and its relationship to observed vegetation. For example, in many systems where climate and biota appear to co-vary, the microbial, plant and animal gene flux entering the system is constant, and the observed biota that survive and reproduce in the systems are dependent (vs. independent) ecosystem variables.

Given the ease with which seeds, microbes and animals are dispersed, it is difficult to make observation comparisons of soil systems or ecosystems that differ only in their biotic factors. To do so must involve sites separated by biotic barriers such as oceans or mountain ranges. Jenny proposed approximate alternatives to this problem by examining soil properties under individual species in areas where the biotic factor is actually constant: 'it is correct to compare soils under two different species and isolate the species factor in soil genesis' (Jenny, 1958). An important implication of Jenny's model, and the definition of the biotic factor, is that the definition of soil does not need to involve plants, or even life (the biotic factor may be nothing). This key distinction now has implications for interplanetary geochemical studies, where it has been argued by some (Markewitz, 1997) that the term soil should not be used on other planets because it implies the past or present presence of plants. This narrow view is obviously inconsistent with our perspective, and one must hope that pedological concepts and knowledge are applied as we explore our solar system.

The factor *topography* ($r$) comprises a number of subvariables that correspond to the configuration of the system. The subvariables of topography are commonly noted to include a system's position along a hillslope complex (shoulder, backslope, etc.), the aspect of the slope, and the proximity to a water table.

Toposequences, sometimes also called *catenas*, have been widely studied. Yet, some challenging questions about sequences remain (e.g. opening 'the black box' of the soil system). For example, how does one determine the age of soils on actively eroding hillslopes? How does one model the inputs and outputs of the system, which obviously include laterally moving sediment? Recent research by geomorphologists, combined with chemically derived rates of geomorphic processes, are providing new and exciting ways of determining soil residence time and a number of other soil attributes on hillslopes (e.g. Heimsath *et al.*, 1997).

*Parent material* ($p$) is defined as the initial state of the sediment, rock, or soil at $t = 0$. For a soil forming on fresh alluvium or a recent volcanic deposit (Dahlgren and Ugolini, 1989b), the parent material is the geological substrate itself. For a soil or ecosystem that is reforming after a disturbance, or major climate change, the parent material is the soil that was present at the beginning of the new state factor assemblage.

The state factor *time* ($t$) is defined as the elapsed time since the system began or was exposed to its present assemblage of state factors. For some systems, this is the starting point immediately after an event such as fluvial or volcanic deposition. In other cases, $t = 0$ may be the point at the end of a major environmental disturbance or change. Many soils older than the Holocene have experienced one or more major climate changes, and possess properties that may be the complex effect of many stages of soil development, and are called *polygenetic* or *relict* soils. However, one of the most important aspects of determining the age of a soil system is that it allows us to determine the rate ($\partial S/\partial t$) of soil property changes.

Determining the age of soils has been, and will continue to be, a challenging research issue. Recent instrumental and theoretical developments have advanced the use of $^{14}C$ (Amundson *et al.*, 1994), U-series isotopes (Sharp *et al.*, 2003), and cosmogenic radionuclides (Goss and Phillips, 2001) in the dating of soils and geomorphic surfaces, but clearly much more research needs to be conducted on this important parameter.

In an increasingly populated planet (Vitousek *et al.*, 1997a), the state factor *humans* takes on increasing significance as an ecosystem variable. Jenny (1941) originally considered humans under the biotic factor because, like other biota, humans contain a genetic component or genotype. However, unlike the other species, humans possess a cultural component that varies from society to society and which operates independently of genotype, arguably making them worthy of a separate factorial treatment (Jenny, 1980; Amundson and Jenny, 1991, 1997).

## 8.3 Importance of State Factor theory

The demands on pedologists, and their activities, are commonly drawn in many pragmatic and applied directions. Included among these activities are soil survey activities, land management and planning needs, soil classification, and the remediation and preservation of wetlands/wildlands. However, it is important for both seasoned and novice pedologists to recognize, and celebrate, that their science has one of the richest theoretical underpinnings in the 'earth' sciences.

The State Factor 'model', 'theory', or 'hypothesis' (whatever term we wish to use) is, in the parlance of the history of science, a 'paradigm' (Kuhn, 1962). It provides the rules for legitimate research and provides a shared language and a focal point for the discipline, and guides us as to the type of questions that should be asked. Kuhn wrote, 'Acquisition of a paradigm and of the more esoteric type of research it permits is a sign of maturity in the development of any given scientific field.' Paradigms never resolve all the problems in a field, and debates about the State Factor theory that occasionally erupt are signs that the field remains vibrant and productive.

From a broader perspective, the State Factor theory of pedology has important implications to sister sciences, particularly ecology. The State Factor model provides a rigorous definition of an ecosystem, and provides a methodogical approach to examining ecosystem properties and processes. 'Gradient analyses' in ecology (Whittaker, 1967) is the intellectual offspring of state factor analysis (Austin *et al.*, 1984), and provides an ideal point of common interest between the earth and ecological sciences. For example, state factor analysis has served as the conceptual framework for the ongoing biogeochemical research on the Hawaiian Islands, work which has revealed much about the fate of soils and ecosystems over long expanses of geological time (Chadwick *et al.*, 1999; Vitousek, 2004) or climate. Natural scientists strive to understand the cause of geographical variations in Earth surface properties and processes, and state factor analysis provides a means to do so. These methods will have increasing importance for the health of the Earth, as scientists endeavour to understand and manage a world that increasingly bears the human footprint.

Finally, while paradigms of science have profound implications for whole fields of study, it is their effect on a personal level that is to me most striking. The historian Thomas Kuhn wrote about the effect of paradigm *change* in a scientific field: 'When paradigms change, the world itself changes with them... It is rather as if the professional community had been suddenly transported to another planet where familiar objects are seen in a different light and are joined by unfamiliar ones as well.' An example that may have some resonance with most earth scientists is the pre and post plate-tectonics field of geology. After adopting plate-tectonic concepts, scientists suddenly saw familiar objects in an unfamiliar way. However, these same ground shifts in perception occur when a student is first introduced to a paradigm, and I remember well how differently the world looked to me following my introduction to the State Factor model as a student. Suddenly, the vast spatial expanses of the Earth no longer seemed baffling, but instead had the look of a large natural experiment that could be unravelled, at my will, through State Factor analyses. Like Archimedes long ago, I experienced my own, albeit modest, 'Eureka' moment, because as Kuhn wrote, 'a paradigm is prerequisite to perception itself'.

# 9

# Factors of soil formation: parent material. As exemplified by a comparison of granitic and basaltic soils

*Michael J. Wilson*

The clearest example of the intimate relationship between the nature of the soil parent material and the properties that the soil eventually acquires through the mediation of soil processes is to be found in soils derived from recent volcanic material. This relationship has been especially well illustrated by a comprehensive review of the topic by Dahlgren *et al.* (2004). This review highlights the fact that volcanic soils possess distinctive physical, chemical and mineralogical characteristics that are usually not found in soils developed upon other parent materials. Many of these characteristics depend on the ease with which volcanic material is weathered to form poorly crystalline clay minerals, such as allophane, imogolite, ferrihydrite, etc. Other examples illustrating the influence of parent material on soil properties could also be cited. Thus, soils derived from carbonate rock may have such distinctive chemical properties that they are described as specific soil types (Rendzinas) in some soil classifications. Again, soils derived from smectitic-rich parent materials may be classed as Vertisols. In addition, it is widely recognized that parent material may be responsible for the origin of some unusual chemical soil properties, such as the high exchangeable Mg/exchangeable Ca ratio or heavy metal content of serpentinite-derived soils, or for the development of particular soil horizons which are used as diagnostic indices in soil taxonomies.

Notwithstanding these examples, however, parent material *per se* cannot be said to be a major criterion for soil classification. For most systems, soil type is based on a number of different criteria related to pedogenesis, stressing particularly the nature and intensity of the process, as controlled largely by combinations of the other soil-forming factors, namely climate, topography, biota, and time. In fact, the influence of parent material is considered to be an inverse

*Soils: Basic Concepts and Future Challenges*, ed. Giacomo Certini and Riccardo Scalenghe.
Published by Cambridge University Press. 

function of time, so that it is important especially in young or relatively immature soils. Thus, Chesworth (1973) concluded that 'as time progresses, the weathered products of rocks as dissimilar as granite and basalt should become more and more alike, to the extent ultimately of becoming indistinguishable'. This may be true from the point of view of an analysis of the total chemical composition of the clay fraction, but such an approach does not provide information on the influence of parent material on the soil properties that are considered to be important to an understanding of soil behaviour in a variety of agricultural and environmental contexts.

The purpose of this contribution, therefore, is to focus on the impact of dissimilar parent materials – in this case granite and basalt, the two most widely occurring igneous rocks on the Earth's surface – on the physical and chemical properties of soils in the context of different weathering intensities to which the soils have been subjected. Is parent material influence ephemeral and if not to what extent can it be detected in soils subjected to long periods of weathering? The initial emphasis will be upon relatively young soils where the influence of the parent material can be most readily discerned, and following this the impact of more intense weathering as well as that of other soil-forming factors will be considered.

## 9.1 Mineralogical properties

### *9.1.1 Granitic soils*

#### *Sand mineralogy*

The mineralogy of the sand fraction of young soils developed on granitic rocks naturally reflects the mineralogy of the parent rock itself. Most frequently granites consist predominantly (>66%) of K-feldspar, mainly orthoclase feldspar and microcline, with subordinate sodic plagioclase feldspar and about 25% quartz. Both biotite and muscovite often occur but hornblende is relatively uncommon. Because of their stability to weathering, persistent accessory minerals in granitic soils include rutile/anatase, zircon, apatite, monazite, tourmaline, garnet and sphene. These minerals represent a store of major nutrients (P in the case of apatite and monazite) or trace elements (for example B in tourmaline) that is not available to plants. The sand fractions of the young granitic soils of Scotland are typically made up K-feldspar and quartz, with subordinate plagioclase feldspar, biotite and muscovite, and a heavy fraction (density $> 2.65$ Mg m$^{-3}$) consisting of many of the accessory minerals listed above, as well as opaque minerals and weathered grains. With intensive weathering, however, the major minerals decompose according to their susceptibility to the process and ultimately, in a tropical environment, the sand fractions of granitic soils may consist very largely (>80%) of quartz with only small amounts (<10%) of feldspar plus micas. Thus the feldspars tend to weather to clay minerals and in fact the sand fractions of intensely weathered granitic soils may contain appreciable amounts of clay minerals such as kaolinite and gibbsite, which may be regarded as undispersed fragments of weathered primary minerals.

### *Clay mineralogy*

Studies of the clay mineralogy of granitic soils in many different countries yield broadly comparable results. The constituents most commonly encountered include kaolinite and halloysite, largely deriving from the weathering of feldspars, and vermiculitic minerals, which are associated with the weathering of micas. The clay fractions of the soils developed upon granitic glacial till in north-east Scotland typically contain kaolinite and/or halloysite and various vermiculitic minerals that may be interstratified with mica or interlayered with non-exchangeable hydroxy-Al, depending on the pH conditions of the soils. It is considered, however, that the kaolinite and halloysite in these soils results from pre-glacially weathered material, of which there is an abundance in north-east Scotland. Dioctahedral illite is also a common constituent of these Scottish soils but this probably represents sericitic or comminuted muscovitic material. The clay fractions of granitic soils developed under more intensive tropical or sub-tropical weathering conditions are usually dominated by kaolin minerals, or kaolinite itself. Gibbsite is also a common constituent of intensely weathered granitic soils, as found in Malaysia, but it also may form rapidly under more temperate climates.

## *9.1.2 Basaltic soils*

### *Sand mineralogy*

Basalt essentially consists of calcic plagioclase feldspar (labradorite to bytownite in composition) and clinopyroxene; olivine may or may not be present. Minerals that occur in small amounts or as accessories include biotite, quartz, apatite and magnetite. Not infrequently the ferromagnesian minerals of basalt are extensively altered to a complex of green minerals often identified optically as 'chloritic or serpentine-like' material, but in reality more likely to be trioctahedral smectitic minerals (Wilson, 1976). The mineralogy of young basaltic soils may reflect that of the parent rock to a large extent, and sand fractions may contain substantial quantities of smectitic aggregates (usually saponite). Frequently quartz is also present, sometimes in considerable amounts as in the basaltic soils of India and

Queensland. There is convincing evidence for an a eolian origin of the quartz in some basaltic soils such as those in Victoria, Australia, but this does not seem to explain the quartz in the Queensland basaltic soils, where it also occurs in coarse sand fractions (Isbell *et al.*, 1976, 1977). These observations are problematical because although the red and brown basaltic soils of Queensland are thought to be residual, if this was the case the amount of quartz that they contain is such that a lowering of the landscape by more than 50 metres would be required. This would appear to be inconsistent with the well-preserved volcanic landforms of the region (Isbell *et al.*, 1976). The major primary minerals in basaltic soils quickly decompose under intensive weathering conditions, the ferromagnesian minerals to goethite and haematite and the plagioclase feldspars to gibbsite and 1:1 clay minerals. Basaltic soils may also contain a substantial content of magnetite or titanomagnetite, which converts to maghaemite or titanomaghaemite under tropical weathering conditions.

*Clay mineralogy*

As mentioned above, basalts often contain significant amounts of expansible minerals which may be directly inherited by the soil and which then occur even in the non-clay fractions. This has been shown for basaltic soils from many countries. The mineral in Scottish basaltic soils was identified as saponite (Wilson, 1976), which is susceptible to weathering and quickly decomposes. Under alkaline conditions, basaltic soils initially become highly smectitic. The smectite formed is often beidellitic and iron-rich as in basaltic soils in India, although aluminous beidellite and low-charge montmorillonite have also been identified. With further weathering the smectite in basaltic soils often becomes interstratified with kaolinite, eventually to become a mixture of kaolin (usually kaolinite but sometimes halloysite) and iron oxide minerals (haematite and goethite). Gibbsite is also a common constituent.

## 9.2 Physical properties

### *9.2.1 Granitic soils*

*Particle size*

Because granite is itself a medium- to coarse-grained rock, with a mean grain size ranging from 1 mm to several centimetres, it might be anticipated that granitic soils would be coarse textured. In relatively young soils this is generally the case, as found, for example, in residual granite soils in Japan and soils developed upon granite-derived glacial till in north-east Scotland with clay contents of $<5\%$ and $<15\%$ respectively. However, granitic soils may acquire a much higher clay content (30–40%) under more intensive weathering conditions, as in tropical climates.

### *Porosity and hydraulic conductivity*

Granitic soils are often characterized by high values for porosity and saturated hydraulic conductivity, although values may vary considerably according to soil type, position in the profile and clay content. For example, in a low clay content (<10%) soil developed upon granitic bedrock in the Sierra Nevada, California, pore space amounted to 49% of soil volume in the AC horizon and to 12% in the Cr horizon (Graham *et al.*, 1997). Average saturated hydraulic conductivity ($K_{sat}$) for the AC horizon was 29 cm $h^{-1}$ and for the Cr horizon it was 3.8 cm $h^{-1}$.

### *Water retention/availability*

Granitic soils often have a very poor capacity to retain water because of their low clay and organic matter contents, combined with high porosity and percentage of macrovoids. Thus, Olowolafe (2002) found that granitic soils in Nigeria had appreciably lower natural water contents and available water capacity than soils developed nearby on basalt, attributing this to a higher clay content of the latter. It should be noted, however, that granitic saprolites (saprolite is the soft, weathered bedrock that lies in its original place) may have the capacity to store large amounts of available water and can be important in sustaining the productivity of commercial forests.

### *Bulk density and compactibility*

The bulk density of granitic saprolites can be very high, even approaching that of the fresh rock itself. For the cultivated granitic soils in Zimbabwe investigated by Burt *et al.* (2001), bulk densities were usually above 1.40 Mg $m^{-3}$ and were often greater than 1.60 Mg $m^{-3}$. According to Vepraskas (1988), bulk density values above this figure virtually stop root elongation due to mechanical impedance and are a feature typical of compacted soils.

### *Structure and aggregate stability*

Residual soils developed on granitic saprolites have only a limited ability to form water-stable aggregates. For example, the granitic Masa soils of Japan lack cohesion to such an extent that they are highly susceptible to erosion (Egashira *et al.*, 1985), as are similar soils in Zimbabwe (Burt *et al.*, 2001). The coarse sandy textures typical of many granitic soils militate against the development of water-stable aggregates and it is only with an increase in clay and organic matter content that water-stable aggregates are able to form.

### 9.2.2 Basaltic soils

*Particle size*

The fine-grained nature of basalt suggests that soils developed from this parent rock might also have a fine-grained texture. This is not always true for young, poorly developed basaltic soils where coarser textures such as loamy sand or sandy loam may be encountered. Higher clay contents are found in soils developed on glacial till derived from Carboniferous basalts in south-west Scotland (Mitchell and Jarvis, 1956). Under free-draining conditions, these soils generally have a loam to clay loam texture with clay contents in the range of 15 to 35%. More poorly drained soils have clay loam to clay textures with clay contents ranging from 25 to 50%. Under conditions of intense weathering the clay contents of basaltic soils may become very high. For example, Isbell *et al.* (1976) found clay contents in the 60 to 70% range for the red basaltic soils of north Queensland. The nearby brown basaltic soils were not quite so clay rich but still had clay contents of between 40 to 60% (Isbell *et al.*, 1977).

*Porosity and hydraulic conductivity*

The high clay contents of basaltic soils would be expected to lead to low porosities and hydraulic conductivities. This is often the case, particularly where there are very large clay contents (50–75%) such as in the basaltic soils of the Golan Heights in the northern Jordan Valley where hydraulic conductivities of 0.1 to 0.3 cm $h^{-1}$ were measured (Singer, 1987). Infiltration of water is thus very slow and decreases from 0.28 cm $h^{-1}$ to 0.1 cm $h^{-1}$ after a short period of time. However, Olowolafe (2002) found that the basaltic soils on the Jos Plateau, Nigeria, had significantly higher infiltration rates and permeabilities (means of 10.3 cm $h^{-1}$ and 1.29 cm $h^{-1}$ respectively) compared with their granitic counterparts (means of 5.3 cm $h^{-1}$ and 0.83 cm $h^{-1}$ respectively), despite containing twice as much clay. The reason for this may be that the clay in the basaltic soils forms water-stable aggregates, thus enabling the soils to behave as if they had a more sandy texture.

*Water retention/availability*

The higher clay content and finer texture of basaltic soils compared with granitic soils suggests that they should be able to retain more water. This is certainly consistent with the findings of Olowolafe (2002) in Nigeria. Here the average natural water content for the basaltic soils that he studied was 40% compared with 17% for nearby granitic soils. However, for plant-available water the difference was not so marked, values of 110 mm $m^{-1}$ and 96 mm $m^{-1}$ being found for basaltic and granitic soils respectively. In clay-rich soils it is often the case that much of the natural water content is held in a form that is not plant-available but it is difficult to generalize because of the wide variations in pore size distribution resulting from differences in the fabric and state of aggregation of the soils.

*Bulk density and compactibility*

Poorly developed basaltic soils may have high bulk densities even in surface horizons but with increased mineral weathering and further pedological development much lower bulk densities are found. Thus the bulk density for a high-clay basaltic saprolite in Malaysia ranged from 1.07 to 1.17 Mg $m^{-3}$ and fell to 0.98 Mg $m^{-3}$ in the Ap horizon (Hamdan *et al.*, 2000). Olowolafe (2002) found values between 0.90 and 1.10 Mg $m^{-3}$ in the clay-textured basaltic soils of Nigeria throughout the profile and down to a depth of 2 m. Bulk densities for adjacent granitic soils were always very much higher.

*Structure and aggregate stability*

Basaltic soils often contain water-stable aggregates, either through direct inheritance from the parent rock (Wilson, 1976) or through an association with the high free sesquioxide content resulting from mineral weathering (Singer, 1977). With further weathering the clay content of basaltic soils may increase to such an extent that they form Vertisols with a high smectitic clay content. This results in swelling and slaking during rainy seasons, with a consequent reduction of infiltration rate, increased run-off (especially on bare slopes) and enhanced soil erosion rates (Singer, 1987). In the dry season shrinking and cracking occur, resulting in the formation of large clods after ploughing, which may eventually break down into stable polyhedral aggregates.

## 9.3 Chemical properties

### *9.3.1 Granitic soils*

*Soil acidity*

The pH ($H_2O$) values in granitic saprolite or in the C horizons of granitic soils are often the highest in the profile and values >7 may be recorded in saprolites (Egashira *et al.*, 1985). This is possibly due to the rapid release of cations from unweathered but strained feldspars in the initial phase of weathering, such as are often found in laboratory experiments. Higher up in the soil profile and with further weathering and pedogenesis, the pH of granitic soils usually falls below 5.5 or even 5.0, especially when they are uncultivated. Thus, in the deep weathering profile on granite studied by Eswaran and Bin (1978) in Malaysia, pH values never exceeded 5.0 at any point. In granitic soils in Nigeria, the mean pH value for surface soils was 4.3 and for subsurface soils 4.7 (Olowolafe, 2002).

*Exchangeable cations and base saturation*

The acidity of granitic soils means that they tend to be characterized by low total exchangeable bases and percentage base saturation values and by high Al saturation. In cultivated granitic soils in Zimbabwe, total exchangeable bases exceeded 3 cmol(+) $kg^{-1}$ in only 2 of the 10 profiles studied by Burt *et al.* (2001). The granitic soils of north-east Scotland show similar low totals of exchangeable bases in a variety of soil types, except where they had been limed and fertilized (Glentworth and Muir, 1963). As might be expected, therefore, percentage base saturation values in granitic soils are typically very low, often being <20% or even <10%. Conversely, percentage Al or H saturation values tend to be very high (80–90%) in strongly acid (pH < 4.5) and moderately acid (pH 4.5–6.0) soils.

*Cation/anion exchange capacity*

Young coarse-textured granitic soils with low contents of clay and organic matter usually have low cation exchange capacities. For example, the C horizons of freely drained granitic soils in north-east Scotland nearly always have CEC values <5.0 cmol(+) $kg^{-1}$ (Glentworth and Muir, 1963). Increasing age and weathering intensity lead to higher clay contents and greater CEC values, although there is considerable variation in this respect. Thus, Volkoff *et al.* (1979) recorded CECs between 6 to 16 cmol(+) $kg^{-1}$ for A and B horizons of granitic soils formed under a subtropical humid climate in south-east Brazil, but for the Zimbabwean granitic soils studied by Burt *et al.* (2001) CECs were <5 cmol(+) $kg^{-1}$ even though there was a wide range of clay contents. Even lower mean CEC values (<3 cmol(+) $kg^{-1}$) were found by Gillman and Sumpter (1986) for intensely weathered granitic soils with clay contents of 20–25% in a high-rainfall tropical area of Queensland. The anion exchange capacity in these soils was also found to be very small, ranging from <0.1 to 0.4 cmol(–) $kg^{-1}$.

*Organic matter status*

The organic matter status of soils is usually considered to be influenced much more strongly by factors such as climate, vegetation, type and degree of soil development, position in the landscape, etc., than the nature of the parent material itself. It would be anticipated, for example, that soils developed on granitic parent materials under a tropical climate would contain much less organic matter than those developed under a cool temperate climate, because conditions in the former instance are so much more favourable for organic matter decomposition. This is generally found to be the case. However, parent material can still be an important factor, particularly with regard to the nature of the organic matter. For example, the humus in Atlantic thin immature soils developed on granite in Galicia, north-west Spain, is thought to be similar to that found in Andosols (Garcia Rodeja *et al.*, 1984) and to be stabilized by Al, resulting in the formation of Al-humus complexes. There is a strict relationship between organic matter and reactive Al, which is presumably derived from weathering minerals, suggesting that the organic matter is stabilized in a similar way to non-allophanic Andosols.

*N status*

The main source of native nitrogen in granitic soils is, of course in the soil organic matter fraction, with little or no N being attributed to the mineral fraction. It is worth noting, however, that a certain amount of native N may be associated with feldspars and micas, although very little of this would be plant-available. Again, some clay minerals occurring in granitic soils, such as dioctahedral vermiculite, may have the capacity to fix ammonium and although this is not immediately plant-available it may become so in time. The nitrogen status of granitic soils is usually low, certainly in terms of plant-available N, although for total N, in temperate climates at least, this is not necessarily true. In humid temperate climates, granitic soils may be high in organic matter and total N, with C/N ratios in the range of 12 to 14 indicating relatively high microbial activity. This is typical of many granitic soils in north-east Scotland (Glentworth and Muir, 1963) and is also found in granitic areas of France and Spain (Carballas *et al.*, 1979). The latter authors concluded that although local climatic factors in Galicia may influence the inhibited mineralization of organic matter in these soils, it was not sufficient to explain the process *in toto*. They showed that organic-rich soils developed on granitic parent material had abnormally low coefficients of endogenous mineralization and consequently showed an enhanced resistance to microbial attack, likely due to the formation of organomineral complexes. Carballas *et al.* (1980) went on to demonstrate the predominance of humus-Al complexes in the Spanish soils and stated that 'in this case parent material is the factor directing humification', although climate and vegetation have influenced the evolution of the soils. Under climatic and other conditions favouring high rates of organic matter mineralization, the nitrogen status of granitic soils is usually low or very low in terms of both total and available N. Thus, the mean total N for surface samples of the Nigerian granitic soils studied by Olowolafe (2002) was 0.07% and was nearly always $<0.1\%$ for similar soils in a variety of topographic situations. The Zimbabwean granitic soils studied by Burt *et al.* (2001) showed similar very low totals.

### *P status*

The mean phosphorus content of granites is 0.08% corresponding to a normative apatite content of 0.4% (Nockolds, 1954), and as the solubility of apatite is low it would be anticipated that granitic soils would be deficient in plant-available phosphate. This is indeed usually found to be the case in granitic soils developed under a range of climatic conditions. For example, Williams (1959) found that the surface horizons of granitic soils in north-east Scotland contained the lowest total, inorganic and organic phosphate amounts compared with surface soils derived from other parent materials (gabbro, slate and sandstone). Organic P (61%) tended to be concentrated in the silt and clay fractions, where it was predominantly Al-bound, whereas inorganic P was concentrated in the sand fractions and was Ca-bound. In both cases, the P was in a non-available form. Granitic soils developed under more intensive weathering conditions are even more likely to be P-deficient. Nigerian granitic soils typically have values of available P below 15 mg $kg^{-1}$ as determined by the Bray method. In fact, the mean value for available P for the granitic soils studied by Olowolafe (2002) was only 4.4 mg $kg^{-1}$ even though total P averaged 4300 mg $kg^{-1}$.

### *K status*

Granites typically contain about 4.6% potassium corresponding to a normative orthoclase content of ~32% (Nockolds, 1954), so that granitic soils should be relatively K-rich, at least in terms of total K content. However, little of this store of K is available to the plant, particularly in relatively young soils. Thus, the soils developed on granitic till in north-east Scotland often have exchangeable K contents between 0.1 to 0.2 cmol $kg^{-1}$ in surface horizons and <0.1 cmol $kg^{-1}$ in the rest of the soil profile (Glentworth and Muir, 1963). With more intensive weathering and with an increase in clay content, however, there may be a significant improvement in the K-status of granitic soils. Thus, some of the Nigerian granitic soils studied by Olowolafe (2002) contain moderate amounts of exchangeable K (>0.4 cmol $kg^{-1}$), and moderately high amounts of exchangeable K were found in granitic soils in other countries in the tropics and subtropics. It is concluded, therefore that the K-status of granitic soils in these countries may be moderately good and is often better than that for granitic soils developed under temperate climates.

### *Trace element status*

Granitic soils may be susceptible to trace element deficiencies, although there is a great deal of variability in this respect. The biologically important trace elements considered here are Co, Cu and Zn, the mean total contents of these elements for granitic rocks being in the ranges 3–5, 8–35 and 30–80 mg $kg^{-1}$ respectively (Wedepohl, 1978). But total content is a poor guide to plant availability, the forms in which the trace elements are held in the soil being the all-important factor. In north-east Scotland, the trace element status of young granitic soils on glacial till is highly variable, although generally they are adequately supplied (Glentworth and Muir, 1963). However, cobalt deficiency in sheep was first recognized in Britain on the granitic soils of Dartmoor, and some granitic soils in Scotland are also cobalt deficient. In strongly weathered granitic soils such as are found in Australia, copper deficiency is more common and copper and zinc deficiencies occur in the granitic soils of the Jos Plateau in Nigeria (Olowolafe, 2002). In summary, therefore, it is concluded that although trace element deficiency does exist on granitic soils, especially when they are strongly weathered, it cannot be said to be a general problem.

### *9.3.2 Basaltic soils*

#### *Soil acidity*

The poorly developed or clay-rich basaltic soils formed under Mediterranean or arid conditions yield pH values in water ranging from 6.5 to 7.5. Lower pH values are found for immature basaltic soils of cool wet climates. The various soils formed on the basaltic glacial till of south-west Scotland generally show pH ($H_2O$) values in surface samples in the range of 5.0 to 6.0, increasing down the profile to values between 6.0 and 7.0 (Mitchell and Jarvis, 1956). Under more intensive weathering regimes, the pH values for basaltic soils may not necessarily fall to particularly low values. For example, the red basaltic soils of north Queensland yield pH ($H_2O$) values between 5.0 to 6.0 up to a depth of 8 m under rainforest, and higher pH values in the range 6.3 to 6.7 are usually found in the thinner basaltic soils under eucalypt woodland (Isbell *et al.*, 1976). The brown basaltic soils of the same region also yielded high pH values in the range 6.4 to 7.0 throughout their profiles (Isbell *et al.*, 1977). However, for the more highly oxidic and presumably more intensely weathered basaltic soils of high-rainfall tropical Queensland, the mean pH was $< 5.0$ in either the virgin or the cultivated state (Gillman and Sumpter, 1986).

#### *Exchangeable cations and base saturation*

Exchangeable cations are usually in the medium to high categories in basaltic soils formed under a variety of climatic conditions. In the basaltic soils in south-west Scotland exchangeable Ca usually falls into these categories (3–8 cmol $kg^{-1}$) as does exchangeable Mg (1–2 cmol $kg^{-1}$), although exchangeable K values usually fall into the medium to low range ($<$ 0.1–1.0 cmol $kg^{-1}$). For the most part these soils are highly base saturated (>60%), although surface horizons often fall into the medium range (20–60%). Low (<20%) to medium base saturation percentages are found in the surface horizons of freely draining soils. The basaltic soils developed under semi-arid conditions on the Golan Heights have high pH values and are fully base saturated (Singer, 1987). Sometimes these soils contain small amounts of free $CaCO_3$. With more intensive weathering conditions it might be anticipated that the percentage base saturation and exchangeable cation contents of basaltic soils would quickly fall to low levels, but this is not always the case. For example, in the brown basaltic soils of north Queensland exchangeable Ca is nearly always high except occasionally towards the base of the profile. Exchangeable Mg is usually very high and exchangeable K falls into the medium category. Base saturation percentages often exceed 80% (Isbell *et al.*, 1977). With further weathering and more intensive drainage, however, as in the red basaltic soils under rainforest (Isbell *et al.*, 1976; Gillman and Sumpter, 1986), values for exchangeable Ca, Mg and K are nearly always in the low category and base saturation percentages are usually low.

### *Cation/anion exchange capacity*

The CEC of immature basaltic soils, even in basal horizons where there is little or no organic matter and apparently only small amounts of clay, can be quite high. Thus in soils developed upon basaltic glacial till in north-east Scotland, the non-clay fractions had high CECs (silt 11–57, fine sand 2–46 and coarse sand 1–26 cmol(+) $kg^{-1}$) and actually contributed more to the total soil CEC than the clay fraction (Wilson and Logan, 1976). Similar findings with respect to basaltic soils have been made in the USA, Australia, Northern Ireland and India. These observations can be rationalized in terms of the parent rock containing significant quantities of hydrothermally formed, high CEC expansible clay minerals, which are directly inherited into the soil as coherent particles that are able to participate in exchange reactions. Coherent high CEC aggregates may also form through pedogenic activity after fairly intensive weathering, as in the brown basaltic soils of Queensland (Isbell *et al.*, 1977). In the even more intensively weathered, strongly kaolinitic basaltic soils in tropical Queensland, very low CECs (<3 cmol (+) $kg^{-1}$) and AECs (<2 cmol(−) $kg^{-1}$) were found whether the soils were in the cultivated or the virgin state (Gillman and Sumpter, 1986). At pH 4 Gillman and Sumpter (1986) found that the preferential occupation of exchange sites by Al in basaltic soils is significantly lower than in granitic soils developed under the same climatic conditions, indicating that the influence of parent material is still apparent despite the intensive weathering to which the soils have been subjected.

### *Organic matter status*

The organic matter content of basaltic mineral soils formed under cool, temperate climatic conditions can be quite high. Thus, the basaltic soils of south-west Scotland usually have organic matter contents between 8 to 17% in surface horizons, decreasing to values between 2 to 6% in the immediate subsoil (Mitchell and Jarvis, 1956). Where conditions favour the development of peaty soils organic matter totals of between 34 to 68% are found in both surface and subsurface horizons. The C/N ratio of the mineral soils usually falls in the range of 10 to 15 whereas that for the peaty soils is between 15 to 30. For tropical basaltic soils organic matter totals are not nearly as low as might be expected from a consideration of climatic factors alone. In the basaltic soils of the Jos Plateau, Nigeria, mean total organic matter was ~4.3% in surface soils, ranging between 2.2 and 6.3%, and ~1.6% in subsoils, ranging between 0 to 4% (Olowolafe, 2002). These basaltic soils had very much higher organic matter contents than the neighbouring granitic soils. The mean C/N ratio of these Nigerian basaltic soils was ~20, showing that humification was still at a relatively early stage. For the red basaltic soils of Queensland, mean organic matter totals were very high, being nearly 12% in the rainforest and nearly 6% under eucalypt woodland (Isbell *et al.*, 1976). In the high rainfall areas of tropical Queensland, Gillman and Sumpter (1986) found that basaltic soils contained more soil organic matter than their granitic counterparts, with mean totals for surface soils and subsoils of 5.6% and 3.2% respectively. In addition, the basaltic soils contained approximately 8 to 10 times more extractable Fe (8 to 10%) than the granitic soils, a clearly discernible parent material effect even in these clay-rich, highly weathered soils. The red basaltic soils studied by Isbell *et al.* (1976) are exceptionally rich in total iron (mean of 21%) and extractable (mean of 17%) and it is reasonable to assume that conditions in these soils would be suitable for the formation of Fe-organic complexes.

### *N status*

As with all mineral soils, basaltic soils are usually N-deficient in terms of available nitrogen, but there may be considerable variation with respect to total N, depending on other soil-forming factors. Thus, in south-west Scotland, of the 21 basaltic soil profiles studied by Mitchell and Jarvis (1956), total N was very high (>1%) in surface horizons in 5 profiles, high (0.5 to 1%) in 6 profiles and medium (0.2 to 0.5%) in 7 profiles. The soils with very high values were of a peaty nature while the high to medium N soils were classed as brown forest soils or non-calcareous gleys. These relatively high total N values may be attributed primarily to the cool wet climate, leading to acidic peaty soils, effectively inhibiting the process of microbial decomposition of organic matter.

For basaltic soils developing under a tropical climate it would be expected that total N contents would fall to lower levels and this is often the case. In Australia, the Euchrozems and Black earths, both of which derive from basaltic parent material, usually contain low to very low levels of total N. However, the red basaltic Krasnozems of northern Queensland are characterized by medium to high total N values (Isbell *et al.*, 1976). This probably relates to their highly oxidic nature, leading to a situation where organic matter is tightly bound to iron oxide minerals or to the formation of iron-organic complexes, both forms being inaccessible to microbial decomposition.

### *P status*

The mean phosphorus content of basaltic rocks is 0.10% for tholeiitic basalts and 0.17% for alkali olivine basalts (Nockolds, 1954) which is 2 to 3 times the P content of granites. As phosphorus is a relatively immobile element in the landscape the P content of soils to a large extent reflects that of the parent material. The chief mineral form of phosphate in basalts is apatite, usually fluorapatite ($Ca_5(PO_4)_3F$) but rare earth phosphate minerals such as monazite and xenotime may also occur. All these minerals are resistant to weathering so that they eventually accumulate in the soil. In the young soils developed upon basaltic till in south-west Scotland total P contents are usually moderately high in the surface horizons, ranging from 0.1% to 0.3%, decreasing to values between 0.04 to 0.1% at the base of the profile (Mitchell and Jarvis, 1956). Results for readily soluble phosphate (in acetic acid) show an inverse relationship with medium to high values (20 to 40 mg $kg^{-1}$) being recorded at the base of the profile and low values (1 to 3 mg $kg^{-1}$) in surface samples. Using direct microprobe methods Norrish and Rosser (1983) found a strong linear relationship between %P in the native form and %Fe in a Queensland Oxisol. Iron was in the form of both goethite and haematite and it was concluded that the relationship was independent of the form of the iron oxide mineral. The influence of the parent material was very evident because the iron oxide minerals are sometimes in the form of microaggregates. These behave optically as single crystals because they are pseudomorphs after olivine which have been inherited from the parent rock.

### *K status*

The mean potassium content of basaltic rocks is about 16% of that for granitic rocks. For tholeiitic basalts the mean K content is 0.68% and for alkali olivine basalts 0.79% (Nockolds, 1954). This corresponds to a normative orthoclase content of 5.0% and 6.1% respectively. Basaltic soils might be expected to be more K-deficient than those derived from granites but this is certainly not always true. Thus the K-status of basaltic soils in south-west Scotland is generally moderate, falling into the 0.2 to 0.6 cmol $kg^{-1}$ range (Mitchell and Jarvis, 1956). Again the brown basaltic soils of Queensland are often high in exchangeable K, ranging between 0.4 to 1.7 cmol $kg^{-1}$, with a mean value of 1.1 cmol $kg^{-1}$ (Isbell *et al.*, 1977). Clay-rich basaltic soils, such as those in India, often show K-deficiency symptoms due to their high K-fixing capacity, relating to the nature of their constituent clay minerals.

### *Trace element status*

With regard to micronutrients, the content for tholeiitic basalts worldwide usually falls into the range of 30–50 mg $kg^{-1}$ for Co, 40–150 mg $kg^{-1}$ for Cu and 100–150 mg $kg^{-1}$ for Zn (Wedepohl, 1978). In the young basaltic soils in Scotland there was no indication of trace element problems (Mitchell and Jarvis, 1956) and this was also the case for the basaltic soils of the Jos Plateau (Olowolafe, 2002). For the more intensively weathered basaltic soils developed upon ancient erosion surfaces in Queensland, trace element problems do occur but they are not universal. Thus the basaltic Krasnozems are associated with Co-deficiency in animals and these soils may also be Cu-deficient. Black earths in Queensland, also derived from basalt, may be Zn-deficient for cereals (Tiller, 1983). However, there is no consistency with respect to soil groups, so that trace element status cannot be reliably predicted from broad-scale surveys. Sometimes basaltic soils may be high in total micronutrients but still show deficiency symptoms (Tiller, 1983).

## 9.4 Conclusions

Many of the characteristic differences between granitic and basaltic soils outlined above and summarized in Table 9.1 can be directly related to the different mineralogies of the parent rocks as well as to the different types of clay minerals that develop through the weathering of these rocks. Thus, granites are coarse-grained and consist predominantly of quartz and K-feldspar, both of which are resistant to weathering, with minor amounts of ferromagnesian minerals other than biotite. Clay mineral weathering products are usually kaolinitic and rarely smectitic. This generally leads to porous soils with a low clay content, a poor capacity to retain water and limited ability to form water-stable aggregates. In addition, parent material mineralogy leads to acid soils, low contents of exchangeable bases and low CECs. Parent material is also implicated in the accumulation and nature of organic matter on and in granitic soils by virtue of the release of aluminium during mineral weathering and the subsequent formation of resistant Al-humus complexes. These complexes also inhibit the release of N and P in a plant-available form. By contrast, basalts are fine-grained, consisting

Table 9.1. *Comparison of general physical properties of granitic and basaltic soils under different weathering conditions*

| Physical property | Granitic soils | | Basaltic soils | |
|---|---|---|---|---|
| | Weak to moderate weathering | High to intense weathering | Weak to moderate weathering | High to intense weathering |
| Particle size distribution | Coarse sandy texture <15% clay | Loam to coarse sandy loam 15–40% clay | Loam, clay loam and clay textures 15–50% clay, except in immature soils | Very high clay contents often in the 50 to 60% range |
| Porosity and hydraulic conductivity | High values at surface, declining with depth | Moderate values at surface declining with depth | Low values due to high clay contents except where clays are aggregated | Higher values than would be expected from clay content |
| Water retention/ availability | Very poor except for saprolites | Improving with increasing clay content | High water content but may not be plant-available in clay soils | Difficult to generalize |
| Bulk density and compactibility | Very high bulk densities | High BDs >1.6 Mg $m^{-3}$: compaction and root impedance problems | High BDs in immature soils | Low BDs in both surface and subsurface horizons |
| Structure and aggregate stability | Coarse, gritty structure. Limited ability to form water-stable aggregates | Increasing clay content improves structure | Good crumb structure due to water stable aggregates but massive structures may form due to high clay content | Good crumb structure at surface due to formation of water-stable aggregates |

predominantly of calcic plagioclase feldspars and ferromagnesian minerals, both of which are susceptible to weathering. Clay mineral weathering products are often smectitic and fresh basalt itself may contain significant quantities of hydrothermally formed smectitic material. This generally leads to soils with a higher clay content and a higher capacity to retain water. Porosity and hydraulic conductivity may be low but some basaltic soils are characterized by a high content of water stable aggregates which makes them more porous than might be expected from their clay contents. Basaltic soils are acidic only under the severe weathering conditions when their clay mineralogy becomes kaolinitic. Usually they are adequately supplied with exchangeable bases that derive directly from mineral weathering. They have high CECs, including sometimes the non-clay fractions, except when they are in a highly weathered state. The accumulation and nature of organic matter in basaltic soils relates to parent material through the formation of Fe-organic complexes or through association with iron oxide minerals, the source of the iron being through weathering of ferromagnesian minerals. Organic matter in this form is resistant to microbial decomposition. This again has implications with respect to the availability to plants of N and P from organic matter in these soils. Finally, the capacity of some basaltic soils to fix K may be related to the beidellitic nature of the smectite component in the clay fraction.

In conclusion, it is evident that the influence of parent material on the properties of soils is important even up to the most severe stages of weathering and is seldom overwhelmed without trace by the other soil-forming factors. Under less extreme circumstances it is crucial to an understanding of soil behaviour, particularly with respect to inherent fertility status, soil management, soil erosion and degradation and soil biodiversity. This conclusion stems from an examination of only two selected types of parent materials but in the writer's opinion had other parent materials been examined in the same way a similar conclusion would have been reached.

# 10

# Factors of soil formation: climate. As exemplified by volcanic ash soils

*Sadao Shoji*
*Masami Nanzyo*
*Tadashi Takahashi*

Global climatic factors, especially temperature and precipitation, most strongly influence soil formation. Temperature acts on the reactions involved in soil processes in a variety of ways. It controls the speed of chemical reactions as described by the well-known Van't Hoff's temperature rule. However, all chemical reactions cease in the absence of soil moisture due to freezing or drying up. Temperature also determines the type and biomass of vegetation closely relating to soil formation. Soil moisture supplied by precipitation is crucial to the forming and functioning of the soil. It contributes to the dissolution, neoformation and translocation of materials and facilitates the growth of vegetation that also acts on soil formation (Buol *et al.*, 1997).

Although there are many treatises dealing with this subject, we will select a conceptual and comprehensive one published by Ugolini and Spaltenstein (1992) and will begin this chapter by introducing it. We will then describe the influence of climatic factors on soil formation, mainly based on our studies of volcanic ash soils which have several distinct advantages for pedogenesis studies. The age, rock types, and chemical and mineralogical properties of volcanic ash as a parent material can be easily determined. Volcanic ash is commonly unconsolidated and comminuted, and is dominated by volcanic glass, which shows the least resistance to chemical weathering (Shoji, 1986). Therefore, volcanic ash rapidly forms a variety of soils in all climate zones, and the soil processes involved are highly accentuated. Intermittent volcanic ash deposition and repeated pedogenesis commonly create sedentary multistoreyed soils. These soils notably resist soil disturbance, truncation, and displacement so that they preserve highly useful information concerning the palaeoclimate, palaeovegetation and palaeopedogenesis.

*Soils: Basic Concepts and Future Challenges*, ed. Giacomo Certini and Riccardo Scalenghe.
Published by Cambridge University Press. 

## 10.1 Global climate and soil formation

Ugolini and Spaltenstein (1992) employed Jenny's approach (Jenny, 1941) as a conceptual framework to relate the soil-forming processes to the global environment. They assumed that soil formation, along a transect from the poles to the equator, occurs under uniform parent material, similar topography, and equivalent time, and the significant variables are climate and biota. Though Jenny considered each of the soil-forming factors as an independent variable, climate and biota are particularly closely linked.

Important classes of soil temperature and soil moisture are cold (cryic), cool (frigid), temperate (mesic), warm (thermic) and hot (hyperthermic), and arid (aridic), semi-arid (ustic), Mediterranean-climate (xeric), and humid (udic), respectively. As shown in Table 10.1, Ugolini and Spaltenstein (1992) divided the global climate into nine zones and employed soil moisture regimes and soil temperature regimes to describe the climatic conditions and soil Orders of Soil Taxonomy that form in each climate zone. Furthermore, they characteristically showed the compartment of each soil Order consisting of a canopy, organic and mineral soil horizons, specific impact of climate and surface conditions on the soil horizons, and speed of soil formation (pedogenesis).

(1) *Cold desert zone.* The cold desert is mostly centred in the ice-free areas of Antarctica and in northern Greenland. This zone has extremely dry and extremely cold climates (ultraxeric soil moisture and ultrapergelic temperature regimes). Physical weathering is mainly caused by frost wedging and insolation. Chemical weathering is restricted by the continuous low temperature and paucity of liquid water. Salts liberated through weathering or accreted by meteoric input are slowly removed in the virtual absence of percolating water. Consequently, chlorides, sulphates, nitrates and other soluble salts are retained and calcium carbonate precipitates in the soil. Soils formed under this environment are extremely slow with respect to the speed of pedogenesis. They were classified as Entisols (soils with very early stage of development) in Soil Taxonomy (Soil Survey Staff, 1975).

(2) *Polar desert zone.* This zone occurs in the northernmost sectors of the Arctic and has extremely dry and very cold climates (ultraxeric moisture and pergelic soil temperature regimes). Though barren conditions prevail, vascular plants are present as small patches. Physical weathering is shown by the existence of frost-shattered rocks. The rate of chemical weathering is low, but higher than in the cold desert zone. The neoformation of Fe-hydroxide and accumulation of secondary calcium carbonates occur in the Bw horizon and in the C horizon, respectively. The soils formed in this climatic zone were also classified as Entisols (Soil Survey Staff, 1975).

(3) *Tundra.* The tundra zone stretches south of the polar desert to the margin of the tree line and has arid and cold climate (aridic soil moisture and cryic soil temperature regimes). Most of the terrain is waterlogged during the growing season because of the

Table 10.1. *Soils and soil-forming processes as summarized by Ugolini and Spaltenstein (1992)*

| | | | | | | Temperate forest | | | | Savanna | | |
|---|---|---|---|---|---|---|---|---|---|---|---|---|
| | Zone | Cold desert | Polar desert | Tundra | Boreal forest | Coniferous | Deciduous | Grassland | Desert | Treeless | Arboreal | Tropical rainforest |
| ST[a] | Soil Order | Entisols (Gelisols) | Entisols (Gelisols) | Inceptisols (Gelisols) | Spodosols | Inceptisols | Alfisols | Mollisols | Aridisols | Vertisols | Ultisols | Oxisols |
| WRB[b] Climate | Soil Group Moisture | Cryosols ultra xeric (very dry) | Cryosols ultra xeric (very dry) | Cryosols aridic (dry) | Podzols udic (humid) | Cambisols | Luvisols udic (humid) | Chernozems ustic (semi-arid) | Solonchaks aridic (arid) | Vertisols ustic (semi-arid) | Acrisols udic (humid) | Ferralsols perudic (very humid) |
| | Temperature | ultra pergelic (extremely cold) | pergelic (very cold) | cryic (cold) | frigid (cool) | mesic (temperate) | | mesic (temperate) thermic (warm) | | isothermic (hot) | | isohyperthermic (very hot) |
| Specific impact of climate and surface conditions on the horizons | O | | | under-saturated non-mobile organic acids | very under-saturated mobile organic acids | under-saturated non-mobile organic acids | | saturated non-mobile organic acids | | saturated non-mobile organic acids | under-saturated organic acids | very under-saturated mobile organic acids |
| | A, E Bh | | | | | | | | | | | |
| | B | | Fe-hydroxide neoformation | smectite, vermiculite formation | non-crystalline aluminosilicates and Fe-hydroxide neoformation | smectite, vermiculite formation by transformation of preexisting phyllosilicates | | Fe-hydroxide neoformation $CaCO_3$ precipitation | Fe-hydroxide neoformation $CaCO_3$ precipitation | smectite neofor-mation | kaolinite neo-formation | Fe- and Al- hydroxide neoformation |
| | C | $CaCO_3$ precipitation | $CaCO_3$ precipitation | Fe-hydroxide neoformation | Fe-hydroxide neoformation | Fe-hydroxide neoformation | | $CaCO_3$ precipitation | $CaCO_3$ precipitation | $CaCO_3$ precipi-tation | Fe-hydroxide neo-formation | Kaolinite neoformation |
| | R | nil | nil | nil | nil | nil | | nil | nil | nil | | Fe-hydroxide neoformation |
| Speed of pedogenesis | | extremely slow | very slow | slow | moderate | moderate | | moderate | slow | fast | | very fast |

[a] ST, Soil Taxonomy.
[b]WRB, World Reference Base for Soil Resources.

permafrost (a soil layer below 0°C for at least two years in succession). A small area of the landscape is well drained, receiving the full impact of the soil-forming factors. Alteration of minerals, such as mica, and the neoformation of Fe-hydroxide take place in the subsoil. These soils consist of A, Bw and C horizons. They were classified as Inceptisols (young, slightly to moderately developed soils) (Soil Survey Staff, 1975).

Ugolini and Spaltenstein described their paper before Gelisols were introduced into Soil Taxonomy. Gelisols are soils that have a permafrost within 100 cm of the soil surface or gelic materials (Appendix Table VI) within 100 cm of the soil surface and permafrost within 200 cm of the soil surface according to the current Soil Taxonomy (Soil Survey Staff, 2006). Therefore, most soils in the cold desert, polar desert and tundra zones are now classified as Gelisols. Gelisols correspond to Cryosols in the World Reference Base for Soil Resources (WRB) (IUSS Working Group WRB, 2006).

(4) *Boreal forest.* This zone occurs south of the Arctic tree line and into the forest belt and has humid and cool climates (udic soil moisture and frigid temperature regimes). There is much more biomass mainly consisting of coniferous trees than in the tundra, and the soil profile differentiation is greatly accentuated, showing a moderate speed of pedogenesis. Organic acids act both as proton donors and chelating agents. They attack the minerals in the upper compartment, resulting in a chromatographic-like redistribution of fulvic acids, iron and aluminium, and distinct differentiation of the soil horizons, such as the O, E, Bh and Bs horizons. Minerals in the E horizon are dissolved while smectite appears to be stable in this environment. In the lower compartment (Bs and BC horizons) which is dominated by carbonic acid, non-crystalline silicates, such as allophane and imogolite, may form as well as the oxides and hydroxides of iron and aluminium. Vermiculite with an Al interlayer is also present. Spodosols (Soil Survey Staff, 2006) or Podzols (IUSS Working Group WRB, 2006) are the soils formed in this zone.

(5) *Temperate deciduous forest.* This zone is found south of the boreal forest belt, and has humid and temperate climates (udic soil moisture and mesic soil temperature regimes). Plants produce litter rich in nitrogen and cations and poor in lignin. The rates of organic matter decomposition are higher than in the boreal forest belt. Earthworms mix the humic substances with the surface mineral soil, and the chromatographic-like redistribution of the organic acids and trivalent metals does not occur. Soils show a moderate speed of pedogenesis and their morphology consists of an A or E, Bw or Bt, and C horizons. Inceptisols and Alfisols (Soil Survey Staff, 2006) or Cambisols and Luvisols (IUSS Working Group WRB, 2006) are the soils developed in this zone.

(6) *Grassland.* The grassland or steppe zone experiences a continental climate with hot, dry summers and cold winters (ustic soil moisture and mesic soil temperature regimes). Rains are prevalent during the spring and autumn. These conditions favour the retention of organic matter in the soil, high $CO_2$ partial pressures, and the formation of carbonic acid. Melanization, the darkening of the soil due to organic matter accumulation, is the prevailing soil process that is well expressed under

grassland vegetation. The soils have a thick dark A horizon with developed structures and plentiful plant roots. The B horizon may be poorly developed, while the C horizon may contain soft concretions of calcium carbonate. The speed of pedogenesis is moderate. Soils common in the grassland zone are called Mollisols (Soil Survey Staff, 2006) or Chernozems, Kastanozems and Phaeozems (IUSS Working Group WRB, 2006).

(7) *Hot desert.* This zone shows scant precipitation and high temperature (aridic soil moisture and thermic soil temperature regimes) so that the rate of pedogenesis is low. These conditions reduce leaching and favour the retention of soluble products of weathering. They also reduce biomass production and soil organic matter content. Therefore, the organic compartment is thin. The formation of 2:1 minerals and Fe-hydroxide and the precipitation of calcium carbonate take place in the soil. The soils of the hot desert are Aridisols (Soil Survey Staff, 2006). They correspond to Solonchaks, Solonetz, Gypsisols, Calcisols and so on in WRB (IUSS Working Group WRB, 2006).

(8) *Savanna.* Savannas occur in tropical or subtropical climates with a marked dry season (ustic/udic soil moisture and isothermic soil temperature regimes). Carbonic acid is the prevailing proton donor. During the wet season, weathering occurs and kaolinite forms in the well-drained soils. The soil solution, charged with $H_4SiO_4$ (silicic acid) and basic cations, migrates to lowlands and depressions where the soil solution is concentrated during the dry season. This leads to an intensive neoformation of expansible 2:1 clay minerals, mostly smectite. Organic matter tends to accumulate all through the soil, forming a thick A horizon, and the clay content markedly increases. These soils are poorly drained during the wet season while they become strongly desiccated during the dry season. Common soils in the savanna are the Vertisols (Soil Survey Staff, 2006; IUSS Working Group WRB, 2006).

(9) *Tropical rainforest.* The increased rainfall and high temperatures (perudic soil moisture and isohyperthermic soil temperature regimes) of the tropical rainforest zone favour a high biomass production, strong weathering, and intense leaching. Though the growth of vegetation is luxurious, mineralization of organic matter is so rapid that the soil organic matter content is low and the A horizon is thin. Intense hydrolysis reactions are favoured by the large volume of water that penetrates the soil, while an aggressive carbonic acid weathering proceeds under a high $CO_2$ partial pressure. Iron becomes segregated as a separate phase and forms intensively red oxides such as haematite, and oxyhydroxides such as goethite. The iron accumulation contributes to the formation of an iron crust at the surface that responds to desiccation caused by changes in the climate or ecological conditions. The weathering profile is very deep and it is difficult to distinguish soil profiles and soil horizons. Oxisols (Soil Survey Staff, 2006) or Ferralsols (IUSS Working Group WRB, 2006) are unique to this environment.

There are considerable differences in climate conditions between soil Orders described by Ugolini and Spaltenstein (Table 10.1) and the equivalent soil groups in the MAP–MAT graph shown in Fig. 10.1. These differences in distribution are considered to be attributable to the assumptions of Ugolini and Spaltenstein

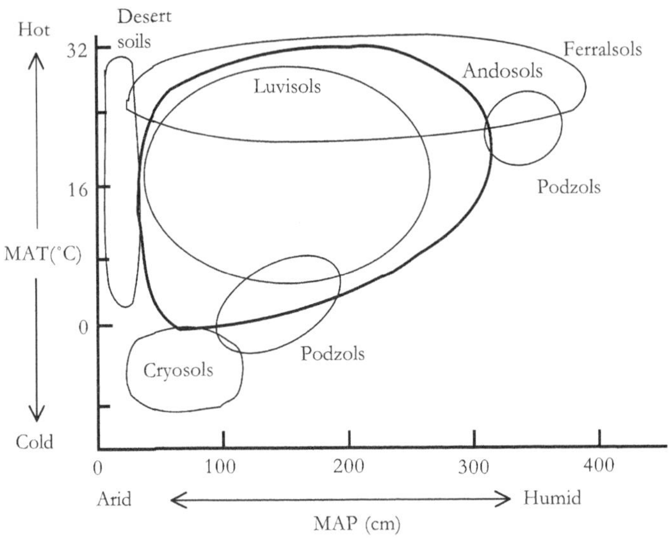


Fig. 10.1. Distribution of selected soil groups in a mean annual precipitation–mean annual temperature (MAP–MAT) graph.

(1992) that the soil formation of a majority of each soil Order is taking place under the present-day climatic conditions and the parent material has not experienced pre-weathering. In fact, Oxisols (Ferralsols) occur not only in tropical soil temperature regimes, but also south of the iso soil temperature regimes. Their distribution tends to be independent of the present rainfall patterns, suggesting that many Oxisols attain their mineralogical characteristics by a process not related to their present location (Buol *et al.*, 1997). Alfisols (Luvisols) are present in humid, subhumid and arid conditions. Under subhumid and arid conditions, Alfisols occupy stable landscapes that have undergone long-term weathering under fairly constant climate conditions or are polygenetic, and have experienced climatic changes during the Pleistocene (Buol *et al.*, 1997). Spodosols (Podzols) are also mainly distributed in the humid boreal climate zone. However, they also occur in the tropical lowlands. This diversity of Spodosol distribution indicates that in addition to the climate, the parent materials with a coarse texture and vegetation supplying mobile and sesquioxide-mobilizing organic compounds contribute greatly to the formation of Spodosols (Buol *et al.*, 1997).

Ugolini and Spaltenstein used nine soil Orders for their discussion on the relationships between climate and soil formation. However, three soil Orders, such as Gelisols (Cryosols), Histosols (Histosols), and Andisols (Andosols), were not described.

A majority of Histosols are typically climate independent while blanket peats and raised bogs are 'climatic': they depend on rainfall for water and nutrients

(Buol *et al.*, 1997). The aclimatic Histosols are formed on the depressions of cold to tropical regions where restricted drainage inhibits the decomposition of plant and animal remains.

Nearly all Andisols are formed from volcanic ash that is unconsolidated, comminuted materials containing a large quantity of most weatherable volcanic glass. Therefore, the development of Andisols is primarily influenced by the properties of volcanic ash, so that all soil temperature regimes except pergelic and all soil moisture regimes are recognized in Andisols. Considering the unique genetic conditions of Histosols and Andisols described above, Ugolini and Spaltenstein would have excluded these orders in their framework (Table 10.1).

## 10.2 Influences of climatic factors on soil formation based on the studies on volcanic ash soils

### *10.2.1 Basic considerations*

#### *Chemical kinetics of weathering*

The kinetic concept is one of the most useful approaches for describing the effects of climate on soil formation (Shoji *et al.*, 1993b). Andosols derived from volcanic ash have some notable advantages because the time zero of soil formation can be clearly set at the time of ash deposition, which is determined with reasonable accuracy by various methods. Tephra or volcanic ash is commonly dominated by volcanic glass that is the most weatherable parent mineral.

The dissolution of fresh volcanic glass consists of two processes. One is the rapid dissolution of the fresh surface of glass. The release of cations such as Na, K, Mg and Ca takes place faster than Si and Al, and the Si-O-Si and Si-O-Al framework remains on the glass surface. The release of Si and Al is a hydrolytic reaction of these chemical bonds and shows almost a linear time-course after the initial very rapid dissolution process. The temperature dependency coefficient ($Q_{10}$) of Al and Si is 1.5 and 1.6–1.7 at ambient temperature, respectively, according to a laboratory dissolution experiment.

Andosols occurring on well-drained sites are subjected to intense leaching under humid climates, and the importance of the leaching process is supported by comparison of the elemental composition between the volcanic glass and allophane–imogolite or Al-humus complex. The Si/Al atomic ratio of non-coloured volcanic glass is around 4.7 whereas the weathering products have the very low atomic ratio of 0.5 for allophane and imogolite and almost zero for the Al-humus complex.

Silicon and basic cations are mostly lost by leaching from Andosols, while Al stays *in situ* mainly as allophane–imogolite and Al-humus complexes. The

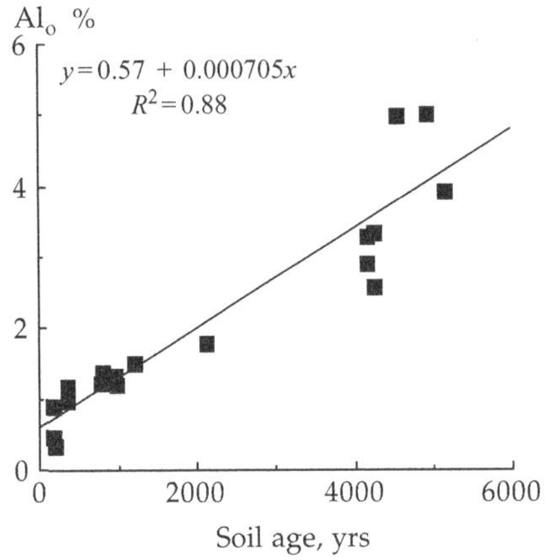


Fig. 10.2. Relationship between soil age normalized to 10 °C and acid oxalate extractable Al in the A horizons of Andosols from north-east Japan. From Shoji *et al.* (1993b).

content of these non-crystalline Al materials can be determined as acid oxalate-extractable Al ($Al_o$). The $Al_o$ content increases almost linearly with the soil age normalized to 10 °C, after a small quick increase during the initial stage (Fig. 10.2). The normalized soil age is converted from the normalized temperature and is useful for determining the temperature effect on chemical weathering of soils with the same soil moisture regime (Shoji *et al.*, 1993b).

During Andosol formation, a huge amount of Si is leached out, resulting in a notably increasing Si concentration of river water in the volcanic ash soil areas. If the weathering product is dominantly Al-humus, the released Si is mostly lost by leaching. The Si incorporated into the weathering product is only about one tenth of the Si released from the volcanic glass even when all the released Al occurs in allophane and imogolite.

According to the equation for the soil age normalized to 10 °C and $Al_o$ ($Al_o + Fe_o/2 = 1.36\ Al_o$), the ($Al_o + Fe_o/2$) of volcanic ash dominated by non-coloured glass attains 2.0% within 1280 years. If the same volcanic ash is subjected to weathering at a higher temperature (such as 25 °C at the same soil moisture) the time required to reach the $Al_o + Fe_o/2$ of 2.0% is 700 years (Shoji *et al.*, 1993b).

### *Database considerations on climate and soil formation*

The Tohoku University World Andosol Database (TUWAD) has been utilized for preparing proposed revisions to the classification of Andosols in the WRB (Shoji *et al.*, 1996; Takahashi *et al.*, 2004). This database is useful for providing the entire picture of Andosols formed under various soil-forming conditions. It can also show the influence of climatic factors on the soil properties and can validate the chemical kinetics studies described in the previous section.

As information on arid Andosols (Torrands in the Soil Taxonomy) is very limited, the arid Andosols are excluded from the database considerations. A total of 1362 samples from 210 pedons have been used to investigate the relationships between soil temperature and soil moisture with soil properties, such as the concentrations of acid oxalate-extractable Al ($Al_o$), acid oxalate-extractable Si ($Si_o$), and organic carbon. A horizons most strongly reflect the soil-forming conditions, so A and buried A horizons were selected to examine the $Si_o$ to $Al_o$ ratios from them. These properties are especially important for characterizing the Andosols.

Aluminium is the least mobile chemical element during the weathering of Andosols, so $Al_o$, commonly called active Al, is useful for determining the weathering degrees of Andosols if the major weathering products are non-crystalline and for estimating the content of Al occurring in allophane–imogolite and Al-humus complexes. The relationship between soil temperature and $Al_o$ values has been studied using humid Andosol samples including very humid (perudic) ones, as presented in Fig. 10.3a. The mean $Al_o$ values are found to be in the order of warm Andosols (4.07%) > temperate Andosols (3.09%) > cool and cold Andosols (2.10%), clearly indicating the thermal effect on the weathering rate of the Andosols. The mean, median and 80th percentile $Al_o$ values of the warm Andosols are 1.8- to 2.2-fold of those of the cool and cold Andosols, while the difference in the mean annual soil temperature between the two types of soil is approximately 20 °C. According to studies on the chemical kinetics of Andosol weathering, the temperature dependency ($Q_{10}$) of the Al dissolution reaction is close to a zero-order reaction. Thus, a temperature rise by 20 °C nearly doubles the dissolution rate, indicating a coincidence with the weathering rate of Andosols obtained by the acid oxalate dissolution measurement. There is no difference in the mean $Al_o$ values between the hot Andosols (3.20%) and temperate Andosols, because a considerable amount of Al released from volcanic ash is incorporated into kaolin minerals under a hot climate.

Soil moisture significantly contributes to the weathering of Andosols compared with soil temperature (Shoji *et al.*, 1993b). According to the frequency distribution of the $Al_o$ values (Fig. 10.3b), the mean, median, and 80th percentile values are 3.06%, 2.47% and 5.03%, respectively, for the humid Andosols, 1.24%, 0.62%, and 1.59% for the semi-arid Andosols, and 1.57%, 1.13%, and 2.55% for the Mediterranean-climate Andosols. Furthermore, the percentages of the soil samples with the least degree of weathering ($Al_o$ values of 1.0% or less) are 60% in the semi-arid Andosols, 42% in the Mediterranean-climate Andosols, and 24% in the humid Andosols. These figures indicate that soil moisture very strongly influences the weathering of Andosols.

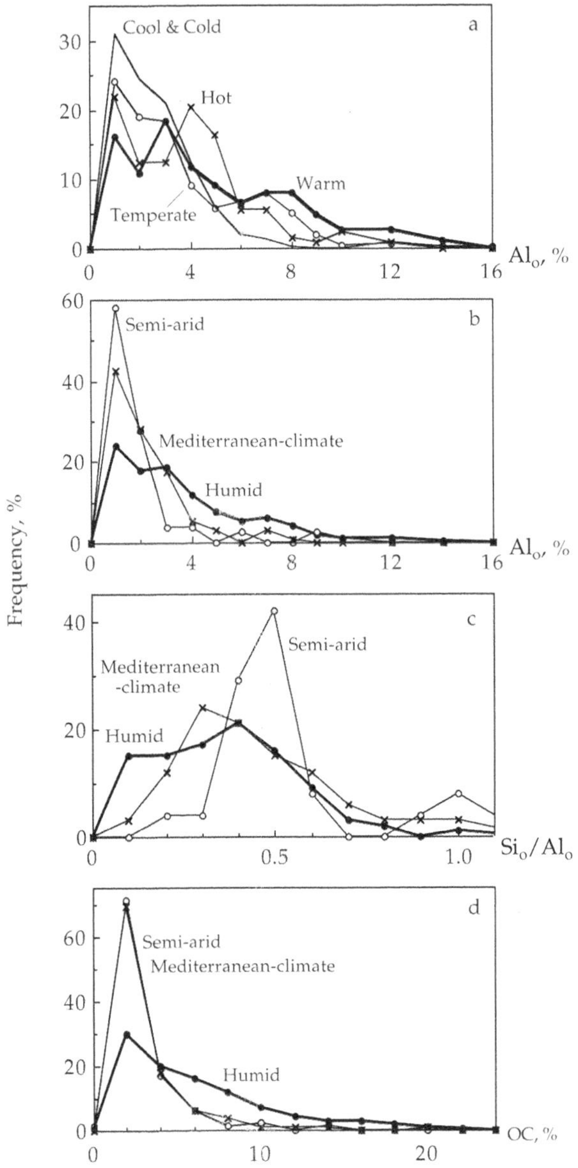


Fig. 10.3. Frequency distribution of $Al_o$ (a and b), $Si_o/Al_o$ ratio (c), and OC (d) of surface horizons of volcanic ash soils with different soil temperature and soil moisture (prepared using TUWAD).

Fresh volcanic ash shows the highest Si content (several-fold above the Al content) and the Si release rate from the parent material is much higher than Al. However, the mean $Si_o$ values of the humid, Mediterranean-climate and semi-arid Andosols are 1.22%, 0.64% and 0.49%, respectively, being approximately 40% of the mean $Al_o$ values. The 80th percentile $Si_o$ values are also notably smaller than the corresponding values of $Al_o$. Thus, a large proportion of the released silica has been subjected to an intense leaching loss under humid climates, while it has been transformed into acid oxalate-insoluble amorphous silica or opaline silica under arid climates.

$Si_o$ mainly occurs in allophane and imogolite and $Al_o$ in allophane, imogolite and Al-humus complexes. Soil allophane often has a Si/Al ratio of 0.5 (Shoji *et al.*, 1993b). Thus, the types of the major weathering product in the Andosols can be determined according to the concentration of $Al_o$ and $Si_o$, and $Si_o/Al_o$ ratios. Most semi-arid Andosols have a $Si_o/Al_o$ ratio of 0.5 to 0.4 (Fig. 10.3c), indicating that the major weathering product except amorphous silica is allophane, though its content is very small, as shown by the low $Al_o$ concentrations (Fig. 10.3b). In contrast, humid Andosols have a bimodal distribution of $Si_o/Al_o$ values. A soil group having a peak at the $Si_o/Al_o$ of 0.1 shows a predominance of Al-humus complexes that is common for present-day surface soils under humid climate. The other soil groups having a peak at the $Si_o/Al_o$ of 0.4 contain many buried A horizons in which intense allophane formation has been taking place without the anti-allophanic reaction by organic matter supplied from vegetation. Most Mediterranean-climate Andosols have $Si_o/Al_o$ ratios of 0.4 to 0.3 and lower $Al_o$ values, so their main weathering products are allophane and Al-humus complexes, though in low amounts.

The accumulation of organic matter in the Andosols is greatest in humid Andosols (Fig. 10.3d) and is primarily attributed to the formation of stable Al-humus complexes. Thus, the frequency distribution of organic carbon (OC) is similar to that of the $Al_o$ content as follows: the mean, 80th percentile and $OC/Al_o$ values are 5.22%, 8.52%, and 1.70% for the humid Andosols, 1.83%, 3.20%, and 1.50% for the semi-arid Andosols and 1.91%, 2.80%, and 1.20% for the Mediterranean-climate Andosols.

### *10.2.2 Soil formation and characteristic weathering products*

Since Andosols or Andisols occur in various environments, the Andisol Order has been provided with eight Suborders. Aquands, Cryands and Udands have many more numbers of Great Groups and Subgroups compared with the other Suborders, indicating that soil moisture regimes most strongly influence the development and diversity of Andisols.

*Soil processes in young volcanic ash soils*

Reflecting the unique properties of the parent material, volcanic ash soils rapidly form and show accentuated soil processes with several characteristic weathering products. Therefore, young volcanic ash soils are the most suitable for understanding the significance of relationships between climatic conditions and soil processes. The soil processes are influenced by the release and activities of the constituent elements, leaching or accumulation of the released elements, soil acidity, and supply, decomposition and humification of organic matter from vegetation.

As confirmed in the previous section, soil moisture influences the weathering of volcanic ash soils more strongly than soil temperature. Thus, the present subject will be discussed according to the relationship of soil moisture with soil processes in the temperate regions where volcanic ash soils are distributed most extensively (Table 10.2).

*Humid (udic) environments* Humid (udic) soil moisture allows various volcanic ash soils to form rapidly. The climosequences of volcanic ash soils also develop during a rather short period of weathering and are useful to explain the climatic effects on pedogenesis as described by the studies on Towada volcanic ash soils, in north-eastern Japan.

Young soils derived from Towada volcanic ash of the rhyolitic type (1000 y BP) typically have a continuous climosequence (Takahashi and Shoji, 1996). The Thornthwaite's or P–E index is helpful to describe closely the climosequence in a small humid area. The study area is divided into three zones based on relationships between the P–E index and soil analytical data as follows (Fig. 10.4):

- Zone 1: P-E index less than 160 and mostly melanic Andosols under Japanese pampas grass
- Zone 2: P-E index between 160 and 320 and melanic Andosols under Japanese pampas grass and fulvic Andosols under beech trees
- Zone 3: P-E index greater than 320 and placic Podzols (Placorthods) under coniferous trees.

The surface soils of zone 1 have a soil pH of 5.2 to 5.8. This soil acidity favours the formation of aluminium hydroxides with a low charge which have a high reactivity with silicic acid ($Si(OH)_4$) and low reactivity with humic acids. Thus active Al (acid oxalate-extractable; $Al_o$) predominantly occurs in allophane (including imogolite), followed by Al complexed with humic acids (pyrophosphate extractable; $Al_p$), as shown by the lower $Al_p/Al_o$ ratios (mean: 0.40). The allophane content (mean: 52.4 g $kg^{-1}$) estimated using acid oxalate-extractable silica ($Si_o$) indicates that the surface soils are allophanic. Common vegetation in

Table 10.2. *Environmental conditions and characteristic weathering products of young volcanic ash soils*

| | Aridic (Arid) | Ustic (Semi-arid) | Udic (Humid) | |
|---|---|---|---|---|
| Soils | Torrands (Entisols) (Vitric Andosols, Regosols) | Ustands (Andosols) | Udands (Spodosols) (Andosols, Podzols) | |
| Leaching | none to very low | low | high | |
| Organic matter supply from vegetation | very small | small | large | |
| Speed of soil development | very low | low | high | |
| *Characteristic weathering products* | | | | |
| Surface soil | amorphous silica | Al-humus | (Allophanic)[a] | (Nonallophanic)[a] |
| | carbonates | allophane amorphous silica | Al-humus allophane | Al-humus Al-interlayered 2:1 min. laminar opalline silica |
| Subsurface soil | amorphous silica carbonates | allophane amorphous silica | allophane | allophane |

this zone is broad-leaved trees and Japanese pampas grass. Plentiful existence of melanic humic acid is attributable to the influence of Japanese pampas grass. The subsurface soils have a limited supply of organic matter from vegetation and slightly higher pH values as compared with those of the surface soils, so active Al effectively reacts with silicic acid, forming an allophanic clay mineralogy.

The surface soils of zone 2 are subjected to strong leaching, so they have a lower base saturation and soil pH values of 4.7 to 4.9, favouring the formation of a high charge Al. Under these conditions, active Al ($Al_o$) is mostly consumed to form Al-humus complexes as shown by higher $Al_p/Al_o$ ratios (mean: 0.82).

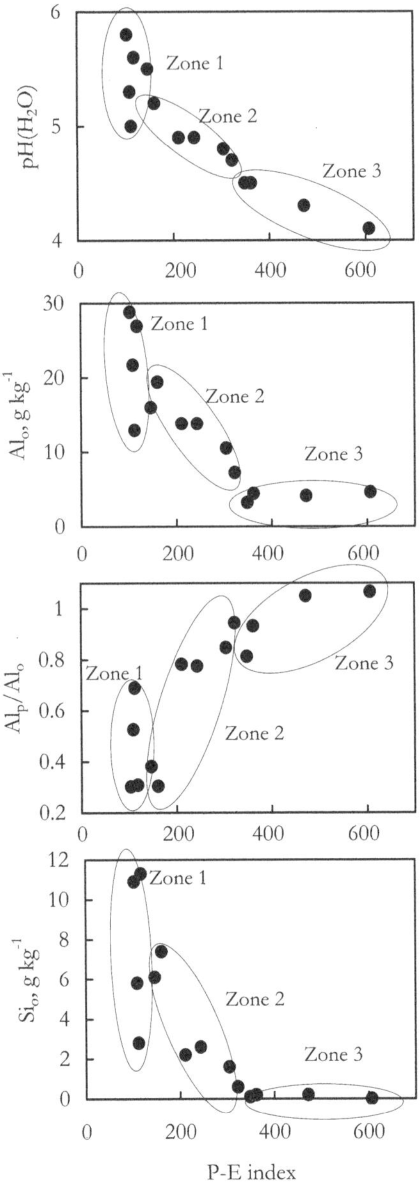


Fig. 10.4. Relationship between precipitation–evaporation (P–E) index and properties of surface soils in the climosequence of Towada volcanic ash soils. From Takahashi and Shoji (1996).

Active Al is also incorporated into hydroxy-Al interlayers of 2:1 layer silicates. These processes inhibit the formation of allophane and imogolite. This inhibition mechanism was termed ''anti-allophanic effect'' (Shoji *et al.*, 1993a). In fact, the allophane content is trace to very small in the surface soils (mean: 12.8 g $kg^{-1}$).

The anti-allophanic effect also contributes to the pedogenic formation of laminar opaline silica. This amorphous silica typically has a coarse clay size and unique morphology such as thin circular and elliptical discs (Shoji and Masui, 1971). It forms in silica-rich surface soils from supersaturated soil solutions, resulting from surface evaporation and freezing. The subsurface soils of Andosols in zone 2 show the same weathering processes as those in the subsurface soils in zone 1.

Vegetational effects on the formation of Andosols are clearly observed in zone 2. Melanic and fulvic Andosols are formed under Japanese pampas grass (*Miscanthus sinensis*) and beech trees (*Fagus crenata*), respectively. *M. sinensis* is the most important andisolizer in north-east Japan (Shoji *et al.*, 1993a). It promotes the accumulation and humification of soil organic matter, resulting in the formation of a melanic epipedon. This epipedon has a cumulative thickness of 30 cm or more within a total thickness of 40 cm and meets the requirements of andic soil properties (Appendix Table VI).

Volcanic ash soils in zone 3 having the highest P-E indices are subjected to podzolization under coniferous vegetation, forming placic Podzols (Placorthods) (Shoji *et al.*, 1993a). The analytical data of the E horizons definitely show intense eluviation, a very low content of acid oxalate-extractable Al, which is totally complexed with humic matter ($Al_p/Al_o = 1$) and the absence of allophane (the content of acid oxalate-extractable Si is zero). In contrast, spodic horizons show the immobilization of organometallic complexes and the intense formation of allophane.

As described above, melanic, fulvic and non-allophanic horizons as well as allophane-rich horizons are the most characteristic for Udands (humid Andosols). Thus, Melanudands, Fulvudands and Hydrudands (allophane-rich Udands) have been established as the most important Great Groups of Andisols in Soil Taxonomy (Soil Survey Staff, 2006).

*Semi-arid (ustic) environments* Semi-arid (ustic) soil moisture regimes conspicuously retard the formation and diversity of volcanic ash soils as indicated by the low acid oxalate-extractable Al ($Al_o$) values in the frequency distribution of the $Al_o$ values of Andisols in semi-arid climates (Ustands) (Fig. 10.3b). The surface soil of Ustands is poorly developed, reflecting the decreased supply of organic matter by vegetation, low leaching intensity and limited weathering of volcanic ash, as compared with those of the Udands. The soil reaction of a weak acid lowers both the Al release from volcanic ash and Al activity, thereby limiting the formation of allophane and Al-humus complexes. In contrast, both the release and activity of Si are independent of the normal range of pH, so amorphous silica accumulates by drying of the soil solutions and forms a cemented horizon (duric horizon or duripan). The subsurface soil has clay mineralogy similar to the surface soil. Thus, the vitric and duric properties are the most important for establishing Great Groups such as Durustands and Ustivitrands. Some Ustands contain considerable amounts of organic matter, so melanic, mollic and umbric epipedons are also used for establishing humic and pachic Subgroups (Soil Survey Staff, 2006).

*Aridic environments* Many soils in aridic environments (Aridisols) have morphologically well-developed soil profiles due to the concentration of the small amount water in a relatively small volume of soil even though leaching of the weathering products is limited (Buol *et al.*, 1997). However, information on volcanic ash soils in aridic environments is scarce, so the general patterns of soil development cannot be described.

Almost all volcanic ash soils at the early stages of weathering are classified into vitric Andosols (Torrands of Andisols) or Regosols (Psamments, Fluvents, or Orthents of Entisols). Since potential evapotranspiration greatly exceeds precipitation during most of the year, the constituent elements with high mobilities in the volcanic ash accumulate, forming carbonate-rich horizons, described as ~~calcic and~~ petrocalcic, as well as silica-cemented horizons (duripans). The origin of carbonates in soil is generally either pedogenic or lithogenic or both. B horizons having both andic soil properties and carbonate accumulation were found in Hawaii under the currently arid climatic conditions (Soil Survey Staff, 2005). The ratio of pedogenic and lithogenic carbonate can be estimated using $\delta^{13}C$ value (Nordt *et al.*, 1996). The silica is amorphous opaline silica as described before.

The duric and vitric properties of Torrands have been used to establish Duritorrand and Vitritorrand Great Groups, and the calcic and petrocalcic properties, for calcic and petrocalcic Subgroups, respectively (Soil Survey Staff, 2006). There is no humic Subgroup in the Torrands because the supply of organic matter by vegetation is extremely small.

*Transformation of Andisols to other soil Orders* As soil age and degree of weathering increase, Andisols transform to other soil Orders in various climatic conditions (Ugolini and Dahlgren, 2002). Although each soil Order may involve one to several soil processes (Arnold and Eswaran, 2003), the transformation certainly involves mineral alterations in the soils, except Spodosols and Histosols, that consume active Al and Fe. In Spodosols acid complexation and removal of Al and Fe with organic matter from the upper part of soil lead to formation of a bleached E horizon and spodic horizon. Thus, volcanic ash-derived Spodosols commonly have the andic soil properties.

Transformations of Andisols to other soil Orders are briefly described below.

Spodosols (Podzols): their formation is favoured by coniferous vegetation and predominance of an organic acid weathering regime under cool to cold humid conditions. Non-crystalline materials such as allophane and imogolite are abundant in the spodic horizon.

Mollisols (Chernozems/Kastanozems/Phaeozems): develop in volcanic material under dry conditions in temperate and tropical environments and under wetter conditions in the presence of basic volcanic ash. Warm/dry conditions promote formation of crystalline layer silicates such as halloysite and maintain high base saturation.

Vertisols (Vertisols): form in warm regions having a distinct dry season as volcanic ash weathers preferentially to smectite and vermiculite.

Inceptisols (Cambisols): under both temperate and tropical environments, non-crystalline materials are gradually consumed by transformation to more stable crystalline minerals such as halloysite, leading to the alteration of Andisols to Inceptisols.

Oxisols (Ferralsols): volcanic ash transforms to Oxisols by alteration of primary weatherable minerals to their oxyhydrate state through extreme weathering on stable landscapes in the perhumid tropics.

Alfisols/Ultisols (Luvisols/Lixisols/Acrisols): their genesis is led by formation of layer silicates and clay translocation in the xeric moisture regime of California.

Histosols (Histosols): under the cool humid climate of north-east Japan blanket peats (climatic Histosols) develop on Andisols which have restricted drainage due to plentiful formation of non-crystalline materials.

### *10.2.3 Palaeoclimates and soil formation*

Global warming started in the late last glacial period (formerly Würm), continued in the early Holocene and peaked at 6000 to 5000 y BP (the Holocene Maximum). Thereafter, the temperature gradually decreased through the cool period of 2500 to 2000 y BP and the Little Ice Age of the sixteenth to nineteenth centuries (Folland *et al.*, 1990).

There are three types of palaeosols, namely buried, exhumed, and relict. Of these soils, buried volcanic ash soils are the most useful for studying the relationship between palaeoclimates and soil formation in Japan (Shoji *et al.*, 1993a) and New Zealand (Newnham *et al.*, 1999). Volcanic ash soils commonly have sedentary multistoreyed profiles formed by intermittent volcanic ash deposition and repeated pedogenesis. They resist soil disturbance, truncation and displacement due to natural and human activities. Therefore, they are able to preserve in their materials and properties highly useful information relating to palaeoenvironment and palaeovegetation and may also include archaeological artefacts (Shoji and Takahashi, 2002).

Soils have been extensively used since human society changed from hunting or nomadism to farming. In north-east Japan, where farming started approximately 2000 y BP, humans strongly impacted the soil environment, as definitely observed for many Holocene volcanic ash soils. Ancient man fired forests for their hunting to flush out wild animals from the forest lands. Farmers also used to burn bushes to develop grazing grasslands consisting of *M. sinensis* until several decades ago. Towada volcanic ash soils clearly show the influences of such climate changes and climate- and human-induced vegetation changes (Shoji *et al.*, 1993a).

Volcanic ash soils developed under the cold climate and coniferous vegetation of the late last glacial period, as shown by phytolith studies (Sase and Hosono, 1996). This environment must have favoured the formation of Podzols from volcanic ash. However, none of the soils has either a visible humus horizon or albic horizon (Hachinohe soil). It is most probable that the organic matter mostly has disappeared by decomposition, and the albic colour has been stained with free iron oxides formed by the weathering of volcanic ash over more than 10 000 years.

Ninokura soil was subjected to a warming climate during 10 000 and 8600 y BP and has a dark-humus horizon. This humus horizon contains humic matter dominated by the A-type humic acid with the highest degree of humification and shows the influence of the *M. sinensis* ecosystem on its formation. The andisolizer *M. sinensis* requires a warm temperature for maintaining the ecosystem in north-east Japan (Shoji *et al.*, 1993a). Thus, the existence of phytoliths of *M. sinensis* indicates the warming climate in the early Holocene.

The warmer climate enabled luxurious plant growth. It also enhanced the activity of humans called the Jomon people (10 000 to 2000 y BP), as shown by a remarkable increase in the number of archaeological remains in north-east Japan. The Jomon people burned woodlands for their hunting, resulting in the extensive development of *M. sinensis* ecosystems. These changes in climate and vegetation favoured the formation of thick dark coloured or melanic Andosols, as observed for Nanbu soil (8600 y BP) and Chuseri soil (5400 y BP), which experienced surface weathering for more than 3000 years. Thus, the age of the most intense formation of melanic soils is coincident not only with the history of active humans, but also with the warmer climate during the Jomon age (8000–2500 y BP), under which *M. sinensis* also could grow luxuriously.

Towada-b soil (2000 y BP) and Towada-a soil (1000 y BP) have continuously been subjected to Andosolization. Though the climate became cool, the thermal condition enabling the development of the *M. sinensis* ecosystem prevailed below

600 m elevation in north-east Japan. Farmers used to burn bushes to establish grazing grasslands and this also helped to maintain the *M. sinensis* ecosystem and accumulate melanic humic matter.

In conclusion, if humans had not lived in north-east Japan, melanic Andosols could not have developed and fulvic Andosols would be more extensively distributed.

# 11

# Factors of soil formation: topography

*Robert C. Graham*

The term *topography* refers to the configuration of the land's surface. The topography of an area incorporates its *relief* (relative differences in elevation), its *aspect* (position with respect to compass coordinates), and the general shape and connectivity of land surfaces. These attributes mediate how external factors, such as solar radiation, precipitation and wind, impinge upon a site. Topographic relief imparts potential energy, by virtue of gravity, that functions to move water and regolith from higher landscape positions to lower ones. The movement of materials, including water and soil materials, on a landscape is influenced by the slope gradient and shape and the degree of connectivity of drainage networks. Thus, from a pedologic perspective, topography is important because it exerts a strong influence on the disposition of energy and matter experienced by soils on the landscape.

The processes that create topography are usually geologic in nature; e.g. tectonic uplift, fluvial erosion and deposition, mass wasting, volcanic activity and glaciation. A landscape produced by these processes is the blank canvas upon which soil patterns are painted by processes that are linked to topography. As time passes, these processes leave characteristic pedogenic imprints on different parts of the landscape, altering the original parent materials and differentiating the physical, chemical and biological nature of the soils by topographic position.

## 11.1 Topographic elements of landscapes

In most landscapes, topography governs the movement of water and is shaped by it. As the name implies, watersheds function to shed meteoric water that falls on upland parts of the landscape; thus they link soils on slopes to fluvial systems. If a watershed drains through an outlet it is termed an *open drainage basin*. This type

*Soils: Basic Concepts and Future Challenges*, ed. Giacomo Certini and Riccardo Scalenghe.
Published by Cambridge University Press. 

of drainage basin is found where the stream network is interconnected, generally formed by headward erosion, with streams becoming larger as they converge downstream.

The place of a drainage in the network can be indicated using an ordering system illustrated in Fig. 11.1a. Zero-order drainages are essentially swales, concave features on hillslopes that lack a defined channel. Down gradient from this, first order drainages have the initial expression of a defined channel, and when first order drainages come together they form a second order drainage. Third order drainages are formed when two second order drainages join, and so on.

The geomorphic components of a watershed are shown in Fig. 11.1b. The *channel* conducts concentrated water flow from the watershed. An *interfluve* is a ridge that is roughly parallel to the channel below. The *divide* is the ridge that separates drainages flowing in roughly opposite directions. The *head slope* is the slope that rises up from the channel toward the divide. The *nose slope* is the end of the interfluve farthest from the divide. The *side slope* is the slope that rises from the channel toward the top of the interfluve. The shapes and positions of these geomorphic components have important impacts on pedogenic processes, as will be shown later in the chapter.

*Closed drainage basins* have no outlet, so any water that falls in them can leave only by deep percolation or evaporation. This type of basin can form by such mechanisms as subsidence, faulting and landslide blockage of valleys. Sediment collects in these basins and, where percolation is slow and evaporation rates are high, the dissolved products of weathering concentrate and precipitate as salts. Again, geomorphology has a major impact on the kinds of pedogenic process.

The slopes within watersheds can be distinguished by their form and relative positions. A slope transect such as that shown in Fig. 11.2 can be established on

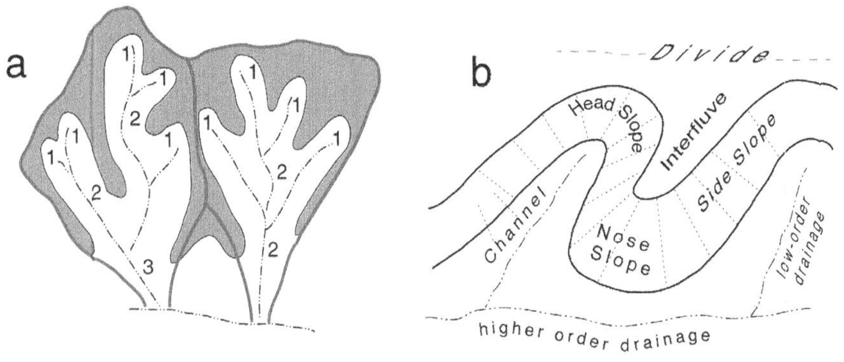


Fig. 11.1. Soils and topography are closely linked to drainage networks. (a) The Strahler method for defining drainage orders; (b) geomorphic components of watersheds. Modified from Schoeneberger *et al.* (1998).

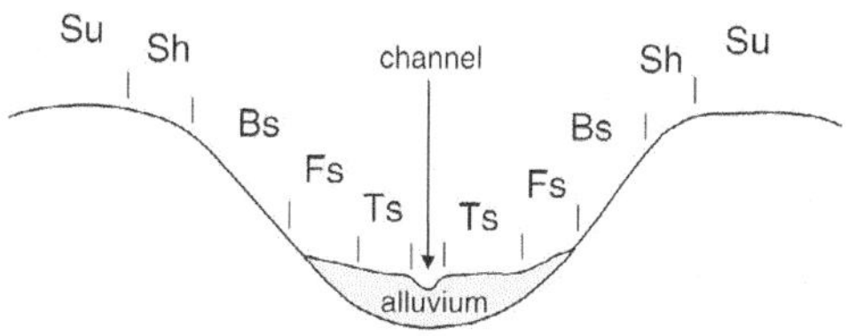


Fig. 11.2. Slope position descriptors for two-dimensional hillslope segments. Su, summit; Sh, shoulder; Bs, back slope; Fs, foot slope; Ts, toe slope. Modified from Schoeneberger *et al.* (1998).

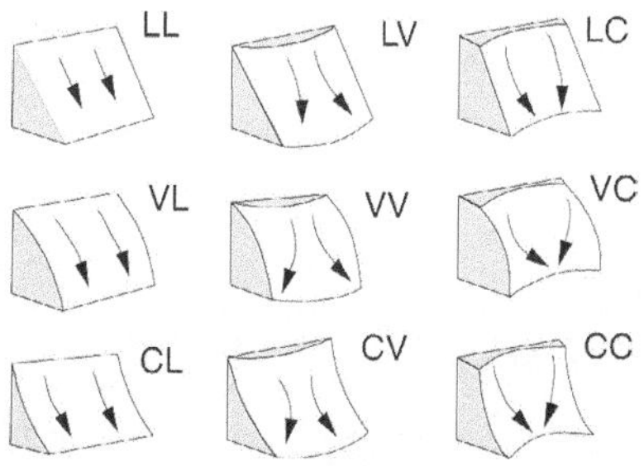


Fig. 11.3. Slope shapes produced by combinations of convexity and concavity in down-slope and across-slope directions. C, concave; V, convex; L, linear. Modified from Schoeneberger *et al.* (1998).

any part of the watershed (nose slope, side slope, or head slope) to illustrate the two-dimensional relationships. The various hillslope positions can be characterized with respect to geomorphic and pedogenic processes (Dalrymple *et al.*, 1968), but three-dimensional models have more predictive value.

Slope shape descriptors, as illustrated in Fig. 11.3, address the three-dimensional character of slopes and have considerable predictive value with respect to water and sediment movement. A slope that has a linear configuration both down slope and across slope neither concentrates nor disperses water flow. A slope that is convex both down and across slope disperses water flow, resulting in soils that are drier. At the other extreme, a slope that is concave both down and across slope converges water flow and has soils that are wetter than the other configurations. A monitoring study of soil water status in a zero-order watershed in southern coastal California found these relationships to hold true, allowing development of a geographic-information-system-based model for predicting soil water dynamics (Chamran *et al.*, 2002).

While the terminology and relationships described above are applicable to landscapes with moderate relief (such as Fig. 11.4a) they are not always appropriate. Some mountainous landscapes have extremely steep slopes hundreds of metres long (Fig. 11.4b). Soil-geomorphic processes in such landscapes are less studied than those on hillslopes, but mass movement is generally a key process in

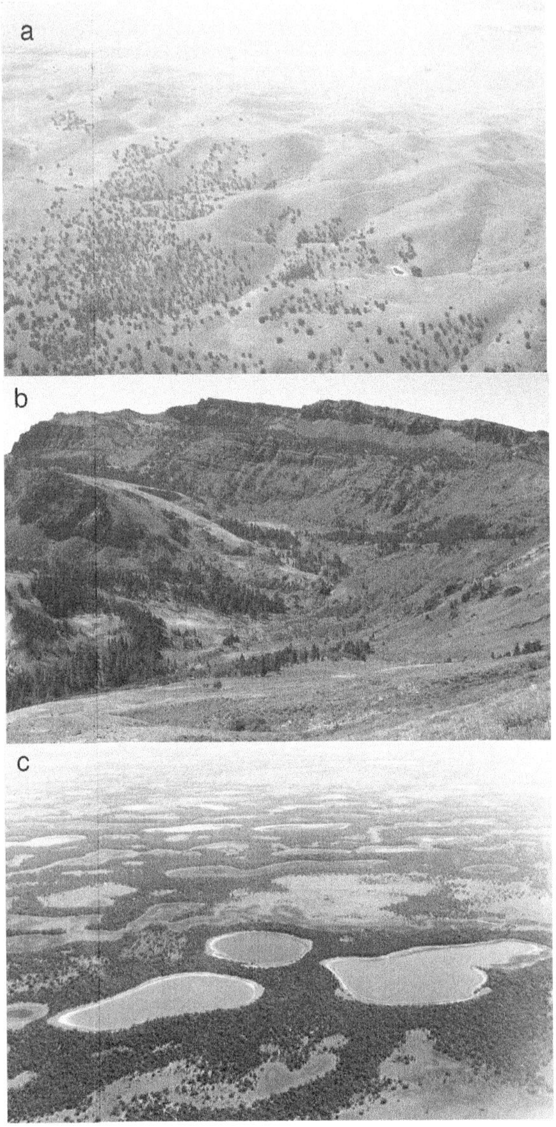


Fig. 11.4. Photographs of landscapes with different relief: (a) hillslopes in central California with about 30 m relief, (b) mountain slopes in north-eastern California with about 300 m relief, and (c) a nearly level landscape in the Pantanal of Mato Grosso in Brazil with about 3 m relief. Photos by R.C. Graham (a,b), and A.Y. Sakamoto (c).

those settings. At the other extreme, some landscapes have very low relief (Fig. 11.4c). In these nearly level landscapes, which make up about 60% of the Earth's land area (Fairbridge, 1968), very subtle changes in topography can strongly influence the soil hydrology and resulting soil properties.

Fine-scale topographic relationships often have important pedological consequences. Microtopographic features, such as the irregular surface of lava flows, bars and swales on stream terraces, coppice dunes and surface clasts, can trap aeolian dust, affect the infiltration pattern of rainwater, alter localized thermal regimes, and provide niches for biological activity, all of which influence the direction and rate of pedogenesis.

## 11.2 External factors mediated by topography

Topographic features affect how water and energy are added to and lost from soils on a landscape. Aspect, elevation and slope position modify the regional climate.

Because of adiabatic cooling and consequent water vapour condensation, rainfall increases with elevation. Colder temperatures may also result in snow rather than rain, in which case the water is stored in solid form on the soil surface for some time, rather than immediately infiltrating or running off, as is the case with rain. But elevation is not the only topographic influence on precipitation. Aspect can be important as well, because the slope facing an incoming storm often receives more precipitation than the leeward side. Substantial aspect-related moisture can even be delivered to a site by fog. Clouds that envelope a mountain may deliver more moisture to the windward side, due to impaction of moisture on the vegetation, than on the leeward side.

In northern and southern midlatitudes, aspect has a very important influence on soil temperature. In northern latitudes, the sun is in the southern sky, so south-facing slopes receive more direct solar radiation than do north-facing slopes. The reverse is true in southern latitudes. In general, the aspect that receives less solar radiation is cooler, has more effective moisture, and therefore more soil water for leaching, as well as more biomass, slower organic matter decomposition rates, and higher soil organic matter contents (Nevo *et al.*, 1998). An example of the aspect effect on soil genesis is found in northern Michigan, USA, where Spodosols are more abundant and strongly developed on north-facing slopes than on south-facing slopes, where Entisols are dominant (Hunckler and Schaetzl, 1997). The soils on north-facing slopes are cooler during the summer, resulting in more effective moisture for leaching (less evapotranspirational loss) and slower organic matter decomposition rates. Significantly thicker snowpacks on north-facing slopes during winter also play a role in that snow cover inhibits soil freezing, allowing at least low levels of microbial activity and water percolation, both essential for podzolization to proceed throughout the year. During spring, snow-melt is more gradual on north-facing slopes, allowing more water to infiltrate than on south-facing slopes where water delivered from rapid melting exceeds infiltration rates and runs off. Again, water infiltration and leaching is a key process in the formation of Spodosols.

The effect of aspect is relatively minimal near the equator, because the sun 'travels' a daily arc perpendicular to the celestial meridian, placing it neither in the southern nor northern sky, but directly overhead at noon. This arc varies little seasonally. The effect of aspect is minimal at polar latitudes as well, where solar radiation is received from many directions during summer and is nil or of low intensity during the winter (Hunckler and Schaetzl, 1997).

The relation of aspect to wind direction is important relative to erosion and deposition at the soil surface. Wind erosion prevails on windward slopes and deposition on leeward slopes. Fine sand- and silt-size particles, including clay aggregates, are most often transported. While we usually think of wind erosion being effective only in arid regions with low vegetation density, this is not entirely the case. In fact, wind erosion in arid regions is usually restricted to certain landscape positions, such as playa margins, that have fresh fluvial or lacustrine sediments or areas otherwise disturbed by humans or natural processes. Wind can remove leaf litter from exposed ridgelines in temperate forests, thereby interrupting the biocycle and draining nutrients from the site. Low-lying branches of shrubs and trees, when blown by wind, can scrape repeatedly across the soil, physically exposing and disrupting the mineral soil so that it is easily entrained by the wind. After severe forest fires, slopes that are normally densely vegetated are left barren. Wind erosion can be very effective in these situations, and prominent, windward slopes are most impacted. Microtopography can be very effective at trapping dust carried by wind. An example is the rough surface of desert pavement, which effectively traps and incorporates dust into the underlying soil so that, over time, the soil actually expands upward (Wood *et al.*, 2005).

Because aspect and slope position affect how much solar radiation and wind impinge on a site, topography also influences evapotranspirational demand. Sites sheltered from direct solar radiation have less evapotranspirational loss. This means that the north-facing slopes in the northern hemisphere (and south-facing slopes in the southern hemisphere) are generally more moist. In at least one case this was found not to be true. In southern California, subsoils on a north-facing slope were drier because greater vegetation density resulted in greater evapotranspiration losses than on the warmer south-facing slope (Poole and Miller, 1975).

## 11.3 Pedogenic processes linked to topography

The relationship of pedogenic processes to topography is expressed in the concept of the *catena* as described by Milne (1935). A catena is a suite of soils that are linked to each other by topographic conditions. The soils at different positions within the catena are often quite different, but are the products of the geomorphic and pedogenic processes that dominate at that particular geomorphic position. The suite of soils within a catena is repeated across a landscape wherever the appropriate topographic conditions are met. The catena concept is a powerful predictive tool because it links extant soil properties and processes that produce them to the readily observable surface topography of a landscape.

### *11.3.1 Erosion and deposition*

Soil erosion is strongly influenced by hillslope position. From a geomorphic standpoint, upper hillslope positions (shoulder, upper back slope) are prone to erosion, whereas lower slope positions (foot slope, toe slope) are sites that favour deposition and accumulation of sediments (Ruhe, 1975). The mechanisms for producing these erosion/deposition trends vary depending on slope steepness, biotic communities, and other factors. On gentle slopes, soil creep and slope wash may be dominant erosion mechanisms, whereas rapid mass movement (i.e. various kinds of landslides) is more important in steeper terrain. Soil properties produced by pedogenesis on the various segments of a catena can also influence erodibility. For example, clayey argillic horizons that form on stable summit positions can be resistant to erosion. Soil organic matter, which tends to accumulate preferentially in lower slope positions, can improve aggregate stability and resistance to erosion. Except on steep slopes where rapid mass movement dominates, slope wash preferentially transports finer particles downslope, contributing to a textural differentiation, with finer-textured soils in the lower landscape positions (Table 11.1). Neoformation of clays in poorly drained soils of lower slope positions is another contributor to this trend, as discussed later.

### *11.3.2 Soil production*

The formation of soil, or regolith, from greywacke bedrock has been studied in relation to topography in northern California (Heimsath *et al.*, 1999). Conversion of bedrock to soil was found to be most rapid where slopes are most convex and soil cover is thinnest. The primary mechanism responsible for the conversion is burrowing rodents. Gophers excavate down into the weathered rock, disrupting its fabric and converting it to soil. They push loose soil material to the surface where it is eroded by rainsplash and slope wash, and is transported downslope off the convexities into the concave parts of the landscape. Over time, the convex parts of the landscape are preferentially worn down.

Table 11.1. *Properties of soils along a catena formed on basalt in southern California*[a]

| | Slope position | | | | |
|---|---|---|---|---|---|
| | Summit | Back slope | Foot slope | Toe slope | Basin |
| Soil Order[b] | Entisols[c] | Entisols[c] | Alfisols | Vertisols | Vertisols |
| Depth to bedrock (cm) | 21 | 40 | 52 | 46 | 98 |
| Maximum clay concentration (%) | 17 | 20 | 55 | 53 | 61 |
| Dominant dry subsoil colour, Munsell | yellowish red 5YR 4/6 | yellowish red 5YR 4/6 | dark brown 7.5YR 3/2 | dark brown 10YR 4/3 | dark grey 10YR 4/1 |
| Clay mineralogy | kaolin >> smectite | kaolin > smectite | not determined | smectite > kaolin | smectite >> kaolin |
| Base saturation (%)[d] | 81 | 74 | 80 | 88 | 92 |
| Fe as pedogenic Fe-oxides (%)[d] | 4.3 | 5.1 | 2.3 | 1.7 | 1.0 |
| Mn as pedogenic Mn-oxides (%)[d] | 0.07 | 0.08 | 0.07 | 0.10 | 0.13 |

[a]Mean annual precipitation is 450 mm, primarily in the winter months. Shallow ponds (vernal pools) form in the basin during the winter and early spring. The summit is ~3 m higher than the basin and about 30 m lateral distance from the basin. Maximum slope gradient is 9%. The vegetation consists of grasses and herbaceous plants.

[b]Soil Taxonomy.

[c]Weak soil development on summit and back slope is due in part to extensive burrowing by gophers, which are absent on the seasonally flooded lower positions.

[d]Base saturation, Fe, and Mn values are weighted means for the profiles. Fe and Mn values are from citrate-dithionite extractions.

*Source:* Weitkamp *et al.*, (1996).

### *11.3.3 Soil organic matter*

Soils of concave, lower landscape positions usually have higher organic matter contents than those upslope or on convex positions. This trend has been observed in the whole range of climatic regions from polar to equatorial (Birkeland, 1999). A number of processes can contribute to this common trend. Lower slope positions generally are the most moist because they receive run-off and/or throughflow from upslope. In regions where water limits plant growth, the higher water status on low slope positions yields more biomass and more incorporation of organic matter into the soil. In humid regions with highly leached soils, plant growth may be limited by nutrient cations, such as calcium. Soils on the lower slopes may have higher base cation status by virtue of downslope leaching (e.g., Graham and Buol, 1990; Weitkamp *et al.*, 1996) (Table 11.1). In such cases, increased fertility may contribute to higher biomass and soil organic matter contents. When lower slopes receive so much water that they are saturated, even during part of the year, microbial decomposition is impeded and organic matter accumulates in the soils. The drainage of cold air to low landscape positions may also contribute to colder temperatures and lower organic matter decomposition rates in soils there. Finally, gravity transports surficial materials downslope, with and without the influence of water. Erosion of O or A horizon material from upslope and deposition of these materials on lower slopes is a common phenomenon.

### *11.3.4 Redox conditions*

The concentration of throughflow water in concave lower slope positions can lead to saturated, anaerobic conditions in the soils there. When soil microbes do not have $O_2$ to use as an electron acceptor in their metabolic processes, they use other compounds. Iron and manganese oxides commonly fill this role and, since they are visually distinct compounds, their presence or absence is obvious and is generally indicative of redox status (Vepraskas, 2001). Manganese oxides, which impart a black colour to soils even at low concentrations, are reduced at higher redox potentials than iron oxides. So manganese is dissolved and transported in solution while iron oxides are still stable. As a result, manganese oxide concentrations are often relatively high in soils on lower slope positions, while highest iron oxide concentrations are on upslope positions (McDaniel and Buol, 1991; Weitkamp *et al.*, 1996) (Table 11.1). In some cases the manganese oxides, in addition to humified organic matter, contribute to the dark colour of soils on toe slopes and in basins.

Iron oxides are more resistant to reduction than manganese oxides and are thus concentrated on upper slopes. Resistance to reduction varies within iron oxide species and is enhanced, at least in part, by increased aluminium substitution for iron within the structure (Bryant and Macedo, 1990). Aluminium substitution in goethite (FeOOH) is often on the order of 33% in soils, whereas it is generally on the order of 15% in pedogenic haematite ($Fe_2O_3$). These two minerals are the most common iron oxide species in soils, with goethite imparting a yellow-brown colour and haematite a red colour. Haematite is such a strong pigmenting agent that it generally overwhelms the colour from goethite even when present in small quantities. Thus, soils that are reddish in colour generally contain both haematite and goethite. If such soils are subjected to a moderate reducing regime, e.g. they are periodically saturated with water, the haematite will dissolve preferentially because it has less aluminium in its structure. The soil will then lose its red pigmentation and become yellow-brown, the colour of the remaining goethite.

The susceptibilities of these iron and manganese oxides to reduction explain the common sequence of soil coloration from red to yellow to grey or black in a transect from the most oxidized to the most reduced parts of the landscape. This coloration sequence is often expressed in hillslope catenas of semi-arid to humid climates (Schwertmann, 1985; Weitkamp *et al.*, 1996; Birkeland, 1999) (Table 11.1), but conditions in low relief landscapes may produce different results. In nearly level landscapes with broad interfluves, the drainage network may be insufficient to rapidly remove water that is delivered by rainfall. In such situations, soils in the centres of interfluves are poorly drained while soils along the terrace edges are well drained because water can readily drain to the adjacent stream. On these landscapes, the reddest soils are the well-drained ones on the edges of the interfluves (Daniels and Gamble, 1967).

### *11.3.5 Leaching and lessivage*

On nearly level upland surfaces, such as interfluves, water infiltrates vertically and the processes of leaching and lessivage reflect the water balance of the regional climate. The soil morphologic features, such as E (eluviated), Bt (enriched in illuvial silicate clay), and Bs (enriched in illuvial complexes of organic matter and Fe- and Al-oxyhydroxides) horizons, produced by these processes on stable geomorphic surfaces are relatively strongly expressed and are also reflective of the regional climate, unless disrupted by bioturbation (e.g. soils on summits in Table 11.1). On slopes, a lateral component is imparted to soil water movement, producing throughflow downslope. The water moving downslope through the soil carries with it dissolved and suspended materials. In humid regions with well-integrated drainage systems, the transported materials may be completely removed from the landscape as they enter the fluvial system. Often, however, the translocated materials have some impact on the downslope soils in terms of chemical, mineralogical and textural properties. The nature of this impact depends to a large degree on the amount of water moving through the soil system; in other words, it depends on the climate.

In arid regions, throughflow leaching moves ions incrementally downslope during rare storm events that wet the soils. As the soils dry, the ions precipitate as evaporite minerals according to their solubility (Boettinger and Richardson, 2001). Calcite is the least soluble and therefore precipitates first. Gypsum is on the order of ten times more soluble than calcite in soils, so it precipitates farther downslope. Salts such as halite, soda and epsomite are a hundred times more soluble than gypsum and are found at the base of slopes or accumulated in closed basins. The depth at which evaporite minerals occur in soils is determined by the water available for effective leaching at each landscape position. Localized surficial conditions can result in microtopographic redistribution of rainwater. For example, desert pavement (Wood et al., 2005) reduces infiltration, resulting in run-off that infiltrates in adjacent soils. Thus, surface conditions can modify expected climatic and topographic effects on leaching.

In temperate to humid regions, the materials mobilized upslope and moved downslope may include reduced Fe and Mn; chelated Fe, Mn and Al; dissolved alkaline earth cations and silicon; anions; colloidal organic matter; and clay. The fate of these substances depends on downslope conditions relative to whether they are sorbed onto the solid phase, precipitated as secondary minerals, or are leached out of the soil system.

The pedochemical processes along a toposequence of soils in Amazonia, Brazil, were studied by placing small bags of cation exchange resin, chelating resin, and vermiculite into major soil horizons along the sequence (Righi *et al.*, 1990). The soils were Oxisols on the upper slope positions, Ultisols on the midslope, and Spodosols on lower slopes. After a year, Fe was minimal on the cation exchange resin on all slope positions, while on the chelating resin it increased about 10-fold from the Oxisols to the wetter, more organic-rich Spodosols downslope. The vermiculite in the upslope positions, particularly in the upper horizons, developed Al-hydroxy interlayering which reduced its cation exchange capacity. Chelation in the Spodosols on the lower slope positions kept the Al in solution and prevented its build-up in the vermiculite interlayer.

Pedogenesis is related to topographic position in ultramafic terrain in the Klamath Mountains of northern California, USA (Lee *et al.*, 2004). Clay contents increased from a maximum of 40% in Bt horizons of the upland soils to a maximum of 60% in the Bg horizons of the soils in the wet meadow at the base of the slope. This was attributed to both erosional transport of clay to the base of the slope and neoformation of smectite there. Neoformation of smectite is favoured by the leaching of Mg and Si weathered from serpentine in the surrounding landscape and their concentration in the wet meadow. Despite the precipitation of smectite, Mg and Si are still the major cations in the stream water draining the wetland. Iron and Mn released by weathering and made soluble by seasonal reducing conditions in upslope positions are concentrated and precipitated as oxides in a well-drained landscape position near the outlet of the wetland.

Because Si is much more soluble than Al under the range of pH values common in soils (pH 4.5–9), it is available for leaching after it is released by weathering, whereas Al precipitates immediately unless chelated. Preferential downslope leaching of Si leads to a common differentiation in soil clay mineralogy along toposequences, with the more Al-rich species in upslope positions and the more Si-rich species on lower slopes. This was illustrated by Tardy *et al.* (1973) for the humid tropics where gibbsite (Si:Al = 0) was found upslope and kaolinite (Si:Al = 1) was found downslope and for the arid and seasonally arid tropics where kaolinite was found upslope and montmorillonite (Si:Al ≈ 2) was found downslope. This latter trend was also observed in an Entisol–Vertisol catena in southern California where the soil moisture regime is xeric (Table 11.1; Weitkamp et al., 1996). Also in southern California, soils along marine terrace edges are cemented by opaline silica while soils of terrace interiors are not (Moody and Graham, 1997). During the rainy season, water flows laterally above a slowly permeable plinthic horizon and, as it reaches the terrace edge, evaporation increases the solution concentration of Si to the point of supersaturation and opaline silica precipitates.

In mountainous terrain, landslides and other kinds of mass movement are important geomorphic processes. While soil development proceeds on all parts of the regolith-covered landscape, it can be interrupted at any stage by mass movement events (Graham and Buol, 1990). This interruption is relatively common on the steepest slopes, so Entisols often predominate there. By contrast, nearly level upland surfaces tend to be very stable sites for soil development. Erosion and deposition is minimal (at least by water), and the vector of water movement for leaching and lessivage is vertical. Materials are translocated downward into the profile rather than downslope. As a result, these soils tend to be more strongly developed than those on steep slopes.

## 11.4 Topography-based models of soil distribution

Because topography so strongly affects pedogenic processes, either directly or indirectly, and is relatively easy to measure, there is considerable interest in using it as the basis for modelling the spatial distribution and behaviour of soils. Spatial topographic data can be collected by traditional land survey techniques or much more rapidly by global positioning systems or LIDAR. Geographic information systems can be used to analyse topographic data in the form of digital elevation models (DEMs) to study terrain attributes that theoretically influence pedogenesis and the development of certain soil characteristics, as described in this chapter. Topographic attributes are considered either *primary*, that is they are calculated directly from the DEM, or *compound* in that primary attributes are combined to predict the distribution of processes on the landscape. Primary attributes that are commonly employed are elevation, slope, aspect, specific catchment area and slope curvature. Compound attributes most relevant to pedogenesis are those that predict the concentration of soil water on the landscape. In some cases these topographically based models have performed well to predict soil features such as organic carbon accumulation, A horizon thickness, pH and soil depth (Moore *et al.*, 1993; Gessler *et al.*, 2000). Less predictive success has been found when topography is complex, subsurface features (e.g. palaeosols, bedrock topography) alter water flow paths, or only coarse-scale DEMs are available. Nevertheless, geographic information systems offer a tremendous advance in the ability to analyse the effect of topography on soil processes and soil formation.

# 12

# Factors of soil formation: biota. As exemplified by case studies on the direct imprint of trees on trace metal concentrations in soils

*François Courchesne*

The role of biota on the genesis and properties of soils was emphasized and thoroughly illustrated in the various models of soil formation published by, among others, Dokuchaev, Joffe, Thorp, Jenny, Wilde, Simonson, Hole, Runge, Johnson, Brimhall, Paton and their collaborators. Indeed, organisms are abundant, with their number reaching up to $10^{12}$ per $m^2$ of soil for bacteria, essentially ubiquitous since they are found even under the most extreme climatic conditions on the Earth surface, and highly active; forests alone assimilate up to 67% of all the $CO_2$ that is removed from the atmosphere by terrestrial ecosystems.

With respect to soil bodies, the activity of the biota contributes to the addition, removal, transformation and translocation of matter (*sensu* Simonson) through a myriad of processes involving organisms such as algae, lichens, mosses, fungi, bacteria, nematodes, acaria, earthworms, termites, ants, beetles, moles, pocket gophers, prairie dogs, birds, badgers, grasses, brushes and trees. The chemical, physical, mineralogical and morphological traits of soil materials are affected by living organisms, either individually or as communities, and key edaphic properties like pH, organic matter quantity and quality, cation exchange capacity, base saturation, nutrient concentration and availability, clay content, colour, porosity, hydraulic conductivity, aggregate formation and stability, oxide content, mineral composition, and horizon diversity have responded to biotic activity since the Carboniferous (Table 12.1).

Some notable field observations illustrating the effects of biota on soils at various spatial scales include the sharp contrasts in soil properties measured at the forest–prairie boundary in Canada and the United States, the singular control exerted by Adelie penguin colonies (*Pygoscelis adeliae*) on the ocean-to-soil fluxes of organic matter in Antarctica, the tight regulation imposed by Kauri pine

*Soils: Basic Concepts and Future Challenges*, ed. Giacomo Certini and Riccardo Scalenghe.
Published by Cambridge University Press. 

Table 12.1. *Potential impacts of soil biota on soil properties and development*

| Biotic components | Selected examples[a] of impact |
|---|---|
| **Flora** | |
| Vegetation formations | Contrasts in horizon sequence, profile depth, pH, clay mineralogy, cation exchange capacity, organic matter and sesquioxide contents, colour and nutrient distribution in profiles at forest–prairie boundary, or under coniferous versus deciduous forests |
| Tree stands | Pits and mounds micromorphology resulting from windthrow with increased water flow and organic matter accumulation in pits |
| Tree canopy | Increased input of matter and water in throughfall, canopy drip (coniferous species), stemflow (*Fagus grandifolia*) or litterfall (*Betula alleghaniensis*, *Agathis australis*) compared with open areas; concentration of elements in surface horizons through biocycling; umbrella effect and soil protection against the impact of rain drops |
| Tree roots | Accumulation of organic matter, acidification/alkalinization, oxidation/reduction, $CO_2$ release and increased mineral weathering in the rhizosphere; preferential hydrologic flow and leaching along root channels; creation and stabilization of soil aggregates and protection of soil surface against erosion |
| Grasses | Secretion of secondary soil minerals (e.g. phytoliths) |
| Lichens and mosses | Role in physical disintegration and chemical weathering of minerals and rocks at early stages of pedogenesis; complexation of metals by excreted organic acids; trapping of atmospheric dusts and accumulation of fines on barren surfaces |
| Fungi | Decomposition of organic residues, litter materials and woody debris; chelation of dissolved metals and accelerated dissolution of soil minerals by exudates; stabilization of soil aggregates |
| Algae, diatoms | Nitrogen fixation; disintegration of rocks and minerals; addition of solid phases in the form of Si-rich cell walls (organic soils) |
| **Fauna** | |
| Mammals (large), e.g. bears, cattle | Excavation of the soil surface; compaction by herds of migrating ungulates or by grazing animals with consequences on accelerated erosion by surface run-off |
| Mammals (small), e.g. moles, rodents | Digging of large burrows in soil with impacts on water infiltration, on the establishment of preferential hydrologic flows, on the formation of *krotovina* (tunnels of rodents back-filled with soil) and on increased mass-wasting; selective transport of soil materials to the surface; disturbance and reworking of surface horizons |
| Birds, e.g. penguins | Cycling of carbon, nitrogen and phosphorus to the soil surface in faeces, egg shells and bone fragments; disturbance of surface horizons by nests; soil acidification by droppings |

Table 12.1 (Cont.)

| Biotic components | Selected examples[a] of impact |
|---|---|
| Reptiles, amphibians | Disturbance and reworking of surface horizons while digging nests; effect of tunnel networks on water infiltration |
| Insects, e.g. ants, termites | Mixing and disturbance of horizons; creation of extended porous networks and increased hydraulic conductivity; increased nutrient (Ca, Mg, K, N, P) content in mounds; cementation of soil materials; incorporation of organic matter; fragmentation of litter materials; sorting and size-dependent segregation of soil particles; selective transport of soil material to the surface; creation of microreliefs |
| Insects, e.g. beetles, spiders | Fragmentation of litter; formation of poorly connected burrows and channels resulting in voids in soils |
| Earthworms | Improvement of soil structure and of ped formation; mixing of soil materials and formation of A horizons; creation of biopores for infiltration of water and aeration; increased nutrient and fine particle contents in casts; decomposition of fresh litter and integration of organic substances; release of calcareous nodules; selective transport of soil material to the surface; nitrogen mineralization |
| Bacteria, Actinomycetes | Decomposition of organic debris and litter materials; weathering of soil minerals; oxidation and reduction of metals; cycling and transformations of macronutrients (C, N, S, P) |

[a]Examples collected from the following sources: Ugolini and Edmonds (1983), Birkeland (1984), Johnson and Watson-Stegner (1987), Paul and Clark (1989), Paton *et al.* (1995), Hinsinger (1998), Kelly *et al.* (1998) and Jobbagy and Jackson (2004).

trees (*Agathis australis*) on the intensity and the spatial distribution of podzolization in New Zealand, the addition as phytoliths of new solid phases to soils by decomposing Si-absorbing plants, and the unique contribution of algae to the disintegration of rocks in arid regions (Ugolini and Edmonds, 1983; Birkeland, 1984).

The contribution of soil processes controlled by biota to the progressive (anisotropy, horizonation and organization of profiles) and regressive (isotropy, haploidization and simplification of profiles) pathways of pedogenesis was synthesized in the evolution model of Johnson and Watson-Stegner (1987). According to this view, soil biota can promote the progressive or regressive vectors by either selectively mixing surface materials to create a biomantle and a stone line or through the biocycling (uptake by root and litterfall) of nutrients like Ca, Mg or P from deep to surface horizons, respectively. Pedogenic thresholds can also be initiated by, for example, invading plant or animal species.

Brimhall *et al.* (1991) developed a view of pedogenesis centred on the volumetric changes that occur during pedogenesis, such as dilation or compaction. To that effect, they confer an essential function to eluvial–illuvial processes but explicitly integrate the key role of bioturbation. For example, the contribution of root development, combined with lessivage, to the deformation of soils is shown. They indeed observe the progressive dilation of soils when translocated materials fill the voids created by successive cycles of root growth and decay. The stress generated by the infillings is converted into three-dimensional displacements, a perturbing effect that can be enhanced by the activity of burrowing animals.

Paton *et al.* (1995) proposed a novel process-oriented model of soil formation that incorporates bioturbation to weathering, creep, rainwash and aeolian processes. In their model, they acknowledge the importance of the biosphere in weathering reactions, new mineral formation and in the process of leaching. Moreover, they include the contribution of invertebrates (worms, ants, termites) and vertebrates (moles, gophers) to bioturbation and on the development of biomantles. Their presentation of the role of plants as a source of bioturbation is, however, more focused on biomechanical processes such as tree fall or root wedging than on biochemical effects. This model is useful to explain the development of texture-contrast soils found in vast areas of Australia, Africa and South America, a type of soil that is composed of a bioturbated topsoil with a basal stonelayer over a saprolite.

However fascinating these results might be, the study of the biotic factor is nonetheless plagued by several conceptual difficulties, a fact that is amply recognized in the soil science literature. This is, for one, because the distribution of biota and of soil properties are mutually linked in space and, consequently, cannot be viewed as independent variables. Moreover, strong feedbacks exist between the biota and climatic conditions. So, all other environmental factors need to be kept constant to assess the specific impact of organisms on soil formation. One potential solution to this dilemma is to use an approach that favours the study of soil-biota relationships at small spatial scales (micrometres to metres) and for short to intermediate time periods (days to years). Under these conditions, the effects of confounding factors like climate are drastically reduced and are considered to be minimal.

## 12.1 Approach

A comprehensive review of the relationships between the biota, soil genesis and soil properties is well beyond the scope of this chapter, as revealed by the introduction. The aim of this contribution is rather to illustrate the impact of the biota on soils through the presentation of field studies documenting the direct imprint of trees on specific soil chemical properties, here the spatial distribution of trace metal concentrations.

The approach is to select two case studies to emphasize the intensity and diversity of the effect of biota on forest soil properties and processes. Each study underlines the role of either the belowground or the aboveground component of tree structures on soils. The first case study focuses on a soil microenvironment, the rhizosphere, where the growth of tree roots and the activity of associated micro-organisms (e.g. mycorrhizal fungi) strongly impact on the distribution of trace metals. Indeed, along the microscale gradient that develops from the surface of fine roots towards the bulk soil, the biota can affect soil properties in a matter of days or weeks. The second case study concerns the relationships between the biocycling of elements through fresh forest litter and the spatial patterns of trace metals in the organic horizons of relatively uncontaminated forest soils. This study targets metre-scale patterns in soil properties evolving within a decadal timeframe.

The emphasis on trace metals like Cd, Cu, Ni, Pb and Zn rests on the fact that their fate in soils, and hence in terrestrial ecosystems, is of crucial relevance to uptake by plants, contamination of surface and ground waters, and to human exposure and health. Indeed, to document quantitatively the role of soils as a key environmental reactor involved in the fate of potentially toxic contaminants has become an important societal issue. Finally, it is of interest to note that forest soils, rather than agricultural, industrial or urban soils, were selected because they are seldom physically disturbed, thus allowing the study of in-place materials that have received and integrated long-term inputs of trace metals.

## 12.2 Case study 1: Trace metal distribution at the soil–root interface

### *12.2.1 Rationale and objective*

The soil is an extremely heterogeneous system and its chemical composition was shown to vary by several orders of magnitude between spatial scales. At the profile scale, the chemistry of soil materials differs between horizons and, for a given horizon, from one site to another. Tonguing created by preferential flow, compound aggregates and localized bioturbations are well-known sources of spatial heterogeneity within soil horizons. Recent work on the rhizosphere, the soil volume (a few millimetres in radius) surrounding living plant roots which is directly influenced by the activity of roots and of their associated micro-organisms, improved our understanding of the microscale heterogeneity of soil bodies. The spatial extent of the rhizosphere is difficult to quantify and it varies as a function of plant type, the texture and mineralogy of parent material, and the soil property of interest (Hinsinger, 1998). The extension of the rhizosphere can further be augmented through mycorrhizal symbiosis to create an environment termed the mycorrhizosphere.

There is a growing body of evidence indicating that the properties of the rhizosphere are in sharp contrast to those of the adjacent bulk soil. This contrast develops in response to the exudation of organic substances by roots (e.g. mucilages, phytosiderophores), the uptake of water and nutrients and following the decomposition of organic debris of plant or microbial origin. For example, the abundance and diversity of microbial populations, the levels of dissolved and solid-phase organic compounds, and the concentrations of macronutrients were frequently shown to decrease along a microscale gradient extending from the root surface to the bulk soil (Marschner and Römheld, 1996). The mineralogy also evolves along the same axis because of the sustained input of $H^+$ ions and, to a lesser extent, of organic acids to the rhizosphere. Yet, important gaps remain with respect to our knowledge of the distribution, concentration and speciation of trace metals at the soil–root interface, notably for chemical species considered to be available for uptake by the biota.

The objective of this research programme was to contrast the concentrations of trace metals, either in the solid or the dissolved form, between the rhizosphere and the proximal bulk components of forest soils that had received a range of metal contamination levels. The approach consisted of comparisons of the materials collected from both components for soils forming under individuals of a variety of tree species to establish the effects of root on soil.

### *12.2.2 Study sites*

The study sites are located in three glaciated forested areas of south-eastern Canada: Rouyn-Noranda (48° 14′N, 79° 01′W), Sudbury (46° 50′N, 81° 04′W) and St-Hippolyte (45° 56′N, 74° 01′W). The sites were selected to cover a range of metal loadings, soil properties and forest canopies. For over a century, the soils of the Rouyn-Noranda and the Sudbury areas received substantial amounts of trace metals from atmospheric depositions related to local Cu and Ni smelting activities. By contrast, the St-Hippolyte area on the Canadian Shield is relatively pristine and the soils received comparatively small atmospheric metal inputs of anthropogenic origin.

At Rouyn-Noranda, the soils were sampled under three young (30 to 50 years) trembling aspens (*Populus tremuloïdes*) at three locations along a metal contamination gradient with sites at 0.5, 2 and 8 km from the Cu smelter. A similar design was used for the Sudbury area, with the three sampling sites being situated at 2.5, 15 and 43 km along the prevailing wind direction from the Cu–Ni smelter. The soils were collected under young (12 to 30 years) naturally regenerating white birch (*Betula papyrifera*). In the St-Hippolyte area, the three sampling locations covered distinct forest canopies and the soils were collected under balsam fir (*Abies balsamea*), sugar maple (*Acer saccharum*) and white birch. To obtain rhizosphere and bulk materials from the upper B horizon, trees were carefully uprooted with the soil attached to the root network. In the field, the adjacent soil material free of roots or falling from the roots after gentle shaking was collected as the bulk soil. In the laboratory, the material adhering to fine roots (diameter $< 2$ mm) after a second shaking was brushed away to collect the rhizosphere. Root fragments were retrieved from samples and materials were air dried and sieved at 0.5 mm.

### *12.2.3 Trace metal distribution at the soil–root interface*

The data from the three sites show that the values of pH in water in the rhizosphere are almost always lower than those of the associated bulk materials (Fig. 12.1). The increased free acidity can reach more than 0.25 pH unit in the pH range of 4.5 to 5.5. The acidification of the rhizosphere is attributed to the excess uptake of cations by roots, compared with anions, with the concomitant increased release of $H^+$ ions to the rhizosphere. The concentration of water-soluble organic

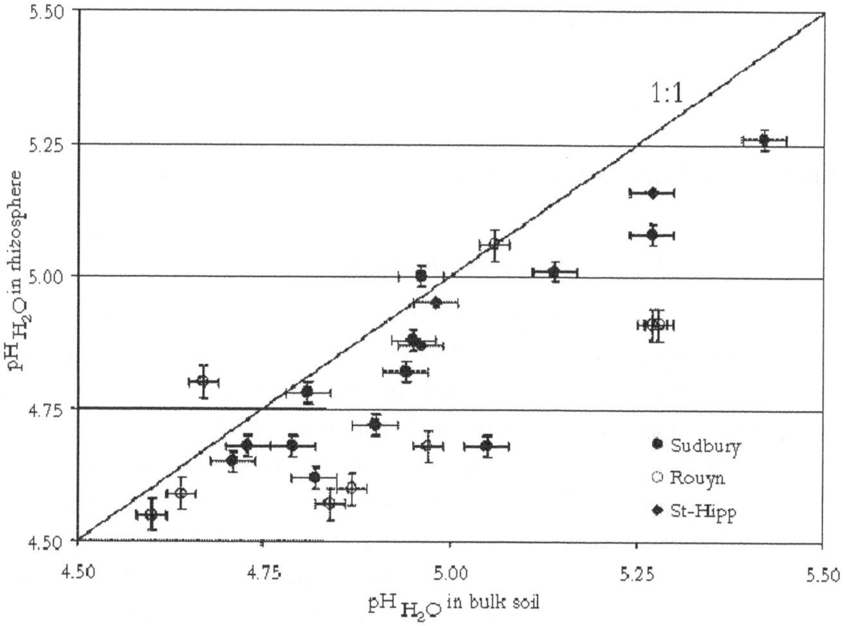


Fig. 12.1. Relationships for pH in water values between the bulk soil and the rhizosphere of three forested sites from southern Québec. Error bars are standard deviation values. Data from Legrand *et al.* (2005) and Séguin *et al.* (2004).

carbon (termed DOC) is systematically higher in the rhizosphere (Fig. 12.2). The DOC levels in the rhizosphere are sometimes twice those of the bulk soil and correlate highly to soil organic carbon contents (Séguin *et al.*, 2004).

For all trace metals except Pb, the water-soluble metal concentrations are significantly ($p = 0.05$) higher in the rhizospheric environment (Figs. 12.3 and 12.4).

The fraction of the acid-extractable metals present in the water-soluble form is generally less than 1% but this fraction is always higher in the rhizosphere component. Metal contents in the rhizosphere of forest soils from several other sites in southern Québec were also shown to be higher than in the bulk soil, for a range of extractants (Courchesne *et al.*, 2000).

Interestingly, the macroscopic observations on trace metal distribution obtained using chemical extractions are strongly corroborated by micrometre-scale investigations of the soil–root interface using synchroton-based technologies. Indeed, metals, in this case Fe, mostly accumulate at the soil–root interface whereas macronutrients like Mg are more evenly distributed and are integrated within the root structure (Fig. 12.5). The spatial pattern observed for Fe was found to be valid for other metals in these soils (Martin *et al.*, 2004).

Only a few reports exist on the chemical composition of the rhizosphere solution because of large technological constraints associated with its acquisition (Dieffenbach and Matzner, 2000). Using water extracts as a surrogate for field solutions, the free $Cu^{2+}$ ion activities and the concentrations of labile Zn ($Zn_L$) were determined to gain insights into this occult soil environment. Results indicate that the rhizosphere is significantly enriched in $Zn_L$ although this trend is not well expressed for $Cu^{2+}$ ions, in part because of the high spatial variability recorded for the latter soil property at the Sudbury site (Fig. 12.4).

### *12.2.4 Contribution to knowledge on soil formation*

The rhizospheric changes documented in this study occur at very small spatial scales and over short periods of time. In the context of soil formation, it is important to estimate the duration and the spatial extent of the impact of the belowground component of trees on soil properties. While performing this generalization exercise, it should be kept in mind (a) that the magnitude of the rhizospheric effect can vary, all other factors being equal, as a function of tree species and soil types and (b) that compound changes are actually measured where an initial rhizosphere change, for example in C sequestration, can subsequently impact on the abundance of microbes which, in turn, may affect the concentration of a nutrient like Fe through the production of microbial siderophores.

The time needed to establish a measurable change in soil properties in the rhizosphere can be as short as several days (Hinsinger, 1998) and can extend to

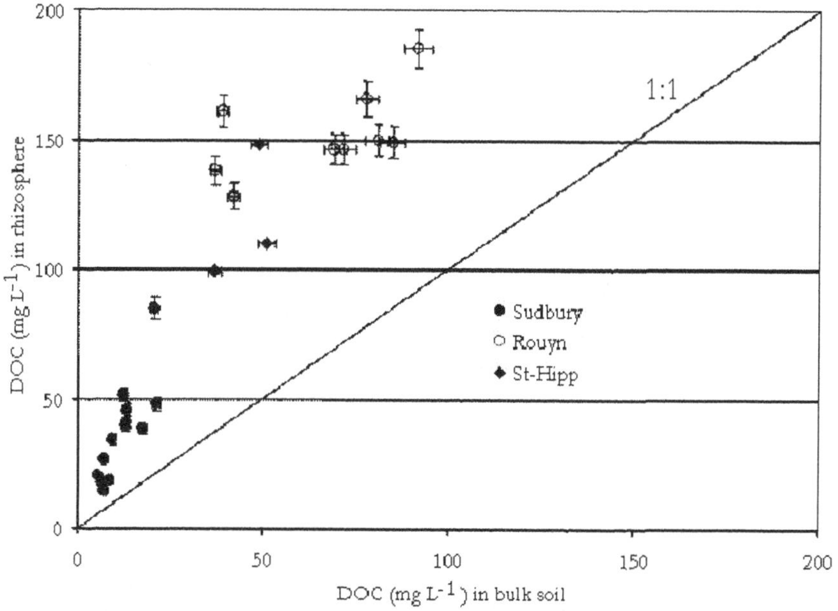


Fig. 12.2. Relationships for dissolved organic carbon (DOC) concentrations between the bulk soil and the rhizosphere of three forested sites from southern Québec. Error bars are standard deviation values. Data from Legrand *et al.* (2005) and Séguin *et al.* (2004).

several months, similar to the life span of fine roots. Very few data are, however, available to establish the stability of these changes through time. It is expected that some rhizosphere changes, such as in nutrient concentrations or microbial counts, can be quickly reversible whereas others will have a much more persistent character, an example being the accumulation of trace elements in combination with Fe-oxides to form sparingly soluble compounds. Changes in soil colour due to root activity may also last for decades or longer. In other instances, the changes are essentially irreversible, as is the case for the selective dissolution of minerals.

The rhizosphere is highly variable in space with soil zones in contact with active root tips being more rapidly and strongly affected than soil materials in contact with the base of mature roots. The spatial extent of the soil zone influenced by root activity is initially of a micrometre to millimetre scale but it can expand to reach a much larger soil volume. This can be explained by the repeated colonization of the same soil channels and voids by successive root generations, notably in nutrient-poor horizons with a low-density root network. A centimetre-scale rhizosphere can then be created. Preferential hydrological flowpaths along root channels can also lead to the formation of centimetre-wide altered and leached soil zones around roots (Certini *et al.*, 1998). Alternatively, in horizons

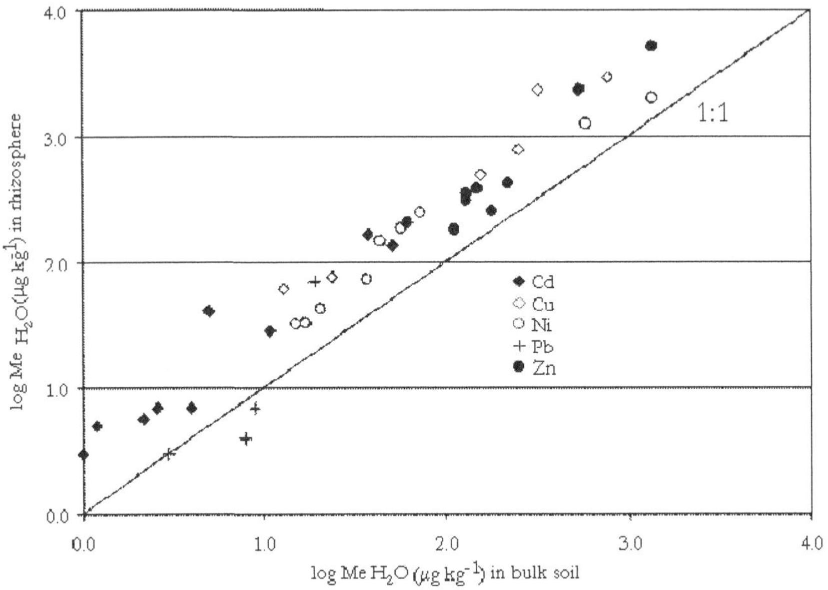


Fig. 12.3. Relationships for the dissolved trace metals Cd, Cu, Ni, Pb and Zn in water extracts, expressed as $\log_{10}$ metal concentration, between the bulk soil and the rhizosphere of three forested sites from southern Québec. Data from Courchesne *et al.* (2006), Legrand *et al.* (2005), and Séguin *et al.* (2004).

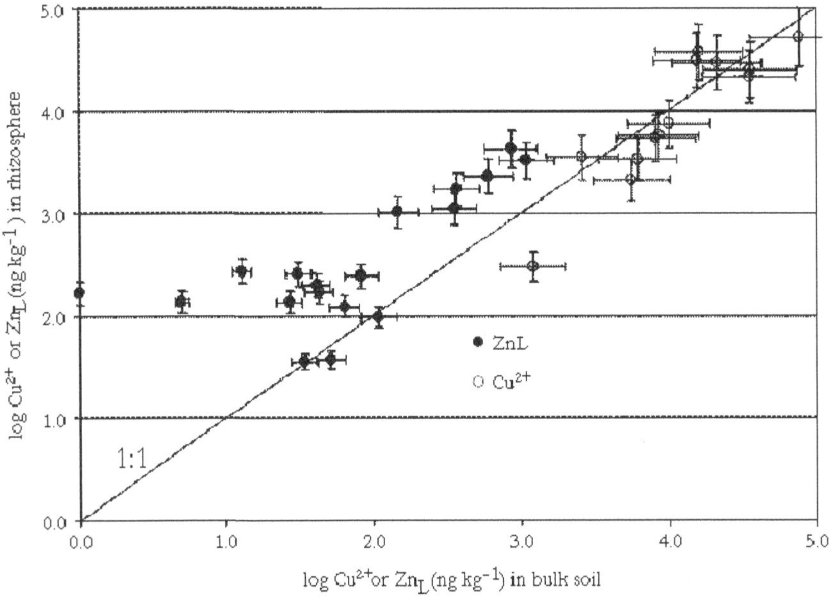


Fig. 12.4. Relationships for free $Cu^{2+}$ ion activities or labile Zn ($Zn_L$) concentrations between the bulk soil and the rhizosphere of three forested sites from southern Québec. Error bars are standard deviation values. Data from Courchesne *et al.* (2006), Legrand *et al.* (2005), and Séguin *et al.* (2004).

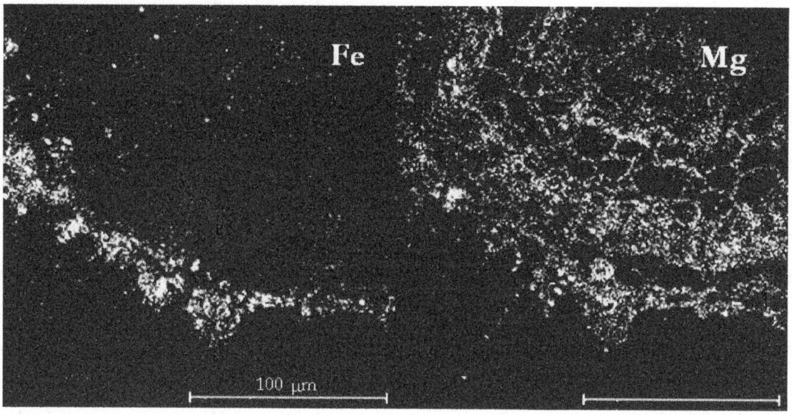


Fig. 12.5. Time of flight secondary ion mass spectrometry (TOF-SIMS) maps for total Fe and Mg. Each image represents the same cross-section of a soil aggregate including a *Populus tremuloïdes* root embedded in the adjacent soil material, here a Luvisol from the Rouyn-Noranda area, southern Québec. The images for a given element are normalized to the total ion yield with lighter colours representing higher relative concentrations. Reprinted from Martin *et al.* (2004), with permission from Elsevier.

bearing a dense root network, like A horizons, the progressive exploration of the soil volume by roots will transform the whole soil materials into a centimetre to decimetre-scale rhizospheric environment.

## 12.3 Case study 2: Trace metal patterns in organic horizons

### *12.3.1 Rationale and objective*

The organic horizons of forest soils are known to behave as an extensive, although heterogeneous, sink for trace metals in areas receiving metals from atmospheric deposition. Indeed, the functional groups (e.g. carboxylic, phenolic) present at the surface of organic substances can interact strongly with dissolved metals in the cationic form and bind them. The strength of the bonds varies as a function of the affinity between the metal and the binding sites, and with ambient conditions such as pH, ionic strength and the presence of competing substrates. These reactions largely determine the subsequent mobility and availability of trace metals in soils containing horizons rich in organic matter.

The sources of spatial variability in trace metal concentrations in the organic horizons of forest soils are varied and include a range of environmental attributes, with each attribute being associated to a specific spatial scale at which its impact reaches a maximum. For example, at the continental scale, the metal content in organic horizons is mostly related to the dominant atmospheric circulation patterns controlling the long-range transport and deposition of pollutants (Reimann *et al.*, 2000). At the regional scale, the spatial patterns are often determined by the presence of metal point sources like smelters and urban centres (McMartin *et al.*, 2002). Within intermediate-scale landscapes, metal patterns in soils have been found to relate to the mineralogical composition of surficial deposits or of bedrock, and to land use. However, little is known of the processes governing the metre-scale mosaic of trace element concentrations observed in surface horizons, notably in relatively pristine terrestrial ecosystems (Jobbagy and Jackson, 2004). Failure to appreciate quantitatively the fine-scale heterogeneity of metal contents in surface soil horizons affects the precision of estimates of metal availability and pools in soils, and precludes a sound understanding of the processes controlling the mobility of potentially toxic substances in forested ecosystems.

In this context, and once the existence of patterns is established, the objective of this study was to identify the environmental controls on metre-scale spatial patterns of trace metal contents in the organic horizons of a podzolic soil forming under a deciduous forest.

### *12.3.2 Study area*

The soils were collected along four 200-m long transects in the Hermine watershed located at St-Hippolyte, 80 km north of Montreal. The catchment is covered by a deciduous forest dominated by sugar and red maples (*Acer saccharum*, *Acer rubrum*) and American beech (*Fagus grandifolia*) with yellow and white birches (*Betula alleghaniensis*, *Betula papyrifera*), and large-toothed aspen (*Populus grandidentata*) as the accompanying species. The forest was either burned or cut some 100 years ago and the age of trees varies from 80 to 120 years. The soils are podzols that formed in a thin anorthositic till.

### *12.3.3 Trace metal patterns in organic horizons*

The leaves of sugar maple and of red maple trees are ubiquitous in the fresh Oi horizons of the Hermine soils (Fig. 12.6) and present strong inverse spatial patterns ($r = -0.52$; $p = 0.05$). The leaves from American beeches contribute a large proportion of the fresh litter dry mass although they are concentrated on the south-facing slope of the catchment. By contrast, the yellow birch leaves mostly accumulated on the north-facing slope, as did leaves of trembling and large-toothed aspens, and white birch (Manna *et al.*, 2002).

Spatial patterns in trace metal concentrations in the Oe horizons clearly exist in the Hermine watershed, as can be seen in Fig. 12.6. Moreover, the results of

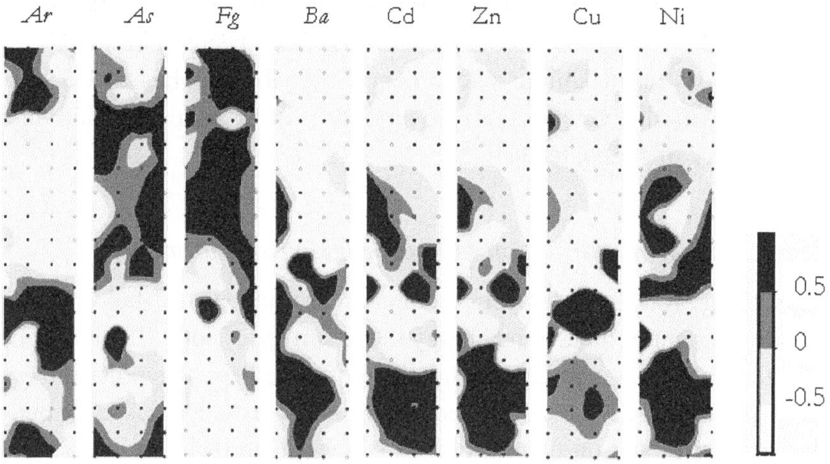


Fig. 12.6. Maps presenting the spatial distribution of leaves from red maple (*Ar, Acer rubrum*), sugar maple (*As, Acer saccharum*), American beech (*Fg, Fagus grandifolia*) and yellow birch (*Ba, Betula alleghaniensis*) in the Oi horizon, and of water-soluble Cd, Zn, Cu and Ni in the Oe horizon of the Hermine watershed. The data are standardized (z-scores). The abundance of leaves vary from 0% to 52, 87, 68 and 48% of total dry litter mass from *Ar, As, Fg* and *Ba*, respectively. The ranges in absolute water-soluble metal concentrations are: Cd = 4–54, Zn = 285–8869, Cu = 70–2156 and Ni = 25–106 $\mu g\ kg^{-1}$. The north is located at the top of the figure. Data from Manna *et al.* (2002).

correlation analyses indicate that the spatial distributions of water-soluble Cd, Cu, Ni and Zn are positively correlated ($r$ = 0.42 to 0.83; $p$ = 0.05). The link between the Cd and Zn patterns is particularly strong at $r$ = 0.83 (Fig. 12.6). Similar spatial structures were detected for acid-extractable and $BaCl_2$-exchangeable metals in the Oe horizons of the Hermine (Manna *et al.*, 2002). Cadmium and Zn concentrations in water and in acid extracts are also well correlated. This suggests that, in some cases, patterns in acid-extractable metals may reflect the potential bioavailability of metals in the surface organic horizons of mineral soils. The distribution of pH values in the Oe horizon is not correlated to metal patterns and, thus, explains only a small component of the overall spatial variability in metal contents.

The spatial patterns for water-soluble and acid-extractable Cd and Zn in the Oe horizons are significantly correlated to the pattern observed for the abundance of yellow birch leaves in the fresh litter horizon ($r$ = 0.46 to 0.60; $p$ = 0.01; Fig. 12.6). The analysis of fresh yellow birch leaves further shows that they are much richer in Cd, Pb and Zn than the leaves from the other tree species of the Hermine. Differences in concentrations sometimes exceed one order of magnitude for Cd and Zn. Data from a similar forested ecosystem at the Hubbard Brook

Experimental Forest also identified yellow birch leaves as a high source of Zn compared with leaves from sugar maple and American beech (Gosz et al., 1973). Our data suggest that a similar phenomenon probably applies to Cd. Finally, metal contents in the Oe horizon are not affected by the pits and mounds microtopography of the watershed.

### *12.3.4 Contribution to knowledge on soil formation*

Under deciduous forests of temperate climates, the annual leaf-fall constitutes a major punctual input of matter and energy to the surface of soils. The biocycling of elements through plant uptake and litter deposition also represents an upward transfer of matter in soils that can subsequently impact on the vertical distribution of elements in profiles and on the creation of surface patterns in horizon properties. The patchy nature of soil organism populations in surface horizons was, for example, found to correlate with variations in litter quality or quantity which are, in turn, linked to canopy composition (Wardle *et al.*, 2004). In the Hermine, the trees affect the shape of trace metal patterns in the Oe horizon and their effect increases with the magnitude of the allocation of a given metal to the aboveground component. Now, how rapidly and to what extent can the imprint of trees on soils develop?

On the basis of age of the individuals studied, the time needed by yellow birch trees to establish a measurable change in metal concentrations in surface horizons is estimated to $10^1$ years. Obviously, this period will vary as a function of metal solubility in the parent material and with the proximity of anthropogenic trace metal sources that can mask the biocycling effect. In general, the influence of trees on soil properties, such as nutrient contents, is thought to last for $10^2$ to $10^3$ years before the succeeding species alters the soil to a point where the imprint of the preceding species is untraceable (Birkeland, 1984). In the case of trace metal patterns in organic horizons, one can submit that the impact of yellow birch, once well established, will probably persist for a period at least equal to $10^2$ to $10^3$ years, because of the strong bonds between metals and organic matter. This is significant when the fate of metals in young terrestrial ecosystems of glaciated areas is a prime concern. The lateral extent of the soil surface affected by the litterfall of tree individuals is metric in scale. This area can slowly expand or contract in time as trees grow or die and may vary between locations with the direction of prevailing winds and topography. The depth of penetration of the effect of yellow birch litter in soil profiles is as yet unknown because only Oe horizons were studied. It is probable that the underlying Oa horizon is similarly enriched in trace metals although a time lag may exist. The impacts on mineral horizons are, at this stage, speculative. Obviously, these hypotheses will need to be tested in the future.

## 12.4 In conclusion

Trees have a direct impact on the spatial distribution of trace metal concentrations in forest soils at spatial scales ranging from micrometres to metres and on timeframes of weeks to decades, centuries or more.

The findings from these case studies are not the first to document the contribution of biota to soil formation, processes and properties and, as such, are not exclusive to forest soils and deciduous trees. In other words, the results presented here can be generalized, although with caution, to other soil–plant associations. The rhizosphere effect on metals indeed occurs in agricultural environments, under grasses and beneath coniferous trees. The generalization can also be extended to soil traits other than metal contents. To that effect, it can reasonably be submitted that a change in dissolved trace metal concentrations in organic horizons or in the rhizosphere will eventually affect microbial activity and thus alter the dynamics of organic matter decomposition and the bioavailability of nutrients. When the results of both case studies are integrated, ideas on how the aboveground and the belowground components of trees can influence soils, either individually or in combination, can be formulated. For example, a marked increase in litterfall Cd content due to the ingress of a new tree species could force other plant species more sensitive to dissolved Cd to develop part of their new roots in deeper horizons, with concomitant effects on organic matter inputs and colour changes in the mineral soil. The aboveground and belowground components of biota have too often been considered separately and feedbacks between both are frequently overlooked. This gap in our approach needs to be filled if we are to increase our understanding of the processes governing the impact of biota on soils.

# 13

# Factors of soil formation: time

*Ewart A. FitzPatrick*

The factors of parent material, climate, topography and organisms are tangible and can be seen and/or touched. By contrast, time is intangible and progressive as the others interact with each other (Fig. 13.1). Time is presented within the Newtonian philosophy which deals with absolute time and space considered as a world framework. Alternatively time can itself be considered as an infinite and ever-increasing universe in which the other factors interact to produce an infinite number of soils, or more precisely soil horizons, the infinite composites of which comprise soils. Perhaps within this context there should be a consideration of the energy in soils (Blum, 1997). There are four main sources of energy – gravity, mineral energy, solar and capillarity (anti-gravity). All processes in soils are dependent upon energy; for example translocation is largely dependent on gravity, while climatic change is largely dependent upon variations of solar radiation.

Climate is not constant and may change to induce different processes, or there may be a change in vegetation as a result of human activity, natural successions or events such as fire. Thus although the progression through time is uninterrupted, the processes may change, leading to a different situation which will show a combination of the properties acquired by the first progression with the imprint of the new properties acquired by the new set of processes. The result will depend upon the strength of the two or more sets of properties. In some situations the new set of properties may completely mask or eliminate the previous set. By contrast, the initial set of properties may remain dominant for a considerable period. In some cases the second set of processes may even preserve the features produced by the first set of processes.

Since soil formation is a very slow process requiring thousands and even millions of years and since this is greater than the lifespan of an individual human

*Soils: Basic Concepts and Future Challenges*, ed. Giacomo Certini and Riccardo Scalenghe.
Published by Cambridge University Press. 

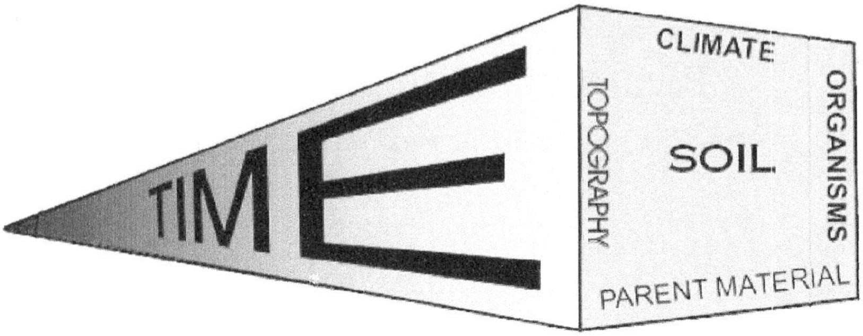


Fig. 13.1. Summary diagram showing climate, parent material, organisms and topography interacting through time to form soil.

it is impossible to make categorical statements about the various stages in the development of soils. A further complication is introduced by the periodic changes of climate and vegetation which often deflect the path of soil formation in one direction or another. Therefore, all that is said about time as a factor in soil development is in part speculation and in part based on botanical, zoological, geological or geomorphological data. The botanical data are usually in the form of pollen analysis from peat and sometimes from the soil itself. The zoological data are from buried fossils. In a number of cases some type of absolute dating such as $^{14}C$ has been used to establish the age of the soil (Pillans, 1998 and 2004).

## 13.1 Time for horizon differentiation

Not all soils have been developing for the same length of time. Most started their development at various points during the last 100 million years. Some horizons differentiate before others, especially those at the surface which may take only a few decades to form in unconsolidated deposits. Middle horizons differentiate more slowly, particularly when a considerable amount of translocation of material or weathering is necessary, some taking 4000 to 5000 years to develop. Some other horizons require even longer periods; for example, a study of the interglacial soils in the central USA has shown that there are strongly weathered horizons known as gumbotil characterized by very high clay content (Kay, 1916). These occur either at the surface or are sandwiched between two glacial deposits and may have formed during a single interglacial period extending over 50 000 to 70 000 years. The rate of development of these horizons can be regarded as very great since they have formed in a zone of permanent saturation where hydrolysis is most vigorous. Weathering of rock to form a Ferralsol may require more than a million years since these situations are found only on very old land surfaces that have been exposed to weathering since at least the Tertiary period (Table 13.1).

Table 13.1. *The time elapsed since 248 million years BP*

| | | |
|---|---|---|
| **Cenozoic Era (from today to 65 MyBP)** | Quaternary (from today to 1.8 MyBP) | Holocene (from today to 10,000 years)<br>Pleistocene (10,000 yrs to 1.8 MyBP) |
| | Tertiary (1.8 to 65 MyBP) | Pliocene (1.8 to 5.3 MyBP)<br>Miocene (5.3 to 23.8 MyBP)<br>Oligocene (23.8 to 33.7 MyBP)<br>Eocene (33.7 to 54.8 MyBP)<br>Paleocene (54.8 to 65 MyBP) |
| **Mesozoic Era (65 to 248 MyBP)** | | Cretaceous (65 to 144 MyBP)<br>Jurassic (144 to 206 MyBP)<br>Triassic (206 to 248 MyBP) |

Also within horizons there are usually different rates of mineral transformation. Quartz and more especially zircon weather slowly and accumulate, while feldspars are completely altered.

## 13.2 Soil development

Initially pedologists tended to interpret most soil features as the result of the interaction of the prevailing environmental conditions at the time when examinations were made. However, it soon became evident that most places have experienced a succession of different climates which induced changes in vegetation and in soil genesis. Therefore most soils are not developed by a single set of processes but undergo successive waves of pedogenesis. Furthermore each wave imparts certain features that are inherited by the succeeding phase or phases. In a number of cases these properties are developed so strongly that they are evident thousands and even millions of years after their formation. The outcome is that soils are regarded as having developmental sequences which manifest not only the present factors and processes of soil formation, but also a varying number of preceding phases.

Although changes in the external environmental factors are usually responsible for progressive changes in the soil, vegetation and landscape, there are a number of cases when the environmental factors remain relatively constant but continuing soil development leads to the formation of new features within the soil. Some of these may become very prominent and may themselves influence the further course of soil formation. For example, the progressive development of an impermeable middle horizon can cause waterlogging at the surface and a change in soil characteristics with corresponding change in the vegetation. These various progressive changes in soils are sometimes known as soil evolution. Examples of these changes have been reported from every major land surface. Many land surfaces in Africa and Australia developed in the mid Tertiary but one or two may date from the earlier part of the Tertiary or even the previous Cretaceous period (Table 13.1). If this is the case the soils occurring on these surfaces may be very old. Consequently, any consideration of soil development and change through time must commence by discussing the events that took place during the early part of the Tertiary period, but the age of the surface may not be the same as the soil, which may be younger. The evidence afforded by plant remains shows clearly that the surface of the Earth at that time did not have the extremes of climate that it experiences at the present. The flora of the late Cretaceous to the Miocene period (Table 13.1) in the northern latitudes grew under warmer conditions, as indicated by the occurrence of palms in Canada and Britain where conditions could be regarded as subtropical. However, full tropical conditions existed in many of the areas where they are found today as well as in many of the present subtropical and arid areas including such places as central Australia and the southern part of the Sahara Desert. These warmer and more humid conditions caused the rocks to undergo profound weathering, by hydrolysis in the case of silicates and solution in the case of limestone and other soluble rocks. Considerable natural erosion also took part during this prolonged period to form the characteristic flat or gently sloping surfaces of planation associated with old landscapes. Where weathering and soil formation kept pace with and even proceeded faster than erosion, great thicknesses of soil and weathered rock are present. The processes of weathering were so complete in many places that rocks of all types were transformed into clay minerals, hydroxides and oxides of iron and aluminium together with the resistant residues such as quartz and zircon, but sometimes even these resistant primary minerals have been weathered. Most of the bauxite accumulations of the world are Tertiary phenomena resulting from profound chemical weathering.

The warm humid tropical conditions were maintained in tropical and subtropical areas throughout most of the Tertiary period, but in the poleward areas the climate became progressively cooler from the end of the Miocene. By the end of the Pliocene (Table 13.1), floras of Eurasia and North America bore many similarities to those of the present but they contained more species, particularly in Western Europe where the present flora is a poor representation of the past. The gradual cooling of the climate culminated in the Pleistocene period (Table 13.1) with its repeated glaciations both in the northern and southern hemispheres but it is the former with its greater circumpolar land surface that the effects on pedology are most significant and widespread. These glaciations together with their associated periglacial conditions effectively removed most of the deep topsoils and weathered rock that formed during the preglacial period. However, there is still sufficient present to leave no doubt about its previous ubiquity or about its contribution to the various superficial deposits of the Pleistocene. Much of the so-called boulder clay (glacial drift) contains fine material derived from the previous weathered soil, mixed with fresh rock fragments. Perhaps the most impressive soil evolution has taken place in Australia and various parts of central Africa, particularly those bordering the deserts where soils that developed under hot humid conditions are now partially fossilized in an arid environment.

Many other tropical and subtropical areas experienced periods of higher rainfall or pluvial conditions during the Pleistocene. It is certain that erosion by water did reduce the thickness of soil and weathered mantle in many tropical and subtropical areas but the effect was small as compared with the widespread and deep striping caused by glacial and periglacial processes.

## 13.3 Holocene soil formation

The rapid climatic amelioration at the end of the Pleistocene possibly induced the most dramatic set of changes experienced by the Earth's surface. This caused wave after wave of vegetation to pass rapidly over wide areas, particularly in Western Europe and North America. These were quickly followed by the influence of humans so that soil development in many places in middle latitudes has been subjected to conditions varying from full glacial to maritime climatic conditions in the relatively short period of 10 000 years. As a result, many of the soils in these areas have diverse properties inherited from the Tertiary and Pleistocene periods.

In historical times many evolutionary paths have been directed by human activity; the polders of the Netherlands, the paddy soils of the eastern countries and the catastrophic erosion in many countries quickly spring to mind. An interesting feature to emerge from recent investigations is that several apparent vegetation climaxes may be due to human activity and are thus plagioclimaxes. Examples are afforded by the ericaceous heaths of Europe and the many savannas of the tropics.

## 13.4 Soil age and progressive change

Undoubtedly the most fascinating relationships of soil properties and soil-forming factors are those that relate to the age of soils. They vary from relatively simple developmental sequences to very complex scenarios involving extremely long periods of time accompanied by changes in climate, vegetation and topography. Fascinating examples include the soil-erosion-landscape development sequence described by Butler (1959) for Australia. The area west of the dividing

Fig. 13.2. Profile of a Podzol from north-east Scotland.

range in Australia and extending to the coast at Perth has a land surface that has experienced a succession of very diverse climates since the Cretaceous period.

Progressive changes will be illustrated in part by the following two soils – one from Scotland and the other from Australia.

The soil from north-east Scotland (Fig. 13.2) is a Podzol with organic horizons (1) underlain by a bleached horizon (2) overlying a strong brown friable fine granular horizon of accumulation of iron and aluminium (3). There is then an abrupt change (4) to a very hard, massive horizon (5) with lenticular structure having silt cappings on the upper surfaces of the lenses and stones.

This soil started as a Tundra soil immediately after deglaciation. Its evolution is shown in Fig. 13.3. The upper mineral horizons (2 and 3) were the active layer that would freeze every winter and thaw every summer and would be mixed by these processes; in addition, any stones that were present would be orientated vertically on flat sites. This orientation is maintained to the present. Beneath the abrupt boundary would have been the previous permanently frozen subsoil or permafrost. The active layer would have shown some evidence of hydromorphism because the permafrost would have been impermeable, producing reducing conditions in the active layer. The permafrost would have had the normal features of lenticular peds delimited by thin, horizontal bifurcating sheets of ice and sheaths of ice around the stones. This structure has been maintained with the addition of silt cappings on the stones that accumulated when the permafrost

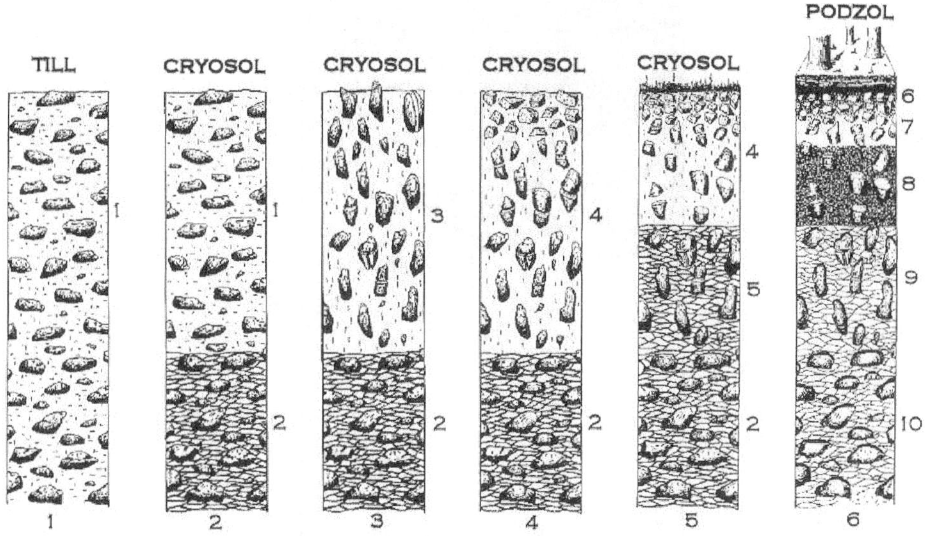


Fig. 13.3. Soil developmental stages from glacial conditions to the present on acid parent material. Stage 1 is the recent deposit of glacial till but it could be glaciofluvial material. Note that the stones are more or less horizontally orientated. Stage 2 shows (1) the active layer that freezes every winter and thaws every summer and the development of permafrost (2) in the lower part of the profile with its characteristic lenticular structure. Stage 3 again shows the active layer (3) and permafrost (2); however, there has been a change in the active layer in that the stones are now vertically orientated due to the freeze–thaw processes. Stage 4 shows a further development in the active layer (4) in the form of much frost shattering of the stones. Stage 5 shows a rise in the permafrost trapping vertically orientated and frost shattered stones in the new permafrost (5). Stage 6 shows the establishment of a coniferous forest and development of a Podzol in the previous active layer with frost shattered and vertically orientated stones. There is an abrupt change (8/9) from the middle to the lower horizon in the Podzol corresponding to the previous change from the active layer to the permafrost. The lenticular structure has been maintained in the lower horizon and silt cappings have formed on the tops of the stones.

melted (FitzPatrick, 1956). In addition, during the Holocene, in the active layer a leached horizon and one of accumulation have formed with the complete elimination of the mottling present during the tundra phase. Colloidal material has been leached into the relict permafrost to cause cementation and hardening and at the same time preserved the structure. This shows clearly how properties develop as a result of the superimposition of processes through time.

The development stages would first be a Regosol followed by a Cambisol and the gradual differentiation of the Podzol on acid material. On basic parent material development would still be at the Cambisol stage. On very acid parent materials, as shown in Fig. 13.2, a thin iron pan has formed.

Fig. 13.4. Australian soil showing (1) loose concretionary layer, (2) layer of breakdown of massive concretionary layer, (3) massive concretionary layer, (4) transition layer from the pallid layer below to the concretionary layer above, (5) pallid zone.

The second example is from Western Australia where the time span is considerably greater, with possible commencement in the Late Cretaceous. This soil, shown in Fig. 13.4, is described by Gilkes et al. (1973) and is summarized as follows. The upper one metre (layers 1 and 2) is composed of loose subspherical concretions in a light grey sandy matrix. The next 50 cm (layer 3) is massive concretionary material that grades through a mottled horizon into a pallid zone (layer 4); and with depth there is a change to hard granite at about 40 m.

Although the field morphology is dramatic it is the micromorphology that is the more revealing, especially the nature of the massive concretionary material as shown in Figs. 13.5 and 13.6. Figure 13.5 shows a hand specimen 15 cm in width while Fig. 13.6 shows a photomicrograph of a single concretion which is extremely complex. On the outside there are conspherical layers of clay coatings composed largely of gibbsite with some boehmite. The individual layers vary in thickness and colour from black to yellowish-brown. It would seem correct to assume that the coatings were the last part of the concretion to form. Precisely how and why the concretion formed around a central core is uncertain. Within the core there are a number of coated quartz grains and coated, rounded fragments of the pallid zone. These two features also occur in the sandy soils in the drier areas to the east and seem to be forming at present or in the recent past. Thus it would

Fig. 13.5. Hand specimen 15 cm wide of concretionary horizon of the soil shown in Fig. 13.4. The arrow indicates a small area where decomposition has started.

seem that there was a phase of deep weathering to form the pallid zone, then an arid phase which eroded the pallid zone forming kaolinite granules by saltation, and then a return to wet conditions with the formation of concretions in wind-transported material. It is not usual to find soil situations that show a change from dry to wet conditions. There are two further stages in the formation of this massive material. There seems to be a return to a dry phase since this material is permeated by opal, which is conspicuous by its fluorescence when the thin sections are examined with incident UV illumination. Finally there is the current wet phase causing this massive material to decompose to form the upper horizon with its sandy matrix surrounding concretions. Thus this soil shows at least five different climatic episodes: (1) deep weathering of the granite to form the pallid zone; (2) erosion of the pallid zone; (3) formation of the concretionary layer; (4) cementation by opal; (5) partial and continuing destruction of the concretionary layer at present.

Such is the contribution of time to soil formation. It is interesting that in both of these soils silica plays an important part in their formation. It would seem that the role of opal in soil formation has been largely overlooked. In most calcite-cemented horizons (calcretes) opal plays an important role in the hardening. After the removal of calcite from thin sections of calcrete there is normally a conspicuous residue of brilliant, white opal.

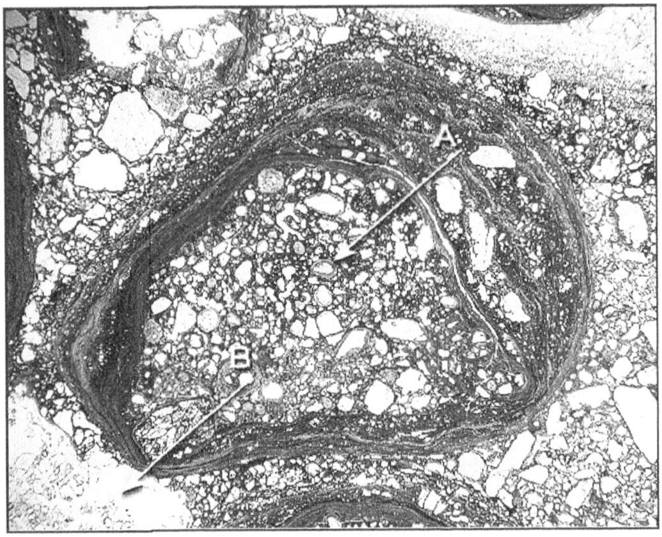


Fig. 13.6. Thin section of a single concretion from the soil horizon shown in Fig. 13.5. Arrow A shows a quartz grain with a coating; arrow B shows an area of decomposition in which all of the fine material has been removed leaving only the quartz grains. Ultimately the whole mass succumbs, leaving the concretions in a loose sandy matrix to form the uppermost layer.

The various soils which comprise any particular area of landscape usually evolve in different directions as determined by differences in the factors of soil formation and variations in the intensities of pedogenic processes. Many relationships are discussed in FitzPatrick (1980).

## 13.5 Time and soil classification

In most classifications the current approach is to use a limited number of the properties with fixed ranges for each property. In practice this has proved very difficult to apply because in many situations a given soil will satisfy all except one of the criteria. A fuzzy system should now be attempted. The suggestion (FitzPatrick, 2002) is to consider that soils have an infinite number of properties and that each property is a coordinate with an infinite number of values along each coordinate. Absolute time is considered as one of the co-ordinates, thus making it possible to classify soils on the basis of their age. These values intersect in space to create an infinite number of points of which 300 have been chosen and regarded as reference points. The results from the analyses of soils (horizons) are compared with the reference points to give a designation. This procedure is carried out with Excel software. There are many virtues of this system; minimum memory is required and numerous classifications can be created to suit various users. Of course these reference points with their myriad of properties are conceptual, as is the corporeal volume they occupy in space. These volumes are not static but are moving along the time co-ordinate. This co-ordinate can be considered as linear when using the accepted units of measurement such as years, millennia, etc.

The two examples of soils in Section 13.4 above show that a given point can be classified in a number of different ways depending on its position along the time co-ordinate. Figure 13.3 shows separate stages in the progressive change as points. Spatially all of these points exist today in different parts of the world. Thus reference points can occur as spatial or geographical realities and as points along the time co-ordinate. Thus, having the concept of reference points brings together within soil classification both spatial and temporal dimensions.

Some aspects of tectonics can also be considered with the time co-ordinate. For example, a given point may be a part of a beach with saline conditions and be gradually uplifted to become dry land that progressively develops a forest with an accompanying soil. So that with time the classification of the point or volume changes. Thus the classification of any particular soil is not sacrosanct but is changing continuously. In pine forests the trees of *Pinus sylvestris* remain constant but the soil beneath is continuously changing. Therefore soil classification must have time as an essential part of its framework.

# 14

# Soil formation on Earth and beyond: the role of additional soil-forming factors

*Giacomo Certini*
*Riccardo Scalenghe*

*Humankind* is an environmental variable whose activities are affecting more and more of the planet and its soils. At present, humans appropriate over 40% of the Earth's net primary productivity (Rojstaczer *et al.*, 2001) and use 60% of freshwater run-off (Postel *et al.*, 1996). It is intuitive that this exploitation of resources results in a massive impact on terrestrial ecosystems. Thus, humankind influences ecological processes on a global scale, sometimes on a par with the role of climate, geological forces and astronomical variations. On this basis, it has been proposed that we are entering a distinct geological era, the 'Anthropocene', a period of intensive human industrialization and land change which started in the late eighteenth century, marked by the increase in atmospheric concentrations of carbon dioxide and methane (Crutzen, 2002). It has been estimated that today 83% of the ice-free land on Earth is affected directly or indirectly by humans (Sanderson *et al.*, 2002). In summary, humans can be viewed as a distinct soil-forming factor, apart from other organisms that are not endowed with the ability to reason, due to the magnitude of their impact on the planet, and their cultural characteristics which drive their decision-making process (Amundson and Jenny, 1991).

In any natural environment, population growth reaches a limit imposed by factors such as light, space, nutrients, or water. Carrying capacity is the maximum number of individuals a particular habitat can support. Commonly, populations exceed the carrying capacity of their ecosystems, and experience a rapid decline until conditions for growth are restored. Mass migrations to sparsely inhabited lands, and the use of progressively higher amounts of energy, have allowed humans to, at least temporarily, modify or circumvent the most limiting population restricting factors. Only a few million humans were sustained by the Earth before

*Soils: Basic Concepts and Future Challenges*, ed. Giacomo Certini and Riccardo Scalenghe.
Published by Cambridge University Press. 

the agricultural revolution, which began about ten thousand years ago. By the beginning of the Christian era, the human population was about 300 million, and there was apparently little increase in the ensuing millennium. Subsequent population growth was slow and fitful, especially due to catastrophic diseases of the Middle Ages. However, since the eighteenth century, human population growth has been uninterruptedly exponential (U.S. Census Bureau, 2005a). As a consequence, the reasonably biologically productive land per capita has decreased continuously. Estimates for the next 50 years indicate that humankind is tending asymptotically to a value of about 9 billion (U.S. Census Bureau, 2005b) that corresponds to a global human density of 1 person for each 0.01 $km^2$ of reasonably biologically productive land (the area of two football fields!). This further increment of population will imply a wave of human expansion and the enhanced exploitation of the environment will have a profound impact on soils (Tilman *et al.*, 2002).

## 14.1 The anthropogenic factor

### *14.1.1 Deforestation and forestation*

In the mid 1990s forests covered about one third of the world's land area, representing about half of their original extent before human intervention. Each year, in underdeveloped countries of the tropics, some 14 million hectares are deforested, a loss counterbalanced to a small degree by a million hectares that are reforested in the developed, northern latitude countries (FAO, 2006).

Soil properties are significantly impacted by the clear-cutting of forests (Hajabbasi *et al.*, 1997). Both the removal of the vegetation mantle and the degradation of soil structure can induce soil erosion and watershed destabilization.

Forestation of virgin substrata or former arable or pasture lands lead to significant modifications of soil morphology and properties, chiefly linked to early pedoturbation, and following decreased erosion and addition of new organic matter (Ritter *et al.*, 2003).

Deforestation, forestation and, more comprehensively, any type of change in the vegetation assemblage impose substantial modifications to moisture and temperature regimes of soil, which in turn affect the soil-dwelling biotic community. The rates of organic and mineral weathering consequently vary.

### *14.1.2 Grazing*

Grazing of domestic animals is the world's most widespread land use, involving more than 30% of the ice-free land (Table 14.1). Grazing leads to changes in vegetation (plant community structure), favouring especially the introduction of

Table 14.1. *Global extent of some types of human-impacted soils*

| | M km$^2$ | % ice-free land[a] |
|---|---|---|
| Pasturelands[b] | 42.26 | 31.77 |
| Croplands[b] | 15.03 | 11.27 |
| Water reservoirs[c] | 0.50 | 0.38 |
| Urbanised areas[d] | 4.74 | 3.56 |
| Roadways[e] | 0.28 | 0.21 |
| Railways[f] | 0.003 | 0.003 |
| Artificial waterways[g] | 0.02 | 0.01 |
| Airports[h] | 0.02 | 0.01 |
| Quarries[i] | 0.24 | 0.18 |
| Conflict areas[j] | 10.75 | 8.08 |

[a]Ice-free land: 133.1 M km$^2$.
[b]FAO (2006).
[c]Shiklomanov (2000).
[d]WRF (2000); CIESIN/IFPRI/World Bank/CIAT (2004). Integrated and recalculated data.
[e]IRF (2003); Lay (1992); CIA (2006). Compared and integrated, then calculated from the length 27.69 M km and multiplied by 10 m (assumed as average roadway width).
[f]CIA (2006). Recalculated from the length 1.13 M km multiplied by 3 m.
[g]CIA (2006). Recalculated from the length 0.64 M km multiplied by 25 m (assumed as average waterway width).
[h]World Watch (1994); Hinkelman (2005); CIA (2006). Integrated and recalculated data.
[i]Moore and Luoma (1990). Data foreseeing the situation in 2000.
[j]Smith (2003). Recalculated by referring to 46 major war zones.

invasive, non-native species. Excessive grazing can impose large modifications on the biogeochemical cycles of carbon and nitrogen. The quantity and quality of soil organic matter can be highly modified, also affecting the pathway of pedogenic processes. Soil is compacted and the roughness coefficients of watersheds reduced, resulting in more surface run-off and soil erosion. Over 50% of the world's pasturelands are overgrazed and subject to erosive degradation, causing high environmental and economic costs (Pimentel *et al.*, 1995).

### *14.1.3 Agriculture*

Most soil disturbance arises from agriculture as a result of tillage, liming, manuring, fertilization, irrigation and drainage. Although the first farmers appeared 12–15 thousand years ago, an evolution of agricultural practices causing widespread soil disturbance began to occur much later. Modern 'intensive' agriculture, based on heavy mechanization and large use of fertilizers and pesticides, was born in the twentieth century. Globally, croplands now amount to 16% of the 93 million km$^2$ of biologically productive land (Sundquist, 2004).

Fig. 14.1. Deep ploughing (1.2 m) for planting a table grape vineyard, Sicily (Italy). A mollic epipedon and a calcic horizon were originally present in this soil.

The impact of agriculture on pedogenesis can be dramatic and soil classification systems have progressively recognized this type of anthropogenic influence. For example, surface soil horizons (epipedons) that are the result of cultivation are now diagnostic for taxonomic purposes.

In most cases, soil perturbation due to ploughing is confined to the top 30–40 cm but site preparation practices such as deep ploughing (subsoiling) and placement of drains involve much deeper disturbance of the soil (possibly down to 1 m or more). The original sequence of horizons is partly or completely destroyed (Fig. 14.1).

Conversion of forests to cultivated land leads to an average loss of 20–30% of soil organic C (Murty *et al.*, 2002). Tillage can even induce complete removal of the fertile surface layer through erosion (Packer and Hamilton, 1993). This represents a serious problem at both local and global scales. Soil conservation measures have spread since the 1930s, especially in North America and Europe. Nonetheless, in the last few decades human-induced soil erosion has reached unprecedented levels. The GLASOD study (at www.isric.nl) estimated that about 8.4% of the Earth's ice-free land is afflicted by water erosion and about 4% by wind erosion. Pimentel *et al.* (1995) report that against an average rate of soil formation of about 1 Mg $ha^{-1}$ $y^{-1}$ soil erosion rates in croplands amount to 20–30 Mg $ha^{-1}$ $y^{-1}$, while in undisturbed forests they range from only 0.0004 to 0.05 Mg $ha^{-1}$ $y^{-1}$. Minimum tillage significantly reduces erosion in croplands by leaving the soil in a more natural state, in particular maintaining soil aggregate integrity.

*Irrigation* is highly impacting on pedogenesis. It is able to convert deserts into fertile fields, changing aridic to udic soil moisture regimes, promoting organic matter accumulation and leaching salts and carbonates. Land levelling can be required for irrigation or, more often, for creation of paddy fields. Otherwise, drainage of peatlands results in the oxidation of organic matter, because of conversion of anaerobic conditions to aerobic ones, and subsidence.

Soil *compaction* is a cause for concern wherever agriculture is based on the use of mechanized implements. It consists of the loss of natural soil structure through excess tillage and/or trafficking – especially damaging in moist to wet soils. The most evident soil effect is the formation of a superficial or subsurface pan, 10 to 60 cm deep, which limits movement of gases and water and, thus, biotic activity in soil. Ploughing can alleviate compaction ephemerally. More effective and durable remedies are by natural means, such as freezing and thawing or wetting and drying, but the time required to completely remove the negative effects of compaction is long, especially when compaction affects the subsoil.

*Manuring* and *fertilization* imply addition of matter to soil. Net primary productivity of crops is increased, but more importantly for pedogenesis, biogeochemical cycles of the main elements can experience an unbalance. Application of pesticides depresses the growth of undesirable living organisms, but also impacts other beneficial soil organisms, organic matter breakdown, and nutrient cycling.

A plethora of once fertile soils have become unsuitable for plant growth due to human-induced ('secondary') *salinization*. Enrichment of soluble salts, such as chlorides, sulphates and bicarbonates can be caused by various anthropogenic practices, such as irrigation mismanagement – in particular use of saline water – poor land levelling, inadequate drainage, and seawater incursion in coastal areas due to soil subsidence caused by depletion of groundwater. Secondary salinization affects pedogenesis through the addition of new minerals and deleterious effects on the biota, and when a strongly dispersing cation such as $Na^+$ dominates the electrolytes, soil structure declines. Globally, more than 0.77 M $km^2$ of agricultural land is affected by secondary salinization: 20% of irrigated land and about 2% of dry land (FAO, 2006).

*Waterlogging* occurs when soil becomes water saturated and loses most or all of the air-filled pores. Apart from natural events, excessive irrigation or high percolation of water from unlined channel beds may cause this condition. In terms of pedogenesis, the main effect of prolonged waterlogging is the establishment of anoxic conditions in the soil, which leads to the formation of methane and unusual minerals. Reduction of Fe(III) to soluble Fe(II) may imply destruction of iron oxides that cement particles in the aggregates, thereby causing soil structure decline.

### 14.1.4 Water reservoirs

An extreme example of waterlogging is represented by the permanent flooding of areas for creating water reservoirs. As long as a thousand years ago, reservoirs were being constructed. However, as objects of global scale, they appeared only in the second half of the twentieth century. Water from reservoirs is used for irrigation and quenching the thirst of people and domestic animals. Many reservoirs are used for fishing and recreation and the largest ones also serve to produce hydroelectric power. Permanent flooding implies a net loss of soils, which are often quite fertile soils occurring on stream terraces and floodplains. In the rare cases of water withdrawal, the original characteristics of the flooded soils may be totally obliterated. In fact, the superimposition of huge water masses exerts elevated pressures on soils causing marked physical modifications, such as compaction and crumbling. Moreover, the anoxic environment leads to accumulation of organic matter and reduced forms of Fe and Mn.

### 14.1.5 Inhabited areas and infrastructures

Semi-permanent human settlements developed during the Neolithic period, roughly ten thousand years ago. Until five thousand years BP, these villages had to be moved whenever the surrounding soils became exhausted. Since then, permanent cities have been established. In the 1800s less than 3% of the world's population lived in cities. This figure increased to about 25% by the mid 1960s and now exceeds 50%. Presently, urban areas occupy almost 3.6% of the ice-free land (Table 14.1). Most of the soils in urban areas have been completely destroyed, buried or built upon. A minor extent is dedicated to lawns and parks, in which soils may be profoundly modified by human activities. Soils used for cities are preferentially floodplain soils that are also prized for farming, so there is augmenting conflict between these different land uses. Urbanization has led to drastically increased surface run-off during storm events and the occurrence of devastating flooding, mudflows and landslides in lower areas.

### 14.1.6 Lines of communication

*Extra-urban roads* have been built since the Sumerians, more than five thousand years BP. The Roman Empire, at its peak, had a road network 80 000 km long, consisting of 29 roads that radiated from the capital and a plethora of minor roads that covered all the provinces. Roman roads had a thickness of 90 to 120 cm, and were formed of three layers of stones with size decreasing with depth. They were amalgamated with mortar and overlaid by a mosaic of slabs. At the beginning of the nineteenth century, new methods were used to construct roads. Those by Telford and McAdam were the most successful because they were particularly able to drain water. Only after the middle of the nineteenth century, roads began to be paved with bricks or covered with an asphalt mantle. Currently, the paved roadways alone cover 0.2% of the global ice-free land (Table 14.1). Impact on pedogenesis due to construction of roads ranges from simple mechanical compaction of soil, such as in the case of forest lanes, to superimposition of inert materials and impermeable bituminous layers, to complete removal of soil material. Soils in the vicinity of roadways can receive de-icing salts or pollutants, such as unburnt hydrocarbons (Yang *et al.*, 1991) and heavy metals (Garcia and Millan, 1998). These soils are also subjected to infestations of invasive weed species that may affect soil properties.

The first steam locomotive appeared at the end of the eighteenth century but only in the twentieth century did a *railway network* develop. Despite the following rapid growth of this network, it is estimated that, at present, railways occupy no more than 0.003% of the ice-free land (Table 14.1). Nevertheless, the soil disturbance their emplacement requires is large and consists of compaction and superimposition of various materials (gravel, asphalt, rails, and ties). On sloping surfaces, the construction of railways and roads causes discontinuities of water run-off and subsurface flow, with implications for pedogenesis such as increased erosion or a reduction in available moisture. Soils adjacent to railways can be affected by trains' wastewaters, and may accumulate excess nutrients and contaminants.

Since the introduction of irrigation in agriculture, portions of cultivated lands have been used for the transfer of fresh water. Later, larger *canals* were required to implement connections between inhabited centres or between the inland and freight ports. In Europe, the construction of navigable canals began in the Middle Ages, but became massive during the Industrial Revolution (by the end of the eighteenth century). At present, artificial waterways cover about 0.01% of the ice-free land (Table 14.1). Preparing canals requires removal of soil and, in some cases, sealing of the bed. The leakage or seepage from canals changes the hydrology of adjacent landscapes and commonly increases the salinity of these soils as the water evaporates.

In the early days of aviation, natural meadows were used as *airports* but as aircraft continued to increase in weight and speed, more and more long paved runways were required. Runways of almost 5 km are now necessary for allowing take-off and landing of long-range aircraft. International airports are wide extra-urban infrastructures sustaining perennially a high population of passengers and workers. Perturbed and sealed soils abound in these areas, but also those apparently undisturbed are human-impacted by mowing and by the addition of a number of pollutants (hydrocarbons from fuels, detergents for cleaning aircraft, de-icing chemicals, etc.). Airports occupy some 0.01% of the ice-free land (Table 14.1).

### *14.1.7 Pipelines*

The global length of pipelines is estimated to amount to almost 1.5 M km (CIA, 2006). Only a part of terrestrial pipelines lies within the soil and has required integral disturbance of the soil profile for their emplacement. However, where not buried, pipes must travel on a high-rise ditch made of vertical supports planted in soil.

The transmission of warm (above 0 °C) gases or oils through buried pipelines alters natural soil temperature regimes. Perennially frozen soils (permafrost) can thaw, with marked negative ecological and economical ramifications (Seligman, 2000). To avoid melting, in some cases refrigerated brine is pumped through additional small pipes.

Most of the gas pipelines are placed in soils already profoundly reworked: for example, more than 90% of the gas network in the current 25 countries of the European Union serves for distribution to customers (Eurogas, 2005) and, thus, runs below roads or other infrastructure. By contrast, oil pipelines are often buried in extra-urban lands not exploited previously.

Serious effects on soil ecology can result from fuel leaks. Soil pollution from oil, for example, is a major problem in Russia, where in 1996 more than 23 000 leaks from pipelines occurred as a result of corrosion (State Committee of the Russian Federation on Environmental Protection, 1997).

### *14.1.8 Electricity transmission*

A little-known anthropogenic influence on soils arises from distribution of electricity to consumers. Electricity transmission is accomplished by an extensive network of cables. A significant pedological impact can be exerted indirectly by the overhead lines (especially the very large transmission lines) in forested environments, where they require removal of a large swathe of vegetation.

Electric current generates an accompanying magnetic field that can be detrimental to human health. For this reason and to safeguard the landscape of some high-value areas, hanging wires can be buried, causing significant soil perturbation. In belowground lines, the magnetic field seems to negatively impact soil-dwelling biota. Preliminary results showed that the microbial activity, measured as ATP content around a buried cable with operative voltages of 220–380 kV, is inversely proportional to the intensity of the magnetic field and, thus, tends to decline toward the cable (R. Scalenghe, unpublished data).

### *14.1.9 Mining*

Mining has occurred since prehistory. Presumably, the first mineral recovered was flint, which, owing to its concoidal-fracturing pattern, could be used to make scrapers, axes and arrowheads. Since the appearance of stable settlements the need for raw materials for construction purposes has required progressively increasing surface and subsurface mining predations. Modern quarries can extend over several square kilometres and reach more than 700 metres in depth. It has been estimated that mining involves almost 0.2% of the ice-free land (Table 14.1). Opening quarries implies not only complete removal of soils but also storage of large volumes of waste spoil material in neighbouring areas. Spoil banks, which can also originate from materials coming from canals, tunnels, foundations of buildings, dredging of hydroelectric basins, etc., in temperate climates experience horizon differentiation in a few decades (Ciolkosz *et al.*, 1985). However, fresh and coherent parent rock exposed to the atmosphere in the excavated areas can require centuries to weather to a few centimetres of soil material.

### *14.1.10 Landfills*

Systematic rubbish collection was born in ancient Rome, although associated only with state-sponsored events, such as parades. Disposal methods were very crude, often consisting of simple accumulations of garbage in the open. A district of the modern city of Rome takes its name, *Testaccio*, from *testae*, the amphoras used since the Augustus epoch to transport olive oil from the provinces to the capital. In fact, in the district there is a hill 30 m high that is composed of 25 million amphoras, broken after a single use, dumped in regular superimposed layers and strewn with lime to stop oil decomposition. Thin soils have formed over time on this unusual substratum. It was only around the end of the fourteenth century that in Rome the scavengers were given the task of carting waste to dumps outside the city walls.

The industrialization of modern societies resulted in a vast increase of rubbish production and at the end of the eighteenth century a capillary municipal collection of garbage started in Boston, New York City, and Philadelphia. The practice of recycling some materials, such as metal, glass and paper began during the Second World War. Nowadays, part of the agricultural, forest, industrial and municipal organic by-products are applied to agricultural and forest soils to improve fertility and this represents a significant anthropogenic influence on pedogenesis. The non-recyclable material is burnt or stored in landfills, impermeable basins whose excavation implies soil removal. Once these basins are full they are covered with a layer of soil material, then revegetated. The soils that form over landfills will be subject to long-term subsidence, gas emissions (*e.g.* $CH_4$), and seepage as the site ages.

### *14.1.11 Pollution*

Soil pollution is particularly severe in densely populated countries because of massive combustion of fossil fuel or inadequate waste disposal, and in industrial areas because of long-term exposure to industrial processes or accidents that cause releases of contaminants. Also cultivated lands can undergo pollution due to excessive applications of fertilizers and pesticides. The plethora of pollutants that can be found in soil include heavy metals, hydrocarbons, organic solvents, tensioactives, synthetic phenols, cyanide, dioxin, etc. The effects of pollution on pedogenesis ranges from simple addition of new materials to serious damage to biota.

Human-induced acidification is a particular type of soil pollution due to the massive supply of proton donors to weakly buffered soils. Using acidifying compounds as fertilizers causes soil acidification in croplands. Moreover, enhanced acidity can arise from the continuous removal of base cations with crops or from $NH_3$ released from animal husbandry. Drainage of waterlogged, weakly buffered soils containing pyrite results in extreme acidification. Exhumed pyrite can also cause appreciable acidification in surface mines and in mine wastes, making revegetation a difficult task. 'Acid rain', those rains containing acids released originally as industrial or transportation N and S emissions, continue to affect areas of the northern hemisphere. In terms of pedogenesis, acidification decreases rates of organic matter decay and increases rates of mineral weathering, leaching of base cations, and solubilization of $Al^{3+}$. Human practices to counteract soil acidity, such as liming (addition of $CaCO_3$ or fly ash), affect further on pedogenesis.

### *14.1.12 Radioactivity*

Since the start of the 'Atomic Energy Era', coinciding with the construction of the first nuclear reactor (in Chicago, 1942), a new type of pollutant began to be introduced in soils: the radionuclide. Nuclear-fuel reprocessing plants are a source of pollution and often contamination occurs accidentally because of uncontrolled leaks (*e.g.* the Chernobyl reactor, in Ukraine). Nuclear reactions were soon also used to prepare devastating weapons (the first atomic bomb test happened in 1945, in New Mexico, USA), which have as an ancillary negative effect the ability to enrich the atmosphere with several types of radionuclides. These latter spread from the point source to wide areas through the fallout. The consequences of high levels of radioactivity on pedogenesis are less intuitive than those of other pollutants. The addition of new materials is a secondary aspect, considering the scarce quantities at stake, while the negative impact on biological properties of soils can be tremendous (Winteringham, 1989). The development of radiation-resistant soil-dwelling micro-organisms has been observed and it suggests a particular ability of this class of organisms to adapt to higher than normal radiation levels. However, genetic mutations are frequent side effects.

The most important radionuclides, from a negative perspective, to soil biota are ruthenium ($^{106}$Ru), caesium ($^{134}$Cs, $^{137}$Cs), strontium ($^{90}$Sr,), and plutonium ($^{238}$Pu, $^{239}$Pu, $^{240}$Pu, $^{241}$Pu, $^{242}$Pu). All these elements have a very low solubility and, thus, their removal from the soil through leaching is negligible.

### *14.1.13 Sports*

*Golf* courses appear as some of the most natural and romantic environments. On the contrary, like gardens created by painstaking owners, they are the expensive products of human efforts. Considerable use of energy and water is required for preparing and maintaining golf courses. Their construction in scenic natural sites often results in the reduction of biodiversity. The mosaic of 'greens' (turfed and carefully mowed areas), 'bunkers' (unvegetated sandy concave areas), thickets, ponds, etc. is obtained through marked modification of the original soils or even their complete removal. The many tonnes of various fertilizers, herbicides and pesticides required every year to keep the greens and fairways healthy, to combat weeds and kill insects can impose serious pollution of soil and drainage waters (Suzuki *et al.*, 1998).

*Skiing* is increasingly impacting soil resources because new ski runs are continuoually being constructed. The first official alpine skiing competition, a primitive downhill, was held in 1767 in Oslo, Norway. Just a few decades later, the sport had spread to the remainder of Europe and to the United States. Nowadays, in the Alpine region of Europe (south-east France, north Italy, south Switzerland, south Austria and west Slovenia), 'ski domains' represent 3.5% of the total area (Hahn, 2004).

Forest clearing, removal of boulders from the topsoil, and levelling required to construct ski runs profoundly affect the original soil properties. Deviations from the original distribution of plant species and the decrease of biodiversity in the machine-graded areas can be considerable (Wipf *et al.*, 2005). Distribution of artificial snow lasting beyond the end of the skiing season in spring can cause substantial modifications to the moisture regime. This and the additional input of nutrients contained in the artificial snow also affect the plant community (Wipf *et al.*, 2005). Salts used as ice nucleants and snow hardeners are leached in spring during snowmelt, and partially accumulate in subsurface horizons. Furthermore, erosion can be easily triggered in the steep and shallow soils of ski runs.

### *14.1.14 Wars*

Land is used by humans for several less noble goals than preserving wildlife, producing food, or even playing golf. The least noble land use is undoubtedly for war. Conflicts have direct effects on soils, such as compaction by tank transit, excavation of trenches, and perturbation by explosions (Fig. 14.2). Pollutants such as depleted uranium and lead are brought to the soil from munitions. Also, the vast amounts of oil that were released into the environment in Kuwait during the Gulf War are a good example of pollution resulting from war. Tremendous soil disturbance is associated with military bases (not necessarily in war zones) where war-games and training are practised. Soils of war zones can be indirectly affected as a consequence, for example, of the laboured search for firewood by

Fig. 14.2. Bomb craters dating back to the First World War (1915–18), Mount Grappa, northern Italy. In the field on the left side the concavities have been filled with additional soil material. Photo courtesy of Alfredo Bini.

civilians or contamination by waste disposal. It has been calculated that the 1991–3 war in Bosnia-Herzegovina indirectly damaged 27% of the total arable land in that region (European Environment Agency, 2003). At the beginning of the third millennium, 8% of the ice-free land is experiencing, directly or indirectly, the effects of war (Table 14.1).

### *14.1.15 Global climate change*

A type of anthropogenic influence on pedogenesis less intuitive than those listed above is the global climate change induced by the recent massive introduction of greenhouse gases into the atmosphere. Ramifications of this unnatural phenomenon are not yet fully understood, but they will most likely concern temperature and moisture regimes, the residence time of soil organic matter and the $pCO_2$ belowground, whose presumable increase due to $CO_2$ fertilization could accelerate mineral weathering.

## 14.2 Other factors of pedogenesis

### *14.2.1 Fire*

Fire is a natural factor of pedogenesis whose occurrence is now often due to human interventions. In this view, human-induced fires must be considered a further way by which humans affect pedogenesis.

In a comprehensive review, Certini (2005) summarizes the effects of forest fires on the various soil properties. These effects depend first upon fire 'severity' that, in turn, is controlled by the fuel amount and type, the moisture content of the live and dead fuel, air temperature and humidity, wind speed, topography of the site and other minor features. Fire severity consists of 'intensity' and 'duration'. Intensity is the rate at which a fire produces thermal energy. Although heat in moist soil is transported faster and penetrates deeper, latent heat of vaporization prevents soil temperatures from exceeding 95 °C until water is absent. Then, once moisture is depleted, 500–800 °C can be reached at the soil surface. However, soil temperature gradients are steep and often no heating occurs below 20–30 cm. This implies that fire-induced modifications are confined mainly to the upper horizons. Duration is perhaps the component of fire severity that results in the greatest damage to biota, particularly when soil temperatures remain elevated for some hours or even days. The removal of the vegetation cover makes soil susceptible to rainsplash detachment and erosion (Fig. 14.3).

The content of organic carbon and its associated nitrogen is the soil property most profoundly affected by fire. It is decreased both directly, through volatilization,

Fig. 14.3. Fires have several immediate negative impacts on soil, such as the removal of the living and dead organic matter and the consequent loss of nutrients through volatilization, leaching, run-off or erosion. However, if plant recolonization is prompt the pre-fire levels of most elements can be recovered and even exceeded in a few vegetative seasons. Photo courtesy of Gianluca Giovannini.

and indirectly, through leaching, run-off, and erosion. However, if plant recolonization is prompt, the original level of soil organic matter can recover in a few years and even exceed original levels after a decade (Johnson and Curtis, 2001). This fact can be explained by the high net productivity during the early stages of secondary succession.

As well as changes in quantity, soil organic matter that experiences heating undergoes structural modifications. Low to moderate temperatures lead to volatilization of hydrophobic compounds, which recondense deeper in the soil to form a more or less continuous impermeable layer; this hydrophobic layer decreases water infiltration, exacerbating run-off and erosion in steep terrains. Moderate to high temperatures induce transformation of non-humic biopolymers, such as carbohydrates and lignin, to humic-like substances. Charred materials, which form as a result of incomplete combustion of wood, may remain in soil for centuries because they are recalcitrant to biochemical attack.

Generally, the inorganic solid phase is not altered to any great extent by fires because the first step of decay of most minerals – dehydroxylation – occurs at temperatures over 500 °C. Among the new minerals that can form in soil through thermal transformation of other minerals are the magnetic Fe-oxide maghaemite and the phyllosilicate illite.

### *14.2.2 Earthquakes*

Earthquakes modify the physical properties of soil. Scalenghe *et al.* (2004) showed that liquefaction induced by vibrations caused disaggregated material of fragipans from seismic areas to redevelop their original bulk density, porosity, and pore-size distribution. This suggests that earthquakes play a role in fragipan formation by providing the necessary close-packing arrangement of soil particles. In sloping areas, a major effect of earthquakes are landslides, which mix horizons in the transported soil masses and cause the burial of lowland soils.

### *14.2.3 Volcanic activity*

Volcanoes are important to pedogenesis by supplying fresh minerals to soils that can rejuvenate soil-forming processes. Depending on the amount of airborne solid ejecta (*tephra*) or lava flows that are added to a landscape, they can even bury pre-existing soils, compelling pedogenesis to begin anew on the fresh parent material (Dahlgren *et al.*, 2004).

Volcanic landslides (debris avalanches) are rapid downslope movements of rock, snow and ice, not necessarily associated with eruptions. They range in size from small movements of loose debris that imply partial removal of material from some soils and addition to others to massive failures of the entire summit or flanks of the volcano, which totally destroy soils.

All magmas contain dissolved gases that are released both during eruptive episodes and between them through small openings called fumaroles. Generally, these gases consist of steam ($H_2O$), followed in abundance by carbon dioxide and compounds of sulphur and chlorine. Minor amounts of carbon monoxide, fluorine and boron can also be present. Distribution of volcanic gases is mostly controlled by the wind. Significant deposition of sulphur dioxide, hydrogen sulphide and ammonia to weakly buffered soils (as often found in young volcanic soils) promotes acidification and accelerates chemical weathering.

### *14.2.4 Meteorites*

The cumulative area of the major terrestrial craters created by the impact of meteorites in the last hundred thousand years – a time interval during which many present soils formed – over the world amounts to just 21 square kilometres! Nonetheless, meteorites have to be considered a soil-forming factor, especially on account of the fact that Earth receives 100–200 thousand Mg $y^{-1}$ of cosmic material, prevalently as very fine particles. Soils previously present in the area of the crater are totally dismantled by the impact of the meteorite and much removed soil and parent material is added to soils in the 'halo' of the crater. In the crater, new soils form in time on a parent material partially modified, both chemically and mineralogically, by the impact. Iridium enrichment and shock metamorphism of rocks are the two most common modifications. Iridium is a stable element, which is present in the Earth's crust in extremely tiny amounts but in much larger quantities in meteorites. Shocked quartz (or 'stishovite') forms under intense pressure and low temperature and has a microscopic structure different from that of normal quartz.

## 14.3 Extraterrestrial soils

In this chapter, we listed and discussed a series of soil-forming factors that can add to the five originally recognized by Dokuchaev. But is soil genesis possible if one of the five 'basic' factors is not present, particularly the contribution of biota? No soil can form without a parent material, and any landscape has a unique topographic position and climate (with or without liquid water). And, of course, it is impossible to stop time!

Abiotic or virtually abiotic conditions are historically present in some zones on Earth, such as the high barren Arctic and the ice-free areas of Antarctica, where average temperatures are well below freezing. Here, ionic migration and weathering of primary and secondary minerals, accompanied by iron oxidation and solute migration, occur in hyperaridic unconsolidated substrata (Ugolini and Anderson, 1973). Several authors (e.g. Ugolini and Edmonds, 1983), recognize these altered materials as soils, and indeed they contain horizonation although lacking the characteristic A horizons often associated with soil formation.

From this Earth-based perspective, extraterrestrial landscapes should be considered as possessing soils. Some who have foreseen that humankind's future is in space have considered whether space-based soils could potentially support future human expansion in the solar system (Ming and Henninger, 1989). Mautner (1997) grew potato and asparagus tissue cultures in ground Martian meteorites that landed on Earth thousands of years ago.

The soils of the Moon have been studied directly by astronauts, and samples have been returned to Earth, while Mars soil chemistry has been measured *in situ* from five different NASA missions and these data now reveal a variety of iron oxides and sulphate and chloride salts.

The Moon's crust is covered by a layer of loose, heterogeneous mineral material that overlies solid rock. Lunar soils, which many authors call the 'lunar regolith', are unevenly distributed, varying in thickness from 3–5 m in the lunar plains called 'maria' (Latin for seas), to 10–20 m in the highlands. Maria are mostly composed of basalts, while the highland rocks are largely anorthosite. Lunar soils have developed under the action of several unique processes which include physical reworking by meteorite impacts, irradiation by solar wind, deposition of impact-generated and sputtered ions, and chemical interactions with the lunar atmosphere. The main chemical effect that has been observed is the formation of a 50–200 nm thick rim of chemical alteration on mineral grains. The rim is composed of material showing a different composition and microstructure from the host grain, not so much in terms of the elements present, but in their relative proportions and oxidation state (Keller and McKay, 1997). For example, ferromagnesian silicates, such as olivine and pyroxenes, have suffered a marked loss of Mg during solar-wind irradiation, and a preferential sputtering of Ca and Al relative to Si and O. Rims on ilmenite grains are not amorphous as in the other lithologies on the Moon, but are microcrystalline and have formed as a consequence of the preferential removal of Fe and O, along with a reduction of $Fe^{2+}$ to $Fe^{0}$ and $Ti^{4+}$ to $Ti^{3+}$.

Lunar soils could represent an archive of the early history of Earth. In fact, because it formed at high temperatures, the Moon as a whole should be depleted in volatile elements such as hydrogen, carbon, nitrogen and the noble gases. But these elements are abundant in lunar soils, pointing to the existence of extralunar sources. According to Ozima *et al.* (2005) these sources are essentially two: (a) the Earth's atmosphere, before the terrestrial dynamo started, and (b) a near-constant flux from the solar wind. The discrimination of the contributions of these two sources through their different isotopic signatures of nitrogen and light noble gases could reveal the epoch of the origination of the geodynamo, as well as the composition of the Earth at that time.

At present Mars has no sea, but has water in the form of ice close to the surface and adjacent to the permanently frozen south polar ice cap, as revealed by NASA's Odyssey spacecraft. There is abundant geomorphic, and now chemical, evidence that liquid water may once have been active at Martian plains, particularly from data provided by the rover Opportunity at Planus Meridiani in 2004. The parent material chemistry is consistent with basalts, with presumed mineralogy being olivine, pyroxenes, plagioclases and accessory Fe-Ti oxides. The ubiquitous presence of such an easily weatherable primary mineral as olivine suggests that physical rather than chemical weathering processes currently dominate (Morris *et al.*, 2004). However, exposed rock surfaces show the presence of a weathering crust. The plains are covered with dust and sediment rich in reddish iron oxides. Gellert *et al.* (2004) suggest that the high Br and low Cl/Br ratios in weathered grains and stones may be indicative of alteration in an aqueous environment. In fact, it is known from terrestrial evaporites found in salt lakes that the precipitation of chloride salts leads to a decrease of the Cl/Br ratio in residual brines. More recently, the unambiguous detection by the OMEGA imaging spectrometer on board the Mars Express spacecraft of a variety of sulphates and phyllosilicates led Poulet *et al.* (2005) to infer that these two main families of hydrated alteration products could trace two different processes separated in time, referring to two major climatic episodes in the history of Mars. The first episode, moist and warm, would have resulted in the formation of hydrated silicates, followed by a more acidic environment, not necessarily implying the long-term presence of liquid water, in which sulphates formed instead of clays.

It is likely that future space research will reveal many more surprising new discoveries on the topic of extraterrestrial soils. More importantly, it is critical that these surface deposits be correctly viewed as soils, and that pedological perspectives and chemistry from Earth guide the sampling, analysis and interpretation of extraterrestrial soils.

# 15

# Soil functions and land use

*Johan Bouma*

The relation between soil functions and land use has been lost to a certain extent in the course of the last century because of technological developments. If certain natural functions were inadequate for certain types of land use, technology was applied to overcome the problem. Soils that were too wet to allow plants to grow were drained, dry soils were irrigated and poor soils were fertilized. In earlier times with less available technology the picture was different as land use was largely determined and restricted by functions that could be performed by the natural soil.

Increasing emphasis on sustainable development during the last few decades has shown that changing the natural functions of soil often comes at a price: drainage may lead to rapid oxidation of peat and generation of greenhouse gases, to acidification in marine soils with pyrite or to drastic changes in natural ecosystems in a broader sense. Irrigation may lead to salinization or erosion and, thereby, also to disturbance of natural ecosystems. Fertilization often results in water pollution when more fertilizer is applied than can be adsorbed by the growing crop, which is common. When designing sustainable land-use systems, in which economic, environmental and social criteria are somehow being balanced, it pays to take natural soil functions into account so as to avoid major deviations of natural processes which are likely to lead to disturbances that may be difficult to correct. Agroecological ecosystems are so complex that changes of one element of the system may have unforeseen consequences for other parts of the system. Emphasis on sustainable land management has, therefore, an implication of 'going back to the future' – looking at natural soil functions as a point of reference, just as in the past, but now for a different reason.

Which soil functions can be distinguished? The following functions are often mentioned: (a) the production function, producing crops; (b) the carrier function, bearing traffic and buildings; (c) the filter, buffer and reactor function, allowing

*Soils: Basic Concepts and Future Challenges*, ed. Giacomo Certini and Riccardo Scalenghe.
Published by Cambridge University Press. 

transformations of solutes passing through; (d) the resource function, providing base materials for industry; (e) the habitat function, providing a living environment for plants and animals, and (f) the cultural and historic function, reflecting past practices. In addressing these functions, soil research often had a rather disciplinary character which has made it rather difficult to communicate with different disciplines. The link with agronomy in the context of function (a) has been quite prominent in terms of soil fertility research and, more recently, in soil physical characterization for dynamic modelling. The engineering literature for function (b) has a separate and quite different character, while function (c) represents a soil-focused, scientific activity emphasizing soil chemistry, soil physics and soil biology. Function (d), on the contrary, has a more applied character when dealing with building materials for example, and has clear links with function (b). The engineering world is quite separate from the other worlds with which soil scientists have to cooperate. Function (e) relates to ecology and biology, again quite different fields in the scientific world as compared with agronomy and engineering. Category (f) has received considerable emphasis in soil science in the past when soil and landscape genesis were prime topics of study as a basis for soil survey. Over time, however, the more utilitarian approaches have become more dominant. Still, recent developments in the European Union (EU), where price support for farmers will gradually be replaced by income support subject to a number of conditions, among which landscape maintenance figures prominently, is likely to make function (f) more important in future.

In this chapter, future demands on the soil and ways in which soil science can contribute to meet these demands, working with various stakeholders, planners and policy makers, will be first analysed. Next, how, in the light of likely future developments, soil concepts should be changed and expanded away from taxonomic based principles will be discussed. Modern techniques can be used to develop a more dynamic approach which fits better with future demands. Soil should not only be characterized in a scientific context but we should also be more receptive to what the soil can tell us: the concept of coproduction.

## 15.1 How to deal with future demands on our soils

Future land use is very difficult to predict (Veldkamp and Lambin, 2001). Higher agricultural production per unit of land increasingly satisfies the worldwide demand for food and a smaller area of land is therefore needed, allowing alternative forms of land use in the remaining land in developed countries where citizens want and need more room for building, nature and recreation. In many developing countries this condition has not been reached as primary food production is still of paramount importance to achieve food security, but eventually they will most likely also proceed to this state of development. Intriguing here is the possibility of a short cut ('leapfrogging') that avoids the pursuit of first cultivating large areas of land on the basis of low-level technology by rapidly adopting high technology, allowing cultivation in smaller areas right away, leaving other areas for alternative forms of land use. This may be very difficult as adoption of high-tech procedures requires a certain level of social and economic organization of society which may not (yet) be present.

In addition, the contrasting tendencies of *globalization* and *regionalization* will lead to land-use patterns that are impossible to predict at this time. Removal of trade barriers in the context of *globalization* could, ideally, have the effect that agricultural crops will only be grown at locations where ecological conditions are optimal. Growing the crops at less suitable sites would cost more and would be associated with higher risks and would, therefore, again ideally, be less competitive in an open market. But successful agriculture is also, and even more, a function of economic, social and marketing conditions. The attractive conceptual ideal of only growing crops under optimal ecological conditions is, therefore, probably unrealistic. *Regionalization*, in contrast, emphasizes local culture and consumption of local products that fit individual tastes in that particular region and is attractive for 'niche' markets elsewhere, providing commercial possibilities that may function independently from global trends. European policies emphasize regionalization and support income – rather than price support for farmers in future, requiring in return attention for product – and environmental quality, including maintenance of cultural landscapes. This is an important development but one with many problems. Farmers are reluctant to accept structural payments for unusual 'products' such as landscape quality because they feel that such payments are politically vulnerable. They are more comfortable with adequate payment for their primary products such as milk, grain and meat. This, however, is increasingly problematic because current prices for agricultural products within many developed countries are well above the level of the world market. Import duties and export subsidies are gradually decreasing under pressure from the World Trade Organization (WTO) in the context of trade liberalization and globalization. This has a negative effect on prices farmers receive. The standard response by economists to this development is cost-cutting and this can be achieved by introduction of more technology and mechanization and, particularly, by making farms much larger. Larger fields make farming operations more effective, but often destroy characteristic features of the landscape such as hedges, ditches, brooks and local microrelief built up over many centuries. These features are as much part of our cultural heritage as are the paintings and sculptures we find in museums. By making farms bigger and more efficient some farmers try to react to the forces of globalization but effects on the landscape are

Fig. 15.1. Characteristic image of the Frisian Woodlands in the northern part of the Netherlands with hedgerows and ponds between relatively small fields. This old landscape is now considered as a National Landscape to be protected. Modern farming would require large fields that would destroy the current characteristic landscape features, built by farmers over many centuries. Soils are described by Sonneveld *et al.* (2002).

sometimes devastating and irreversible. Solutions for land use have, therefore, to be found that allow economically feasible farming while at the same time preserving our cultural heritage. In the Netherlands some twenty National Landscapes, which are culturally significant, are now distinguished and they are to be preserved (Fig. 15.1). Here, the Government is prepared to partly finance farming operations as these landscapes are considered to be made by humans and preserving them for the future means that farming has to continue. In areas beyond National Landscapes, market forces will more strongly determine future land-use patterns.

Different philosophies are followed when considering the feasibility of affecting land use in a given region. The demise of communism has led to the realization that rigid, central guidance of land use by government does not work. Different countries try different forms of milder guidance that all try to avoid a 'free-for-all' attitude that tends to be determined by short-term market forces and business interests leading to haphazard development. This way, valuable landscapes (and soils) can be destroyed and, once gone, they are gone forever. In the last decade, various forms of mild guidance have taken the form of participatory approaches where policy makers, scientists and land users try together to define and design forms of land use that satisfy the often quite contrary demands of the many groups of people involved (Bouma, 2001a).

How does soil fit into this complex and murky future picture? Unfortunately, it hardly does. Land use has historically been interpreted from a dominantly economic and sometimes aesthetic point of view: most planners and politicians use blank sheets of paper for their plans, not soil maps. It is difficult enough to add the ecological and sociocultural dimensions to the economic dimension when trying to deal with the sustainability concept. Soil has the major handicap of being invisible to the eye of the observer walking over the land. Landscape features are visible and so is biodiversity. The soil is not. Even characterizing land use is difficult as it is increasingly derived from land cover observations obtained by remote sensing. But land use is different from land cover and relates to what is being done with the land during the year; this cannot be deduced from a single remote-sensing image and requires extensive field surveys and interviews with land users. Again, the soil tends to get lost in the process. What should we do about this, aside from complaining?

I believe that soil by itself has not enough pushing power to get back into the picture. This is a hard conclusion for a soil scientist but a conclusion that is difficult to avoid. It is better to face this fact head on than to keep complaining about the neglect of soils in modern planning. So what can we do?

One approach, advocated here, is to not only talk about soils but to include water as well at all times. As has been said: 'Water in soil is like blood in man'. Looking at water regimes in individual soils, but also in landscapes (the 'throbbing landscape'), provides a key to getting soil back into the land-use debate (Lin *et al.*, 2004). Conditions in different soils lead to quite different hydraulic processes in soils and landscapes in any given climate and these processes are crucial to either understand current land use problems or to design new forms of land use. Looking at water regimes in soils and landscapes requires a systems analysis including at least soils, hydraulic, meteorological and agronomic expertise, and there are by now a variety of computer simulation models as well as modern monitoring techniques that allow validation at field or regional level (e.g. Lin *et al.*, 2004). We will now discuss what this structural 'marriage' of soil and water could mean for soil science in future, considering the relation between soil functions and land use.

## 15.2 To characterize soil functions better

Here, the question to be explored is: how can an integrated systems approach, including soil science and hydrology/soil physics, contribute towards new products that are more attractive for modern users of soil information than soil products by themselves? Emphasis will be on the particular contributions by soil science in this broader context. The six soil functions will be briefly discussed.

### *15.2.1 The production function*

Major advances have been made in the area of simulation modelling of crop growth (e.g. Kropff *et al.*, 2001). Also, these modelling activities have been interpreted in terms of their significance for policy analysis (Kuyvenhoven *et al.*, 1998). When considering the production function, soil input is important to represent soil processes in terms of infiltration and transport of water in the soil, moisture and nutrient supply to the crop and groundwater regimes. It is a continuous struggle to make sure that the level of soil input into the models corresponds with the overall level of the models, We often see that the physiological detail in which plant processes are described in models is continuously increasing, because plant scientists are strongly involved with model use and development, while soil input does not improve as soil scientists are not part of the team. The result is an unbalanced model. Of course, this phenomenon also relates to other disciplines, such as meteorology and climatology, entomology, soil fertility, engineering, etc. An example are the DSSAT models (see Bouma and Jones, 2001) where ever more highly detailed plant submodules are combined with a highly generalized 'tipping-bucket' representation of the soil module. The latter does not allow adequate representation of soil processes (Bouma and Droogers, 1999). Even when more sophisticated soil physical modules are included, they may not work well because soils are still represented as isotropic-homogeneous media, which they are not. Swelling and shrinkage and occurrence of macropores lead to entirely different processes in soils which can be represented in such a way that they fit well in the type of complex simulation models for crop growth that are considered here. Of particular concern is bypass flow, which is free water flowing rapidly downwards along air-filled macropores, such as animal burrows or shrinking cracks through an unsaturated soil matrix. This process has long been ignored by soil physicists who only considered flow to be saturated or unsaturated at any given depth, not both at the same time (Bouma, 1989; Booltink and Bouma, 2002; Crescimanno and De Santis, 2004).

The central problem here is the fact that non-soil-scientists have a feeling that they can easily integrate soil information into their models and that they do not really need soil scientists to help. Our development of widely accessible *pedotransferfunctions*, which relate existing soil data such as bulk density, organic matter content and texture to hydraulic parameters needed for simulation, may unintentionally hurt our case (Wosten *et al.*, 2001). Such functions should never automatically be applied by non-soil-scientists who do not realize their limitations, but this is often not understood nor communicated well enough by the developers of these functions.

Surprisingly, soil scientists usually present their data such as pedotransferfunctions in a generic manner, lumping all soils together. Would it not be wiser to stratify information by soil type or region, avoiding the mixing of 'apples and oranges'? One reason may be that soil survey and classification has always been somewhat apart from other subdisciplines of soil science and that at least part of pedological thinking has not been adopted by other subdisciplines of soil science. Maybe it is time for a change, as will be discussed later. Soil scientists can also play a key role in fine-tuning the production function by making sure that agricultural crops do not receive more manure than they can utilize. This principle of precision agriculture has a strong soil character (Bouma *et al.*, 1999). Strangely enough, manure legislation in Western Europe emphasizes fertilization rates when trying to reduce the impact of fertilization on the quality of surface water and groundwater. One rate for the entire European Union area was defined in the nitrate guideline of 1991: 170 kg N from organic manure per hectare, even though soils and climates are highly different (Sonneveld and Bouma, 2003). The nitrification process is the key to nitrate pollution of groundwater. It is a soil process but soils are not, or only in a very general way, taken into account when developing environmental guidelines relating to water quality. It would be more realistic to fine-tune fertilization rates for crops grown on given types of soils in given climate zones by using the actual soil moisture regime as a guideline (Bouma *et al.*, 1999). Perhaps new regulations will try to do this and soil expertise is essential to achieve this.

### 15.2.2 The carrier function

If there is one area where combining soil science and hydrology is necessary, it is here. Soil stability is a direct function of water content. Characterizing soil water regimes either by monitoring with new automated techniques or by prediction using simulation models allows predictions of the water content of surface soil as a function of time. Defining critical water contents ('threshold values') below which compaction is unlikely, allows estimates to be made of 'trafficable' or 'workable' days for any given soil. Many examples are available in literature (e.g. Bouma, 1989). Defining such 'threshold' values for different soils would be important to obtain key values to predict soil behaviour. Field studies are important here. Rather than exert a range of pressures on different cylinders filled with aggregates with different moisture contents, soils in the field can be wetted to represent different relevant moisture contents, and effects of pressure can be obtained by measuring the compacting effects of tractors or other types of farm equipment driving over the land. One way to do that is by measuring the infiltration rate before and after the driving. This is a more realistic measure than a bulk density value, because one concern about compaction centres on its effect on infiltration. Moisture contents of the soil associated with unacceptably low infiltration rates are then considered to represent a 'threshold'.

### *15.2.3 The filter, buffering and reactor function*

Natural soils can be very effective in filtering, adsorbing and otherwise removing pollutants. Examples were given for septic tank effluent by Bouma (1979) but many other examples are to be found in the literature, also for heavy metals and organic pollutants. The flow regime is very important here because high fluxes may be associated with relatively low retention times in the soil which do not allow adequate contact between solutes and the soil surface. This, in turn, leads to inadequate filtering. Rapid flow often follows the larger pores, while water in finer pores remains virtually stagnant (Bouma, 1991). An application of this understanding of flow processes in structured soil is the manipulation of the flow regime by defining for each particular soil structure a flow rate (at a given initial moisture content) that leads to a sufficient retention time allowing adequate filtering and adsorption, while at the same time an acceptable flow rate is obtained. The latter is important for waste disposal on land where available space is limited. The reactor function being considered here also includes decomposition of dead organic matter, transforming it in its mineral components which, in turn, can be taken up by plants and re-enter the life cycle. Soil acting as a 'reactor' includes transformations by countless soil organisms.

### *15.2.4 The resource function for industry*

In many countries the soil definition is not restricted to the upper metre of the land but extends much deeper. The resource function for industry often derives its significance from the occurrence of gravel, sand, clay, peat, coal and various minerals which may be found at deeper layers. Here, geological expertise is important to identify deposits that are worth developing. The resource function for industry is less relevant when the soil concept is restricted to the upper metre of the land, except, perhaps, when using clay for baking bricks.

### *15.2.5 The habitat function*

The hydrological and chemical properties of the non-biotic environment are of crucial importance to natural plants and animals. They can be strongly affected by agriculture, and precision agriculture can help to reduce loads of excess fertilizer to acceptable levels (Bouma *et al.*, 1999, 2002). But the habitat function relates particularly to characteristics of the natural environment at any given location that follow from the genesis of the area or from active soil-forming processes that may be affected by humans. Oxidation of pyrites in marine areas after drainage of agricultural land may, for example, lead to excessive acidification that is likely also to affect adjacent nature areas. Soil scientists can recognize this in time, avoiding possible questions and problems before they arise. Clearly, the habitat function of soils requires consideration of hydraulic regimes in soils and its surrounding areas in order to explain movement of solutes from the soil and its geologic substrate into surface water and groundwater, thereby determining its quality for plants and animals.

### *15.2.6 The cultural and historic function*

Soil survey reports published before the 1990s paid considerable attention to landscape genesis, its geologic and geomorphological characteristics and the role of humankind in shaping the land. Now that many soil surveys have either been completed or terminated because of lack of funds, soil scientists are increasingly engaged in short-term topical research being tightly funded by outside funding agents. Broader landscape studies do not fit in this context. This is unfortunate and needs to be redressed. We need to go back to the older type of work to address the increasing number of questions about the cultural and historic functions of soils in days past. For example, the European Union pays much attention to what is called the cultural heritage and special funding is provided to regions that have a particular cultural identity which needs preservation. Soil scientists can play a role in discussions about the cultural and historic role of soils in pointing out that soil morphology often reflects past human practices, such as evidenced by thickness of plaggen epipedons, evidence of soil tillage in early periods, dissolution of certain soil horizons following changes in human practices, etc. The rejuvenation of the cultural and historic function may serve to encourage soil scientists to return to the old emphasis on the relation between soils and landscapes. Modern monitoring and simulation techniques can help to quantify such relations more than was possible in the past, thereby providing specific contributions to planners and politicians when they try to preserve certain features of landscapes that are of cultural value.

## 15.3 Storylines: what can the soil tell us when we listen?

By not only talking about soil in isolation, but to fully incorporate it when dealing with moisture regimes in soils and hydrology of landscapes, we can make soil's input more effective for society when defining and working with the six functions described above. Is that all there is? I believe not. Earlier this chapter discussed the need for science, including soil and water science, to be more focused on the user, be it a farmer or a planner or a politician – not with the objective to follow blindly what they have in mind but to facilitate the process of interaction between various government officials and planners on the one hand and soil users on the other. How to do this? One way is to inject the right knowledge about soil behaviour into the discussions at the right time or by initiating new research at the right time and in the right place. Rather than provide final judgements about soil behaviour and firm conclusions as to what should be done (as was common in earlier times), soil scientists are likely to be more effective when they become proactive partners and participants in the negotiation process (Bouma, 2001a, 2001b).

So what does appeal to this highly heterogeneous group of users? Rather than only present sterile scientific characterizations of soils and model predictions of expected future behaviour, as seen from a scientific point of view, soil scientists should also refine their somewhat forgotten ability to 'read' the soil profile to see how any given soil has reacted to a wide variety of (mis)treatments. This requires 'listening to the soil', by interpreting soil morphology and where, again, new monitoring and simulation techniques can provide intriguing dimensions and meaning to observations alone. In this context, sociologists refer to this two-way interaction between soil and men as 'coproduction'.

There is still much to be learned by studying soil profiles in the field, even now that most soil maps have been completed, and by talking to soil users. Replacing field work by exclusive laboratory analyses and computer simulations is unwise because there is still very much to be learned out there. Of course, these modern techniques are, in turn, essential to better interpret field observations and far-ranging opinions of soil users. But in this context they are used to serve a purpose and they are not a purpose in themselves. Sonneveld *et al.* (2002) and Pulleman *et al.* (2000) showed, for example, that organic matter contents of surface soil in a given soil type could be well predicted by regressing it with data on past land use. This presents a clear alternative to mechanistic simulations of organic matter contents as a function of land use for which complex models are available. But lack of basic data or the sketchy, approximate character of some of these models makes simulation modelling a shaky proposition.

Sonneveld *et al.* (2002) presented the following equation for a major sandy soil in the Netherlands:

$$\text{C org}(\%) = 3.40 - 1.54 \times \text{Maize} + 0.19 \times \text{Old} + 0.55 \times \text{GWC} \quad [R^2 = 0.75]$$

where Maize equals 1 if maize growing was continuous, 0 if non-continuous; Old equals 1 in case of old grassland value, 0 otherwise; GWC (groundwater depth-class) equals 1 for class Vb, 0 for class VI.

In this work, the authors focus on a particular soil type and try to tell its 'story' by describing soil genesis and effects of human impact which have either a short-range or a long-range character. The first *short-range* category includes the compaction when driving over the soil when it is too wet, excessive leaching of nitrate when more fertilizers have been applied than can be absorbed by the crop or stored by the soil, breakthough of heavy metals when the adsorptive capacity of the soil is exceeded or loss of agricultural production in a dry year. The second *long-range* category includes the gradual increase of organic matter content as a function of green manuring and application of organic manure or a decrease of vulnerability to degradation by various soil conservation measures. The 'story' for each type of soil may help to explain and predict future behaviour and shows in which areas any particular soil is more vulnerable and where the 'niches' for success are to be found. It also may function to illustrate soil resilience, which is the capacity of a disturbed soil to recover its original quality, an important but underdeveloped aspect of soil behaviour. This appeals to many users, particularly when new forms are found for communication.

Such an approach contrasts with the one where soils are treated as belonging to one big group, making it much more difficult to generate specific conclusions for any particular type of soil.

## 15.4 In conclusion

The time when soil scientists could restrict themselves arbitrarily to only some soil functions and define land use in terms of arbitrary, descriptive soil suitabilities for a wide variety of uses, is over (Bouma, 2000). Land users are more critical and not only want specific answers to their questions but also help in their discussions with colleagues and politicians and planners. Here, soil scientists enter an inter-disciplinary arena where they are not perceived as key players. To be better equipped for this arena, this chapter suggests: (a) a strategic combination of soil and water expertise when discussing soil and landscape behaviour; (b) a focus on defining realistic options for land use, each one with all their limitations and advantages to be expressed in terms of risks using simulation modelling, rather than single recommendations for soil suitability; (c) emphasis on recognition and use of user knowledge and expertise that is checked and validated by innovative field monitoring techniques and computer simulation models; (d) development of 'storylines' for major types of soil, which not only sketch their genesis but also the effects of use by humans in a way that appeals to users and policy makers alike, and which requires much new field work and a fresh vision.

# 16

# Physical degradation of soils

*Michael J. Singer*

Organic and inorganic solids, liquids of various kinds and gases in intimate contact with the solid phase comprise the three-phase system that is soil. Soil management includes management of all three phases but emphasis is typically on the solid and liquid phases. The liquids and gases occupy the soil pores, which are the spaces between the solids. Soil management is challenging because of the great spatial and temporal variability of the liquid and gas phases that are in constant flux. Unless disturbed by growing roots, burrowing animals, or humans, the temporal changes to the solid phase are slow to occur but spatial variability is often great. Physical degradation of soil most often reflects a change in the total volume or size distribution of the pores brought about by reorganization of the solid phase. When reorganization occurs in the subsoil, it is most often called compaction and when it occurs on the surface it is often referred to as soil sealing or crusting. Compaction, sealing and crusting are significant soil physical management problems worldwide. The objective of this chapter is to describe some of these problems, the processes of soil physical degradation, their impacts on wind and water erosion and control of the degradation.

Soil hydrology is controlled for the most part by the total volume of pores and the pore size distribution, hence a decrease in either of these has the primary effect of reducing the rate of fluid entry into the soil and the rate of fluid flow and chemicals through the soil. Ultimately, these processes influence crop growth and environmental quality. Pore continuity and tortuosity also affect the rate of fluid flow through soil. Soil management can significantly affect these properties and the processes that they control (Green *et al.*, 2003). A secondary effect of a reduced rate of fluid entry into soil is increased overland flow and the potential for an increase in soil erosion by the overland flow. 'Fluid' is used here rather

*Soils: Basic Concepts and Future Challenges*, ed. Giacomo Certini and Riccardo Scalenghe.
Published by Cambridge University Press. 

than 'water' to emphasize that soil is often the medium in which disposal of wastes of various types occurs and where the unintentional entry of various liquids is a part of modern life.

## 16.1 Soil compaction

Soil compaction is the process that rearranges the soil solid phase into a more compact arrangement, resulting in higher bulk or apparent density, increased soil strength, smaller total pore volume and decreased macropores (Logsdon and Karlen, 2004; McNabb *et al.*, 2001). Soil compaction decreases the amount of air in the soil without a significant change in the amount of water. Although the problem is well known and in many ways there is much knowledge about how the problem occurs, it continues to be a serious problem in agriculture, forestry and recreation (e.g. Van den Akker *et al.*, 2003). The term soil compaction is also used to describe the result of the process and it implies an increased density of the soil mass. Bulk density or apparent density is the mass of solid particles per unit volume (Eq. 16.1). Units for bulk density are megagrams (Mg) of oven-dry soil per cubic metre.

$$\text{Bulk density} = \frac{\text{Mass (Mg)}}{\text{Volume (m}^3\text{)}} \qquad (16.1)$$

Density increases when a load on the soil surface exceeds the strength of the soil and causes particles to slide or roll into new more dense arrangements. As the load per unit area of soil increases, the bulk density of the soil increases up to a maximum, which is a function of the load, soil properties and the soil water content. Bulk density increases as water content increases because the water on particle surfaces lubricates the particles and reduces friction, enabling rearrangement. After the maximum bulk density is reached, more and more of the pore volume is filled with water. The water exerts a positive pressure against the mineral grains that forces the particles apart, thus reducing bulk density. The curve in Fig. 16.1 is an example of a Proctor curve, named after the engineer who developed the test. Engineers use the Proctor test to determine the water content at which a load will produce the maximum soil density. The engineers will adjust the soil to this water content (termed 'the optimum water content') when compacting the soil for construction purposes. The optimum water content is often near the field capacity water content and is the least desirable soil water content for manipulating the soil (e.g. cultivation), driving or walking over the soil if the objective is to maintain the natural soil density.

Porosity is the fraction of the soil volume occupied by pores (Eq. 16.2) and is a convenient measure of total pore volume.

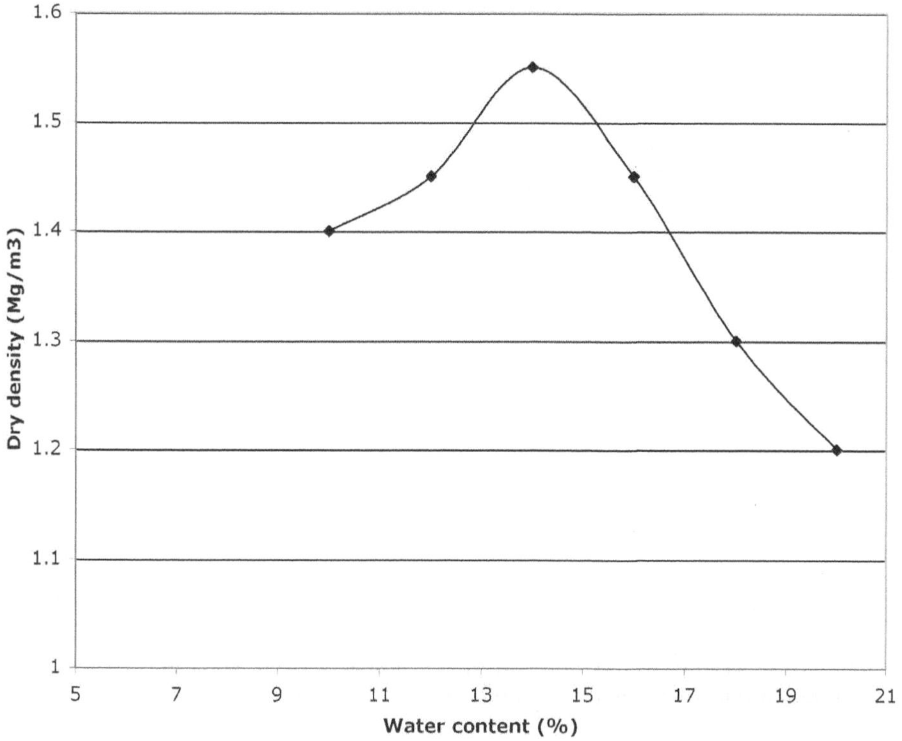


Fig. 16.1. A hypothetical Proctor curve for a cohesive soil illustrating the optimum soil water content to achieve the maximum density at the specified compactive effort.

$$\text{Porosity} = 1 - \left(\frac{\text{bulk density}}{\text{particle density}}\right) \tag{16.2}$$

Percentage pore space is the porosity multiplied by 100. Another measure of soil porosity is the void ratio ($e$), which is defined as the ratio of the volume of pores or voids ($V_{\mathrm{v}}$) to the volume of solid particles ($V_{\mathrm{s}}$) (Eq. 16.3). Void ratio also decreases as soil density increases.

$$e = \frac{V_{\mathrm{v}}}{V_{\mathrm{s}}} \tag{16.3}$$

A useful reference for this and other soil science definitions is the Glossary of Soil Science terms: www.soils.org/sssagloss.

In cultivated soils, tractors, harvesters and implements produce the compaction that results in a dense, low porosity layer that is referred to as a traffic or plough pan. It is found below the maximum depth of tillage. This high-density layer is delineated in soil profile descriptions in the USA by the lowercase 'd' after the

Fig. 16.2. Traffic in this orchard has compacted the soil and decreased infiltration rate significantly, resulting in standing water in the wheel tracks.

master horizon designation (e.g. Ad). Similar problems are produced during forest harvesting operations due to the weight of the harvesting equipment. This is most often reflected in lower tree survival rates or growth rates on compacted skid trails. Soil compaction on trails or camping areas where frequent foot traffic occurs often results in loss of vegetation and subsequent severe water erosion. Vegetation is either killed outright by foot traffic, or is killed by lack of water due to reduced porosity and water-holding capacity. High soil strength also reduces root growth, further stressing plants.

In addition to the resistance a pan presents to roots, it is a management problem because of the slow rate of water flow through this induced soil layer. Fig. 16.2 illustrates a common soil compaction problem in California orchards where late winter rains and warming temperatures stimulate plant disease. Growers respond by spraying the trees with chemicals using tractor-drawn spray equipment. Often this is done when the soil is near the Proctor optimum and severe compaction results. In Fig. 16.2, the severely compacted soil in the wheel track has reduced infiltration rates, resulting in standing water in the wheel tracks.

At any water content, air-filled porosity is reduced in compacted soils compared with non-compacted soils. In the orchard illustrated in Fig. 16.2, the slower infiltration and percolation rates lead to high soil water contents above the compacted zone and the potential for reducing plant vigour when the trees break dormancy if the condition persists. Effects of soil compaction include changes in the soil temperature regime and changes in biological processes including increased denitrification and loss of mycorrhizal fungi (Ellis, 1998). Other consequences of reduced porosity, decreased macroporosity and increased microporosity are that soil organic matter may be physically protected against microbial decay, and predation of micro-organisms by protozoa and nematodes may be restricted (Breland and Hansen, 1996).

Increased soil strength typically results in reduced crop growth, which is most often related to the detrimental effect of higher strength on root growth. Lipiec *et al.* (2003) reviewed models of root growth as a function of soil strength and found that root growth is reduced when penetrometer measurements of soil strength vary between 1.0 and 1.7 MPa. The range is due to the differences in root strength that are a function of root morphology and soil properties.

Decreased infiltration rate due to fewer pores and reduced macropores results in a higher proportion of rainfall running off the soil surface and an increase in the potential for soil erosion and pollution of receiving waters with sediment and the nutrients and pesticides carried by the soil particles.

These pans should not be confused with natural subsurface horizons of high density and low permeability such as claypans, fragipans, duripans, petrocalcic, petroferric and petrogypsic horizons. Claypans are natural subsurface horizons produced by the accumulation of translocated clay. They are characterized by an abrupt increase in clay content of the subsoil compared with the clay content of surface horizons. Claypans will slake (fall apart) in water. Fragipans are subsurface zones of high density that are brittle when dry. Neither claypans nor fragipans are cemented, unlike duripans that are subsurface horizons cemented with silica, petrocalcic horizons that are cemented with calcium carbonate, petroferric horizons that are cemented with iron, and petrogypsic horizons that are cemented with gypsum. Cemented layers do not slake when shaken in water.

## 16.2 Sealing and crusting

Unlike compaction, sealing and crusting are soil surface conditions. Seals are wet soil crusts. Compacted zones in soil may be tens of centimetres thick. Seals and crusts tend to be significantly thinner. Structural crusts may be as thin as a few millimetres. Depositional crusts may be a centimetre or two thick. Seals and crusts have higher density, higher shear strength, finer pore size distribution, and lower saturated hydraulic conductivity than uncrusted soils (Shainberg, 1992). Crusts and seals have greatly reduced infiltration rates because of their low porosity and lack of macropores. In addition, seals and crusts have slow gas exchange. Because of their high shear strength, crusts reduce emergence of

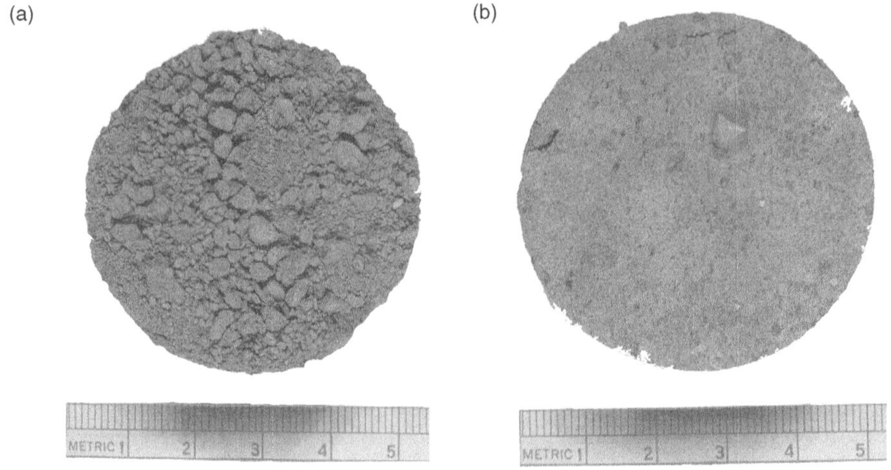


Fig. 16.3. (a) An uncrusted soil and (b) a structural soil crust formed after 90 minutes of simulated rainfall of 42.5 mm $h^{-1}$.

fine-seeded crops. Fig. 16.3 illustrates the effects of rainfall on soil surface conditions of a soil with 47% sand, 34% silt and 19% clay, 1.1% organic carbon, an exchangeable sodium percentage of 2.5 and 39% water stable aggregates. The uncrusted soil sample on the left has large aggregates and obvious macroporosity. On the right is the same soil sample after 90 minutes of 42.5 mm $h^{-1}$ rainfall from a rainfall simulator. The aggregation has been destroyed and the macropores are no longer present. The result is a much reduced infiltration rate and increase in volume of soil splash and overland flow.

Two different mechanisms create either depositional crusts or structural crusts. These are different from naturally occurring chemical and biological crusts that are common in arid areas (Eldridge, 2003; Belnap, 2003). Neither chemical (salt) crusts nor biological crusts are the subjects of this chapter. Physical processes, erosion, sorting and deposition form depositional crusts. Depositional crusts form when soil suspended in water is deposited on the soil surface as the water infiltrates. This often occurs during surface irrigation when the sides of furrows are eroded and the eroded sediment is deposited in the furrow bottom. Crusts up to several centimetres thick can be formed in this way, depending on the soil susceptibility to erosion and irrigation method. Depositional crusts are typically layered because the suspended sediment is sorted by size as it settles on the furrow bottom. The sorting also leads to low total porosity due to a dense packing of grains. Subsequent applications of irrigation water result in more surface run-off because of the very slow infiltration rate of the depositional crust.

Structural crusts form when soil structure is destroyed by the combined action of swelling and dispersion of aggregates and raindrop or irrigation drop impact.

The combination of chemical and physical processes that produce structural crusts differentiates structural crusts from depositional crusts. The beating action of water drops breaks down aggregates and reduces the pore size distribution of the surface few millimetres of the soil. Rapid wetting of aggregates causes aggregate slaking (breakdown) because air trapped in pores pushes against particles as it is replaced by water entering the aggregates. Under conditions of low electrolyte soil solution, clay dispersion further facilitates structural crust formation. Crust formation of this type is a significant problem in soils with low aggregate stability. Aggregate stability is low when a soil has little organic matter, clay or iron oxides. Sodium on the soil exchange complex further reduces aggregate stability. Soils with less than 15% clay and 2% organic matter tend to be susceptible to sealing and crusting because of low percentage of water stable aggregates. In arid and semi-arid regions, cultivated soils tend to have low organic matter contents, and they are more likely to have higher exchangeable sodium percentage (ESP) than soils of more humid regions. This combination decreases aggregate stability and increases the potential for sealing and crusting, wind and water erosion.

## 16.3 Physical soil management

### *16.3.1 Compaction*

Avoiding compaction is the least costly and most effective management choice because there is little natural reduction in soil compaction from shrinking and swelling or freezing and thawing of soils (Alakukku *et al.*, 2003). Reducing the number of trips across a field, decreasing the size of equipment, increasing tyre size, decreasing inflation pressure or replacing tyres with tractor treads reduces compaction effects by reducing the total load per unit area of soil. Many farmers in the United States, particularly in the Midwestern corn and soybean growing areas, have adopted conservation tillage as a means to reduce the number of tractor trips across a field. Conservation tillage not only reduces the number of mechanical operations, but it increases the amount of organic materials returned to the soil. Reducing the number of trips across a field through conservation tillage also reduces the total load on the soil surface, but a disadvantage is that the lack of cultivation allows dense soil horizons, once formed, to persist. Occasional ploughing or ripping may be necessary to disrupt dense horizons in conservation tillage systems. According to Green *et al.* (2003), long-term management using a no-tillage or minimum-tillage approach increases macropore connectivity without changing either total porosity or soil bulk density compared with conventional tillage. Results from their work were inconsistent across soils, climates and conservation tillage practices and it is not possible to predict the exact response of soils to conservation tillage without detailed knowledge of the soils, climate and practices at individual locations.

Improved infiltration of naturally occurring soil horizons and induced pans typically requires physical disruption such as slip ploughing, deep ploughing and ripping. All three of these options require significant mechanical power and are unlikely to be useful in developing countries. These tend to be temporary solutions to soil compaction, which recurs with tillage and other traffic on the soil surface.

Cemented horizons require much more power to disrupt than do compacted layers but present another dilemma. Soils with naturally occurring slowly permeable horizons like clay pans and hard pans are typically intensively weathered and once physically disrupted are not likely to reform. This is exactly what is desirable in agriculture, but such activity destroys the natural soil. Amundson *et al.* (2003) argue that these soils have become rare and endangered because of the deliberate destruction of impeding horizons. They state that 508 soil series, occupying 1.8 million hectares in the United States, are in danger of being lost forever. They define endangered as $< 50\%$ of the total area of the soil remaining.

### *16.3.2 Crusting and sealing*

Depositional crust formation is minimized by improving soil structure, and by reducing erosion by reducing the slope of irrigation furrows and rate of water application. Shallow tillage disrupts crusts and returns infiltration rates to pre-crusted levels (Singer and Warrington, 1992). Addition of polyacrylamide to irrigation water has been shown to be effective in reducing depositional crust formation and maintaining an uncrusted infiltration rate. Polyacrylamide flocculates the suspended sediment. The large flocs, when deposited on the furrow bottom, help to maintain the soil porosity. Gypsum can similarly help to flocculate the clay. Flocculated clay forms a more porous fabric when deposited on the furrow bottom, thus helping to maintain infiltration rates. Amendments such as these are practical in irrigated agriculture, but impractical when crops are rainfed (Singer and Shainberg, 2004).

Structural crust formation can be reduced by chemical amendments that favour clay flocculation and by mulching that reduces direct raindrop impact on the soil surface. Clay dispersion occurs when high quality irrigation water is applied to soils, regardless of the soil exchangeable sodium content. The potential for swelling and dispersion increases as the exchangeable sodium content increases and the electrical conductivity of the soil solution decreases. Gypsum has been used in California to maintain the electrical conductivity of the soil solution with some success. If exchangeable sodium levels are high, removal of the sodium through standard reclamation techniques is required. Gypsum is often an essential ingredient in reclamation of sodium-affected soils.

## 16.4 Secondary effects

Among the most important soil management problems worldwide is soil erosion by wind and water. Two complementary processes, detachment and transport, produce wind and water erosion. Soil physical and chemical properties determine the resistance of soil particles to detachment, the soil erodibility. Detachment of soil particles from the main soil body is accomplished when the wind velocity is sufficient or when raindrop energy is sufficient to dislodge particles. Transport of particles from the site of detachment is primarily a function of the flow of wind and water over the soil surface carrying particles. If the overland flow becomes concentrated in rills, the velocity can increase sufficiently to cause further detachment of particles.

Compaction, sealing and crusting affect the erodibility of soils in different ways. Compaction reduces the permeability of near surface and subsoil horizons and limits the rate at which water flows through the soil. In many cases, water infiltration is reduced significantly, time to ponding decreases, run-off occurs more rapidly on compacted soils and soil erosion is accelerated. Sealing and crusting have a similar effect in that infiltration rates are greatly reduced, reducing the time to ponding and increasing the volume of run-off and potential for soil erosion (Singer and Le Bissonnais, 1998). Fig. 16.4 illustrates the effect of different sealing potentials on run-off. The Holland soil, which is found under mixed conifer forest in California, has 4.5% organic carbon and 1.4% dithionite extractable iron (iron in pedogenic oxide minerals). The stable aggregates in this soil resist destruction, and under the simulated rainfall of this experiment produced no run-off. In contrast, the San Andreas soil series has aggregates with low water stability because it has only 1.2% organic carbon and 0.25% dithionite extractable iron. The aggregates quickly break down under the $40\,\mathrm{mm\,h^{-1}}$ simulated rainfall, rapidly forming a seal that lowers infiltration rate and increases run-off. The Salinas series has properties intermediate between the Holland and San Andreas soils. It has 2.1% organic carbon, twice the clay (35% compared with 17 and 13% respectively) and 0.66% dithionite extractable iron. After one hour of simulated rainfall, run-off is continuing to increase, suggesting that the seal has not completely formed. It is clear from these data that the rate of seal formation contributes to the amount of run-off. The erosion rate of the three soils

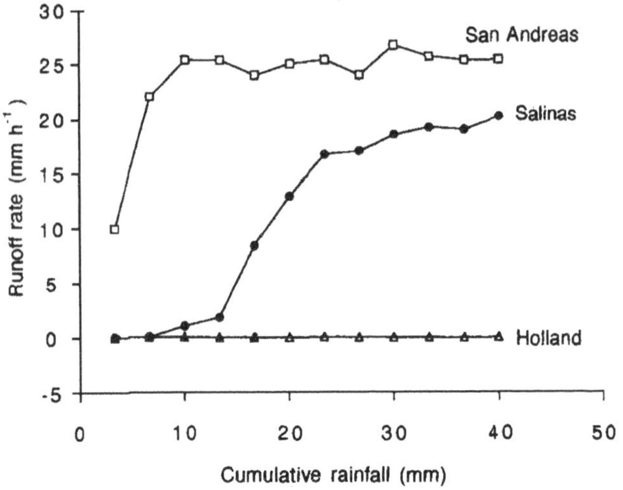


Fig. 16.4. Run-off response of soils with different crust-forming potential.

follows the same trend. The Holland soil had no soil loss. The Salinas soil final erosion rate (not including splash) was 0.46 kg $h^{-1}$ $m^{-2}$ and the final erosion rate of the San Andreas soil was 0.71 kg $h^{-1}$ $m^{-2}$ (Singer and Le Bissonnais, 1998).

Crusting also increases soil strength, which may reduce detachment, counteracting the effect of slow infiltration rates early in a storm, but the net effect of crusting is to increase the volume of run-off and the potential for soil erosion.

Compaction has little effect on wind erosion unless it is at the very surface of the soil, in which case it may reduce wind erodibility by increasing soil strength and the threshold wind velocity required to detach the soil. Surface crusting may influence wind erosion in two opposite ways. It increases soil strength and resistance to wind erosion but it also reduces surface roughness, which tends to increase the potential for wind erosion. Therefore, the net effect of crusting is to decrease wind erosion.

## 16.5 Conclusions

Soil physical condition directly affects the hydrology of a soil profile and the potential for wind and water erosion. In addition, soil physical condition influences the volume of soil available for plant roots to explore for nutrients and water and the ease with which seedlings can emerge from the soil. Soil compaction and surface crusting and sealing are processes that increase soil strength, decrease total porosity and change pore size distribution from a normal to a skewed distribution in which micropores dominate. The net effect of these three processes is to decrease water infiltration rate, increase run-off volume and increase the potential for water erosion. Sealing and crusting, in contrast, tend to decrease wind erosion. Strategies for management of soil compaction include reducing the impact of traffic on the soil surface through fewer trips, and reduced load per unit area. Soil sealing and crusting can be reduced through organic matter management and the addition of amendments that help to stabilize soil aggregates.

# 17

# Chemical degradation of soils

*Peter Blaser*

Soils can be considered as a finite non-renewable resource. A resource is any material that is of benefit to human life. The formation of soils through weathering of the underlying parent rock, the formation of humus, and the development of a soil structure all require a long time. The time span for soil development depends on the intensity of the soil forming factors. On average, the formation of a soil layer a few centimetres thick in a humid climate takes several hundred years (Jenny, 1980; Tutzing Project, 1998). In relation to the time span of a human life, any loss of soil is to be considered permanent.

Soil degradation as defined for the Global Assessment of Soil Degradation (ISRIC, 1990) is 'a process that describes human-induced phenomena which lower the current and/or future capacity of the soil to support human life'. In other words, soil degradation can be defined as human-induced deterioration of its quality, which means the partial or entire loss of one or more functions of soil (Blum, 1988). Soil quality then should be related to the potential socioeconomical and ecological soil functions. Important ecological soil functions include the ability to produce biomass, to store nutrients and water, to transform plant residues to soil humus (humification), and to release organically bound elements by mineralization (nutrient cycling). Furthermore, soils operate as a buffer against acidification and as a filter for pollutants protecting ground and surface waters. Soils are also the habitat for plant roots and countless soil organisms. Soils act therefore as a genetic reservoir.

The function of soil is generally threatened by the increasing and very often conflicting demands of a constantly growing human population and its activities, as well as by land use and climate change. This leads to a number of physical and chemical degradation processes that affect the sustainable functioning of soils (EEA/UNEP, 2000).

*Soils: Basic Concepts and Future Challenges*, ed. Giacomo Certini and Riccardo Scalenghe.
Published by Cambridge University Press. 

Among the most prominent chemical threats affecting soil functions, those of acidification, contamination with heavy metals, organic pollutants, radionuclides and salinity/sodicity are of major concern. In addition, soil eutrophication, primarily by excessive N and P additions, may also have detrimental or at least negative effects on the proper functioning of the soil–plant system. Other serious problems arise from mismanagement practices such as inappropriate irrigation, overexploitation of agricultural land or the introduction of plant species that affect soil properties in a negative way. There is also a clear link between climate change, sustainable development, environmental quality and soil degradation (EEA, 1999). In this chapter, it is shown in which ways the various chemical degradation processes act on soil functioning.

## 17.1 Chemical soil degradation processes

### *17.1.1 Soil degradation by acidification*

In the humid and temperate climate zones, soil acidification is a natural process that is controlled primarily by the chemistry of the soil mineral phase, the vegetation and the climate. This natural soil acidification is part of soil development, and is therefore not considered as a process of degradation. Soil degradation by acidification refers only to that part of acidification caused by anthropogenic processes, such as the deposition of acidifying substances, mismanagement, land-use change and global climate change.

#### *Causes*

- Emission of acidifying substances such as $NO_x$, $NH_3$ or $SO_x$ by industry, traffic, agriculture and households to the environment. $NO_x$, and $SO_x$ are transformed in the atmosphere to strong acids, transported over short to long distances and deposited onto the soil by wet or dry deposition. $NH_3$ transforms in the air to $NH_4^+$ which in soil may be nitrified ($NH_4^+ + 2O_2 = 2H^+ + NO_3^- + H_2O$), releasing protons into the soil solution. This type of soil acidification generally originates from diffuse contamination of the environment making it difficult to localize.
- Management practices: intensive use of acidifying fertilizers, converting broadleaved forest to coniferous forest, and other land-use changes, particularly the transformation of agricultural land to fallow land.

#### *Direct effects on soil functions*

Soil acidification is followed by a number of buffer reactions, which depend on the acidity status of the soil. Regardless of the buffer substance involved, soil acidification generally decreases the soil buffer capacity.

The most important buffer reactions are: weathering of the parent rock, cation exchange, and the dissolution of pedogenic Al- and Fe-oxides and hydroxides. As long as a soil contains carbonate minerals in the parent rock or in the fine-earth fraction, protons are buffered efficiently by the dissolution of the calcium carbonate. Under such conditions, the pH value remains in the slightly basic range and no soil acidification can be observed. A pH decrease below 7 is followed by a constant decrease of the cation exchange capacity due to the protonation of functional groups with variable charges. Consequently, fewer nutrient cations can be held in the soil. Soil acidification therefore lowers the soil's nutrient storage function. In relation to plant nutrition, the buffer reactions become important once the pH drops below approximately 5.5. In this pH range, acidity starts to dissolve pedogenic aluminium compounds and the proton buffering is followed by a release of free aluminium cations into the soil solution. Due to the strong selectivity of the cation exchanger for trivalent cations, this aluminium gradually replaces the divalent and monovalent nutrient cations ($Ca^{2+}$, $Mg^{2+}$ and $K^{+}$) from their exchange sites. These elements are leached to the subsoil or out of the rooting zone to the ground and surface waters. The base saturation decreases and the soil becomes depleted in nutrient cations. Indirectly soil acidification affects the biomass production function by nutrient depletion. Once the base saturation drops below roughly 15%, the concentration of free aluminium in the soil solution may reach levels that are potentially phytotoxic. At this stage of acidification, there is a potential risk for detrimental effects of free aluminium to plants (Cronan and Grigal, 1995).

Soil acidification also affects the transformation function and the filter function of soils. Although soil organisms, especially micro-organisms, may have a remarkable potential to adapt to changes in the soil environment, their activity is limited at low pH values. Consequently, turnover rates and the biodegradation of harmful organic substances are reduced under such conditions. The soil macrofauna, particularly earthworms, are even more affected by low pH values. In soils with pH below 4.5, earthworms are scarce and once the pH drops below 3.5 they are generally absent (Curry, 2004). Because of this, litter decomposition and the incorporation of organic matter in the mineral soil are inhibited in acidic soils. The typical humus form of strongly acid soils is raw humus with large amounts of fibrous plant residues that are not entirely decomposed and remain at the soil surface. The absence of an active earthworm population also affects the soil structure and consequently the soil aeration and the soil water-holding capacity.

Under acidic conditions, most heavy metals of natural or anthropogenic origin become increasingly soluble and can be leached to ground and surface waters. In addition, at higher heavy metal concentrations many of these metals are toxic to plants and soil organisms, and directly decrease biomass production.

To demonstrate the potential risk of increasing Al-concentration in the soil solution to plants at low pH values, the results of a laboratory study are given below.

### *Example 1: Effects of aluminium on tree fine-roots*

In recent decades, forest soil acidification has been accelerated by acidic deposition at sensitive sites in forested landscapes (Driscoll *et al.*, 2001). Although atmospheric deposition of sulphur has decreased over the last 30 years due to reduced emissions of sulphur dioxide (Palmer and Driscoll, 2002), the remaining acid deposition still leads to a depletion of base cations and to accelerated soil acidification coupled with an increase in aluminium concentrations (Blaser *et al.*, 1999). The consequences can be an increase in the incidence and severity of Al-toxicity, Al-antagonism, and nutrient imbalances in forest trees (Cronan and Grigal, 1995), as observed with red spruce and sugar maple in the north-eastern United States (Driscoll *et al.*, 2001).

Based on the concept of critical loads of acidity, indicators of soil acidification are, amongst others, the base saturation (BS) of the soil matrix (threshold 15%) and the molar Ca/Al ratio of fine-roots (threshold 0.2) (Cronan and Grigal, 1995). At soil BS below 15% and soil pH below 5, phytotoxic forms of Al are solubilized into the soil solution, inhibiting root growth and function, and thus reducing plant yields. Aluminium interferes with a wide range of physical and cellular processes. As Al is highly reactive, there are many potential sites for injury, including cell walls, plasma membranes, and the cytosol. One of the earliest responses to Al-toxicity is the inhibition of root growth. Aluminium, after entering into the cytosol, quickly disrupts root cell expansion and elongation. In addition, Al can strongly interact with the negatively charged plasma-membrane surface, and thus for example affect the activity of the mitochondrial respiration chain (Fig. 17.1, Ruf and Brunner, 2003). It has been suggested that the displacement of Ca from the membrane surface by Al may increase the apoplastic Ca pool. Such an increase of apoplastic Ca, however, stimulates callose synthesis. In Norway spruce seedlings, callose synthesis is strongly induced by Al at concentrations as low as 84 $\mu$M (Fig. 17.2, Hirano *et al.*, 2004). Callose, thus, can be regarded as a sensitive physiological indicator for Al-toxicity. The development of molecular tools may facilitate and improve the detection of Al-toxicity in the near future.

### *Indirect effects to the environment*

Soil acidification is followed by changes in vegetation, with plant species more tolerant of acidic conditions replacing more sensitive ones. For forests, such a transition may temporarily lead to less stable stands which are more sensitive to

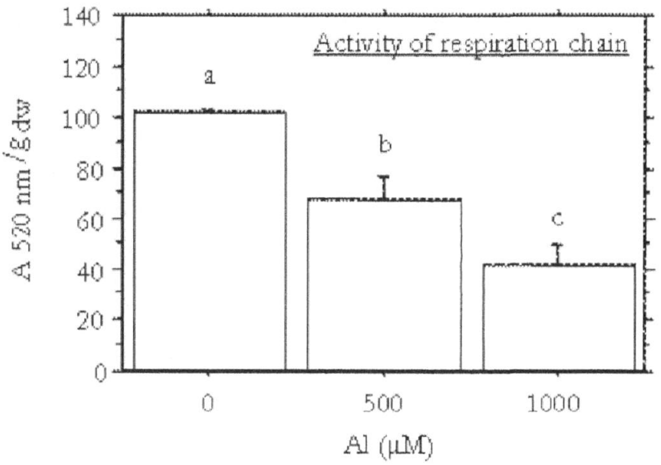


Fig. 17.1. Activity of the mitochondrial respiration chain of roots of Norway spruce seedlings after treatment with various concentrations of Al. From Ruf and Brunner (2003).

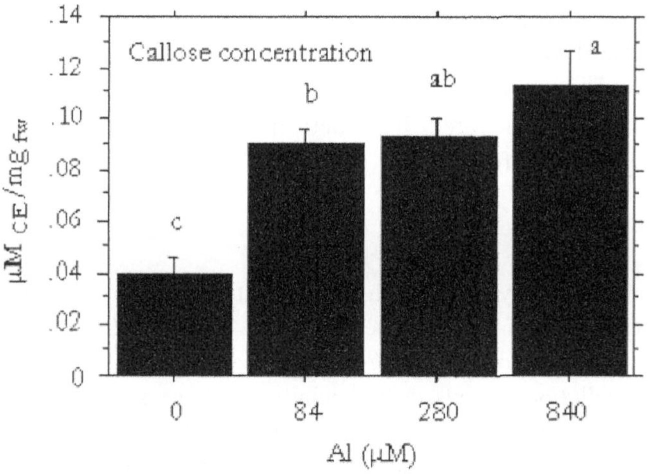


Fig. 17.2. Callose concentration of roots of Norway spruce seedlings after treatment with various concentrations of Al. From Hirano *et al.* (2004).

additional stresses such as drought or pest attack. Furthermore, changes in plant community composition may lead to lower ecosystem biodiversity.

Heavy metals are more mobile under acid conditions. Heavy metals enter the biosphere and accumulate in the food chain through groundwater pollution and enhanced uptake by plants and soil organisms. Soil acidification may lead to severe stream and lake acidification (low pH and elevated $Al^{3+}$) that has been shown to adversely impact lower food-chain transfers (i.e. phytoplankton–zooplankton) that ultimately impact the higher components of the food chain (i.e. fish).

*Risk assessment*

Sites where acid deposition exceeds the rate of proton buffering over the long term are vulnerable to acidification. Acidification is favoured by acid parent rock such as granites with low weathering rates. It is also favoured by shallow soils, coarse soil texture, low humus content, high porosity of the mineral soil and by high precipitation. Under such conditions the residence time of the acid soil solution in the mineral soil is short and results in inefficient buffering.

Sites on calcareous parent material, in contrast, are marginally threatened by all acidifying factors.

### *17.1.2 Soil degradation by heavy metals*

Heavy metals are of special concern with respect to their toxic effects on biology.

Soil degradation by heavy metals depends on the concentration of the various elements in soil and soil solution, and on the specific sensitivity of the soil flora and fauna. Some heavy metals are essential for soil organisms, plants, animals and humans, and have adverse effects if their concentration is either too low or too high. Moreover, some heavy metals are of concern for micro-organisms, animals and humans, but rarely have detrimental effects on plants; some others are of primary concern for plants.

*Causes*

- Soil pollution by heavy metals originates most often from local sources such as industry, incineration of waste, waste deposition, landfill and agriculture.
- There are many well-known examples of industrial deserts being created as a result of heavy metal pollution of soils in the vicinities of large smelters. Examples include around Nikel and Monchegorsk in Russia, and around Trail and Sudbury in Canada.
- Environmental disasters are another important source of pollution. For example the failure of the tailing lagoon of the Boliden zinc mine in Aznacóllar (Seville, Spain) in 1998 caused enormous socioeconomical and ecological consequences for the Doñana wetland ecosystem. In another case, a mining waste spill near Baja Mare, Romania, in 2000 flushed a huge amount of mineral waste into the Viseu river and contaminated large parts of the Tisza and Danube rivers.
- Besides local sources, there are also diffuse sources such as traffic, which may contribute substantially to the total load of some heavy metals in soils. The contamination originates from the dry and wet deposition of pollutants, often after long-distance atmospheric transport.

### *Direct effects on soil functions*

Heavy metal pollution is of primary concern for the transformation and the biological filter functions of soils because both functions depend on the integrated activity of soil organisms. There are strong indications that high contents of heavy metals such as lead, cadmium, copper and mercury inhibit the decomposition of soil organic matter and consequently have adverse effects on nutrient cycling.

Heavy metals also directly affect the biomass production function. Some heavy metals have both beneficial and deleterious effects on living organisms, depending on the concentration and on the chemical form in which they occur in the soil environment. Heavy metals, such as chromium (Cr(III)), copper, zinc, iron, and manganese, are essential for the health of plants and animals. They are required in very low quantities and are considered to be essential trace elements. In their absence or if they are not available in sufficient amounts, specific deficiency symptoms can appear. The same elements, however, are toxic once their concentrations exceed a certain threshold value. The critical concentrations of various trace metals are specific for each element and for each living species.

For other heavy metals such as antimony or silver little is known about their beneficial value for living organisms, whereas heavy metals such as arsenic, mercury, chromium (Cr(VI)), cadmium or lead are well known as being toxic at even very low concentrations. Elements such as Hg, Pb, Cd and As are especially toxic to animals and humans while Cu, Ni, Co or B are toxic primarily to plants (Pierzynski *et al.*, 2000).

### *Indirect effects to the environment*

Most heavy metals are almost insoluble under near neutral or only weakly acid conditions. In addition, heavy metals are likely to be associated with soil organic matter (SOM) due to strong chemical binding. Consequently, heavy metals originating from pollution tend to accumulate first and foremost in the humus-rich upper horizons of soils and may be stored there for a very long time. This part of the soil usually is most intensively rooted and is therefore of special concern with regard to heavy metal uptake by plants and the enrichment of potentially toxic trace elements in the biosphere. However, because of the strong association of heavy metals with SOM, heavy metals are unlikely to adversely affect plant growth or enter the biosphere in harmful concentrations under near neutral pH conditions. In acid soils, in contrast, the solubility of these elements is much higher. They are more mobile, and subject to leaching or uptake by plants and soil organisms. Under such conditions, potentially harmful heavy metals may contaminate ground and surface waters and enter the food web of the biosphere. Once they accumulate in the food chain they are a potential risk to the health of plants, animals and humans.

When heavy metals occur at critical concentrations for the biota, specific stress symptoms can be detected using appropriate biomarkers, as shown in the next example.

*Example 2: Biomarkers for heavy metal exposure and effects*

An important goal of heavy metal research is the development of biomarkers to assess environmental risks. Biomarkers of heavy metal stress can generally be subdivided into biomarkers of exposure, biomarkers of effects and biomarkers of susceptibility. Receptors in terrestrial ecosystems can be decomposers, comprising micro-organisms, consumers, such as invertebrates, and primary producers, specifically plants.

The impacts of elevated heavy metal pollution on soil microbial parameters such as soil enzymes, litter decomposition, and nitrification have been well documented. Besides measuring a single microbial parameter, the effect of heavy metals on the structure and composition of the microbial community has been analysed. The microbial community can be altered without resulting in changes in the overall performance of the soil system. In fact, an increasing body of evidence suggests that heavy metals have a strong influence on the structure of bacterial and fungal communities (Sandaa *et al.*, 2001), although the bacteria can develop tolerances against heavy metal contamination. Many soil micro-organisms cannot be cultivated, which impedes our knowledge of microbial diversity in soil. Culture-independent DNA-based methods have been developed that have aided in the analysis of microbial community diversity. Terminal Restriction Fragment Length Polymorphism (T-RFLP) is a recently introduced polymerase chain reaction (PCR) based tool for studying the genetic diversity of bacterial communities (Blackwood *et al.*, 2003). T-RFLP analysis is based on the detection of a single restriction fragment in each sequence amplified directly from the environmental sample of DNA. Fig. 17.3 shows electropherograms of a mercury-spiked forest soil and a control, non-spiked soil. Peaks are considered as operational taxonomic units (OTU). Comparison of the two spectra reveals that various OTU disappear while new OTU appear, indicating that mercury pollution in the soil caused a shift in the microbial community composition. Some taxa may be eliminated whilst others increase in abundance.

*Risk assessment*

The risk of detrimental effects of heavy metals on living organisms is highest in heavily polluted areas with acid, shallow soils, low in clay and humus content. A special threat arises from agricultural soils on acidic parent material that have been abandoned, a situation that is not uncommon in developed countries. Such

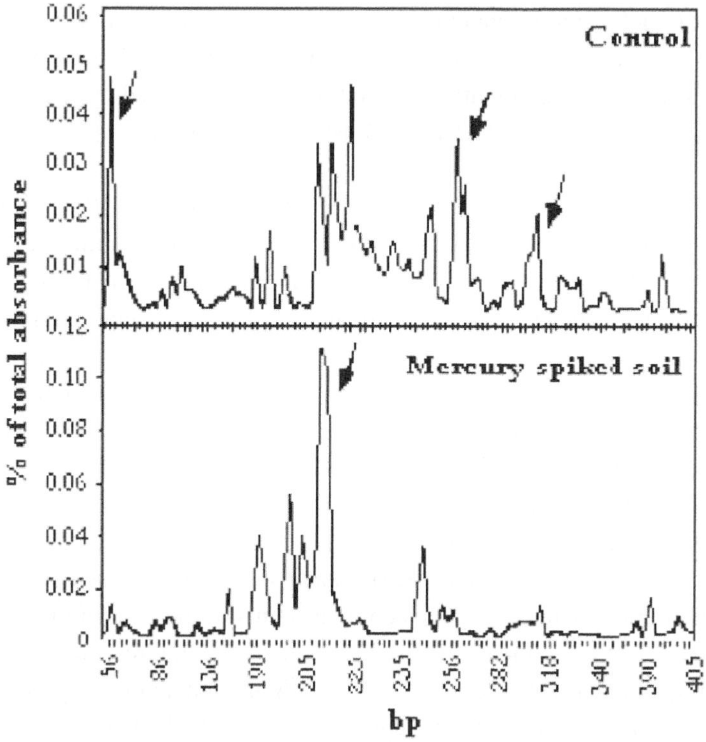


Fig. 17.3. T-RFLP community fingerprint pattern produced by HaeIII digestion of 16S rRNA gene amplicons of (a) a natural forest soil and (b) the same forest soil spiked with 100 μM $HgCl_2$. Arrows indicate major changes in the bacterial community. From Frey and Sperisen, unpublished.

soils tend to acidify because they are no longer limed and the plant cover may slowly change to some form of forest that accelerates soil acidification. Agricultural soils usually contain much higher heavy metal concentrations than wildland soils because of the use of heavy metal containing mineral fertilizers. Lowering the pH by land-use change may trigger the mobilization of these potentially harmful elements, a process referred to as a 'chemical time bomb' (Stigliani *et al.*, 1991).

Environmental pollution with heavy metals and abatement strategies based on the precautionary principle offer a special problem. All heavy metals considered can also be of natural origin and may be liberated by weathering of the parent rock. For this, soils act not only as sinks but due to the weathering of rocks and minerals they represent at the same time the dominant natural source of these elements. A high or even harmful concentration of a given heavy metal in the soil is therefore not necessarily caused by pollution but might represent the natural geochemical background concentration. Abatement strategies based on emission control are therefore only justified for harmful elements of anthropogenic origin.

A way in which this problem can be addressed is revealed by the results of the following case study in Switzerland (Blaser *et al.*, 2000).

*Example 3: The problem with the natural background concentration*

Every soil contains a natural background concentration of almost all trace elements. Chemically, these elements are indistinguishable from the corresponding contaminants, and for most trace elements of environmental interest the concentrations in soils may exceed the input concentrations of the pollutants by many orders of magnitude, depending on the parent material from which the soil was derived. There is still a question, however, of whether metallic pollutants can be distinguished from elements derived from natural sources and how they can be quantified. One way to overcome this problem is the estimation of enrichment factors (EF) that quantify the extent to which soils may have become contaminated by human activities. Enrichment factors have been widely used to quantify the enrichment of trace metals in various compartments of the environment. Generally, the EF considers the abundance of a heavy metal of interest, relative to a conservative lithogenic tracer element with no significant anthropogenic source. In addition, conservative elements that can be used to calculate enrichment factors in soils should not be part of the nutrient cycle, meaning that they should not be taken up by plants and soil organisms, and they should be insoluble and immobile in the chemical environment of soils. The elements that largely fulfil these requirements are zirconium and titanium. To estimate enrichment or depletion of a trace metal in a soil profile, soil samples have to be collected at various depths with special emphasis being given to the deepest soil horizon containing negligibly weathered parent material. The ratio of the trace metal of interest and the conservative lithogenic tracer is then calculated for each soil horizon and normalized either to the corresponding ratio in the parent material or to the ratio determined in the deepest soil horizon. Provided the parent rock is the same throughout the entire soil profile, the enrichment factors in all horizons should be equal to 1. Enrichment factors $>1$ indicate a trace metal concentration in the corresponding soil horizon that cannot be explained by weathering of the parent rock whereas an enrichment factor $< 1$ indicates depletion, probably by leaching. In polluted soils the enrichment factors of the pollutants considerably exceed the theoretical value of 1.

Fig. 17.4 shows the lead distribution in two soil profiles, calculated with the enrichment factors where zirconium has been used as a conservative trace metal.

Profile 1 has developed on gneiss. The natural background concentration is on the order of 25 mg $kg^{-1}$. The much higher concentrations in the upper horizons cannot be explained by weathering and natural enrichment and point to severe lead

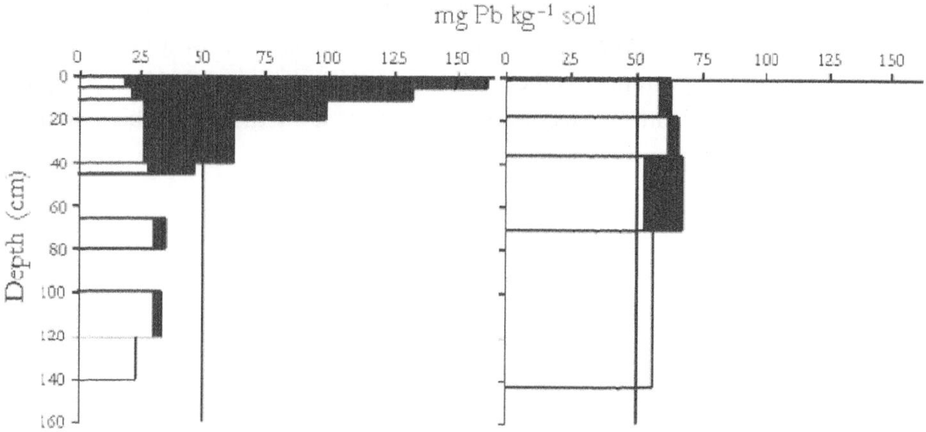


Fig. 17.4. Lead distribution in two soil profiles, calculated with the enrichment factors where zirconium has been used as a conservative trace metal. The natural background concentrations in the various soil depths are marked with open bars whereas the black areas on the corresponding bars represent the amount of lead of non-natural origin. The vertical line marks the guide level for Pb according to the environmental legislation of Switzerland.

pollution. In profile 2, developed on calcareous rock, the natural background concentration is much higher but enrichment by pollution is almost negligible (Blaser *et al.*, 2000).

The cases shown in Fig. 17.4 demonstrate the dilemma of legislating against heavy metal pollution by emission control based on fixed threshold values for the individual heavy metals. An abatement strategy might be successful in case 1 but would not consider the naturally high background concentrations seen in case 2. The same consideration holds for the approach of critical limits as defined by the protocol of heavy metals to the Convention on Long-range Transboundary Air Pollution (UN-ECE, 1998). In areas with high natural background concentrations of heavy metals, the critical limit may be reached even if the level of pollution is low. Such areas are vulnerable to heavy metal pollution and need to be monitored more intensely and managed with special care. The variable abundances of trace metals, their diverse chemical behaviours, and the broad spectrum of soil physical and chemical properties mean that soil monitoring requires clear and comprehensive understanding of the soils themselves.

### *17.1.3 Soil degradation by pesticides and organic contaminants*

Pesticides are substances or mixtures of substances used to prevent, destroy or mitigate a pest. They refer not only to insecticides but also include herbicides, fungicides, nematocides and rodenticides. Pesticides and organic pollutants which are not easily decomposed remain in the environment for a long time and are considered as persistent organic pollutants (POP). Through biological uptake they accumulate and are subsequently concentrated in the food chain and are therefore of particular concern.

*Causes*

- Use of biocides for pest control in agriculture.
- Drift of polychlorinated biphenyls (PCB) used as insulating, cooling or hydraulic fluids.
- Byproducts such as dioxins, furans, polycyclic aromatic hydrocarbons (PAH) from industrial processes or from combustion of petroleum products.
- Released to the environment by accidents such as that in Seveso, Italy (1976), where highly toxic dioxins were released from a reactor of a chemical plant, or that in Schweizerhalle, Switzerland (1986), where the storehouse of a chemical factory burned down and a large quantity of chemicals was flushed into the Rhine river.

*Direct effects on soil functions*

Biocides (*nomen est omen*) that are not quickly degraded after their application have detrimental effects on the soil flora and fauna and in turn they may severely affect the transformation and the biological filter function of soils. Consequently, the natural turnover of soil organic matter, humification and mineralization may be inhibited. Because of their toxicity, such substances may irreversibly eliminate some vulnerable species among the soil organisms and lower the biodiversity of the soil flora and fauna (reduction of the gene pool). Persistent herbicides can reduce crop yields of sensitive cultures and directly affect the biomass production function of soils.

*Indirect effects on the environment*

Degradation of pesticides in soils can occur not only through the activity of soil organisms but also by hydrolysis and photochemical reactions. All degradation reactions can generate degradation products, which may be even more toxic than the parent compound itself. In some cases, such degradation products are significantly more resistant to further decomposition.

Entering the biosphere by digestion of polluted food or drinking water, these highly persistent and toxic substances may accumulate in the food chain and exert detrimental effects on wildlife and humans.

*Risk assessment*

The solubility, mobility and bioavailability of organic pollutants determine the impact of this group of pollutants on the environment. Chemical, physical and biological processes control the ultimate degradation of such substances. In general, the biodegradation of pesticides is favoured by all environmental factors that stimulate microbial activity as a whole. Degradation is promoted by warm temperatures, sufficient soil humidity and by easily decomposable organic matter serving as source of energy. However, recalcitrant organic pollutants may persist for long periods of time in the soil when the chemical and biological degradation is suppressed by the toxicity of these chemicals.

Biocides and persistent organic pollutants have detrimental effects in all kinds of ecosystems, in particular when they reach ground and surface waters and enter the food chain. They threaten life, because they can affect the fertility of animal species and contribute to their diminution or extinction. Surface run-off loaded with pesticides and POPs endanger ground and surface waters.

In soils, significant amounts of contaminants are transported through preferential flow which leads to a rapid movement of solutes and compounds through the soil system which in turn makes risk assessments difficult. The following case study demonstrates the effects of preferential flowpaths on solute transport at a forest site in Switzerland (Bundt *et al.*, 2000, 2001).

### *Example 4: The preferential flow phenomenon and its implications*

'Preferential flow' is the rapid movement of water and solutes through the soil that bypasses a large portion of the soil matrix. As a consequence, the residence time of incoming solutes is shorter than expected under the assumption that solutes are moving through the whole soil volume. This also implies that only part of the soil is responsible for sorbing, buffering and degrading incoming elemental loads and contaminants. The physical phenomenon of preferential flow has been intensively studied since the end of the 1980s. Field studies have shown that surface-applied herbicides or fertilizers, heavy metals, radionuclides and micro-organisms are transported faster and to greater depths than predicted from bulk soil analysis and the assumption of uniform water movement through soils (Bundt *et al.*, 2000, 2001). Preferential flow is caused by a variety of different mechanisms, such as flow along macropores or roots, inhomogeneous infiltration, unstable wetting fronts, or water repellency of soil material.

Preferential flow can be visualized by staining the flowpaths with different colours or fluorescent dyes, and by studying the flow patterns in soil profiles (Fig. 17.5). However, it is nearly impossible to predict these flow patterns without disturbing soils. In addition, there is no practical way to collect the soil solution *in situ* separately from the flowpaths and the surrounding matrix. The only possible way to quantify solution transport is the use of large lysimeters in the field or the study of whole catchments.

Fig. 17.5. Preferential flowpaths visualized (in black) in a well-drained forest soil after homogeneous irrigation with 45 mm of dye tracer solution for 6 hours. Units at the four sides are 10 cm. From Bundt *et al.* (2000).

### *17.1.4 Soil degradation through eutrophication by nitrogen and phosphorus*

Elements such as N and P are not pollutants *sensu stricto*, but rather are essential for plant growth and soil organisms. However, they may have a serious impact on the environment, if their concentrations exceed the requirement of the vegetation. Eutrophication can be understood as 'nutrient pollution'. It causes an increase in the nutrient status of a soil that in turn stimulates microbial activity and provokes excessive plant growth. In addition, it may also lead to nutrient imbalances. A driving force for excessive use of N- and P-fertilizers is the pressure to produce increasing amounts of food on a limited area of arable land (Pierzynski *et al.*, 2000).

*Causes*

- Emissions of N-containing compounds to the atmosphere by traffic.
- Diffuse sources facilitated by long-distance atmospheric transport and deposition.

- Local sources: overapplication of livestock manure and mineral fertilizers in agriculture as well as intensive animal husbandry.
- Soil application of sewage sludge from wastewater treatment.

### *Direct effects on soil functions*

The stimulating effect of N and P on plant species primarily affects the soil flora and as a consequence the transformation function of the soil. Accelerated microbial activity favours the mineralization of labile organic substances, delays humus formation and changes the nature of the soil humus (specifically the C/N ratio). Eutrophication may therefore also affect the storage function of soils.

### *Indirect effects on the environment*

Nitrogen and phosphorus that exceed vegetation needs are leached to ground and surface waters and affect drinking water quality (Hagedorn *et al.*, 2001). Groundwater contamination is common in areas with high N-deposition and is a problem worldwide.

Many natural ecosystems are N-limited. As a consequence, increasing N-supply stimulates plant growth and induces nutrient imbalances. Given that the availability of other essential elements (or trace elements) is limited, various deficiency symptoms may develop in the plants. An excessive supply of N may have various physiological implications for plants and changes in the community structure can be observed. It may also lower the resistance of plants to frost or pest attack and hence weaken the overall stability of forest stands. Increased availability of nitrogen in terrestrial ecosystems also affects species diversity, often promoting an increase of invasive species (Tilman, 1987). Coastal eutrophication is becoming common in regions with elevated nitrogen deposition leading to excessive production of algal biomass, blooms of toxic algal species, hypoxia, fish kills, and loss of important plant and animal diversity (Jaworski *et al.*, 1997).

### *Risk assessment*

In contrast to other pollutants described above, the adverse impacts of elevated N- and P-concentrations in soils are not directly detectable and the effective risk is unclear. Areas with intensive agriculture and concentrated animal production, with inappropriate fertilization and organic waste management, are prone to eutrophication. If the soils have inefficient filter functions (e.g. shallow soils with coarse texture) ground and surface water pollution is likely to occur.

The devastating effects of storm events like Vivian in 1990 or Lothar in 2000 in central Europe promote the mineralization of soil organic matter. This is followed by a rapid release to surface waters and groundwaters of N which had been stored in soils for centuries.

### *17.1.5 Soil degradation by radionuclides*

Understanding of the nature of radioactivity, of the effects of high radiation doses on living tissues, and of the behaviour of radionuclides in the soil–plant system are crucial for the assessment of potential harmful impacts on ecosystems and, ultimately, human health.

#### *Causes*

Although radioactive elements occur naturally, harmful effects derive through:

- fall-out from nuclear weapons testing
- deposition of radioactive waste
- accidents from nuclear power plants such as Chernobyl, Ukraine, in 1986.

#### *Direct effects on soil functions*

Comparable to similar non-radioactive trace metals, radioactive elements having a long half-life are either taken up by soil micro-organisms or strongly bound by soil organic matter. In both cases, they accumulate in the rooting zone of the surface horizons. Their vertical movement is generally slow and depends on the physical–chemical properties in the soil environment (e.g. Riesen *et al.*, 1999). At high doses of radioactivity, there may be detrimental effects to soil organisms and, thus, ramifications on the transformation and the biological filter functions of soils.

#### *Indirect effects on the environment*

Indirect effects on the environment are of general concern because high doses of radioactivity are a threat to all living species. The low mobility of radioactive elements in soils together with the long half-life of the most prominent radioactive pollutants may render large areas hostile to human settlements or at least useless for agricultural purpose. Caesium and strontium behave similarly to the nutrient elements potassium and calcium. They are easily taken up by plants and soil organisms and hence enter the biosphere.

#### *Risk assessment*

The risk emanating from radioactive pollution depends on the intensity of radioactivity that is not modified by environmental conditions. Radionuclides bound in exchangeable forms may undergo exchange reactions by acidification or through the input of competing cations. Such processes render them available for biological uptake or vulnerable to leaching. Special attention needs to be given to the phenomenon of preferential flow, leading to the transport of radionuclides to greater soil depths than otherwise would have been predicted (Bundt *et al.*, 2000). Uptake will tend to decrease with increasing clay and organic matter content of soils.

### *17.1.6 Soil degradation by salinization*

In climate zones with an excess of evapotranspiration over rainfall during a substantial part of the year, the soil solution moves upwards and evaporates at the surface, thereby enriching the uppermost soil horizons with salts. Salinization eventually renders soils completely unproductive. Irrigation in arid and semi-arid regions can induce salinization. Approximately one third of the irrigated lands in the world are adversely affected by excess salt.

#### *Causes*

- Inappropriate irrigation practices or use of saline groundwater.
- Lack of properly working drainage systems.

#### *Direct effects on soil functions*

Salinization affects biomass production directly through decreasing the water availability for plants at high salt concentration and through the specific toxicity of some ions such as chloride. Trace element deficiencies are also common in many saline soils due to high pH. As a result, the growth of plants intolerant to salt is drastically restricted. Sodic soils (high sodium concentrations) have an additional problem in that $Na^+$ disperses the soil resulting in breakdown of soil structure, which in turn greatly reduces water infiltration and permeability. Thus, it is difficult to leach salts from sodic soils.

#### *Indirect effects on the environment*

Salinization has a major impact on the economy of a region and may force people to abandon their homeland. It has become a serious problem that has contributed to the desertification of large areas of land. As side effect, eroded dusts loaded with salts may be carried by wind and pollute other areas. Drainage waters from irrigated agriculture may also contain high concentrations of potentially toxic trace elements, such as selenium, arsenic, boron, molybdenum and uranium, that makes their environmentally safe disposal difficult.

*Risk assessment*

Soils in arid zones are generally sensitive to salinization. Regions under high population pressure are often subject to overuse and inappropriate irrigation practices.

### *17.1.7 Soil degradation by global climate change*

Soil processes are controlled by climatic conditions. As a consequence, the climatic changes expected in the near future could have the potential to induce soil degradation (EEA/UNEP, 2000), but the magnitude of these changes is unknown. Climate change is a natural process that is difficult to predict. Over the last 100 years, concentrations of greenhouse gases in the atmosphere such as $CO_2$, $N_2O$ and $CH_4$ have increased significantly. As these gases absorb the radiant energy of the atmosphere, increases in their concentrations are thought to be mainly responsible for the observed warming during recent decades. Some of these gases are naturally occurring whereas others are of anthropogenic origin.

*Causes*

- Excessive $CO_2$ release to the atmosphere by fossil fuel combustion and land use changes.
- Release of $CH_4$ and $N_2O$ from agriculture, livestock and landfills.

*Direct effects on soil functions*

Global warming directly affects soils and soil functioning. It will enhance desertification in some regions, but the extent of these changes is debatable. The current increase in mean annual temperature potentially accelerates all chemical and biological processes. Consequently, the transformation function and the biological filter function, both governed largely by the activity of soil organisms, might be affected. Given that soil organic matter dynamics strongly depend on moisture and temperature, changes in humus forms and the amount of soil organic matter are expected. As a consequence, the storage function of soils will also be affected. In soils, dissolved organic matter (DOM) is mobile and this phase largely determines the transport of pollutants, such as heavy metals and persistent organic pollutants. As DOM dynamics are sensitive to climatic conditions, we can expect changes in transport phenomena facilitated by DOM in the near future. The increasing DOM concentrations in surface waters of northern Europe are already a cause of concern for water suppliers.

*Indirect effects on the environment*

Elevated $CO_2$ concentrations and temperatures stimulate plant growth and lead to long-term changes in natural vegetation and biomass production. Soils may play a key role in buffering or accelerating climatic change as they contain the largest fraction of C in terrestrial ecosystems and about three times as much as the atmospheric reservoir of $CO_2$. However, soil organic matter can either be a source or sink for greenhouse gases and there are still uncertainties whether changes in soil organic matter will lead to a positive or negative feedback between elevated $CO_2$, climate, and soils (Hagedorn *et al.*, 2002, 2003).

*Risk assessment*

Global climate change will lead to unknown, irreversible and unpredictable changes in terrestrial ecosystems because of the many physical, chemical and biological functions that respond primarily to temperature changes. Considerable uncertainty arises over the predominance of feedback mechanisms in the environment; these make any analyses difficult.

## 17.2 Our duty

The degradation of soils through human activities is a severe threat to one of the most vital natural resources for sustaining life on Earth. As a result of the resilience of soils and because chemical soil degradation happens so slowly, the disturbance of soil functions by chemical pollution is difficult to detect and may only be noticed once the damages are well advanced and difficult to reverse (EEA/UNEP, 2000). The risk of soil degradation calls for a co-ordinated international effort for a sustainable use of soils in order to protect them from erosion, pollution, overexploitation and mismanagement. To protect soils, reliable indicators are needed and intelligent monitoring is required. There is justification for internationally binding agreements as proposed by the Convention on Soil (Tutzing Project, 1998). In addition, there is a need for scientific programmes that will improve our understanding of degradation processes, for scientifically appropriate risk assessment, and for a strong scientific basis for the decisions taken by policy makers and decision makers.

# 18

# The future of soil research

*Anthony C. Edwards*

A rather pragmatic view of soil is adopted in this final chapter which is based on three broad categories of land use/function: soil that is (a) managed for production (food, fibre or energy) purposes, (b) either used directly by humans (urban, recreation and transport) or provides necessary raw materials, (c) not directly managed for human purposes. The first two categories when combined represent more than half the available global land resource. The overwhelming significance and impact of anthropogenically related factors on the properties of atmosphere, biosphere, hydrosphere and pedosphere are becoming increasingly evident. The increased exploitation and utilization of soil with time has seen a change in the geographical scale over which impacts have become detectable. While local and geographically isolated issues of environmental degradation have occurred throughout recent history (see Hutchison, 1970, who describes the early influences of a Roman road, Via Cassia, on sediment delivery and subsequent eutrophication) there is increasing evidence of regional and global scale impacts. It has been well argued that probably none of the Earth's ecosystems remains in a 'pristine' condition. This situation is most easily demonstrated using the example of the nitrogen cycle and aspects of this relevant to soil are described later. Conventional ecosystem level description of nutrient cycles now require an upscaling component that includes the transport of food products from rural to urban areas of high population densities and movement of livestock and human waste products.

The growing awareness of the need to manage and protect soil as a finite and essentially non-renewable resource is currently emphasized by the development and adoption of national and international soil strategies for soil management. In situations where erosion and degradation is an issue, rates of soil formation are slower than rates of loss. The timescales over which soil becomes degraded and basic functions are lost vary greatly.

*Soils: Basic Concepts and Future Challenges*, ed. Giacomo Certini and Riccardo Scalenghe.
Published by Cambridge University Press. 

Table 18.1. *Estimated land area (in thousands km²) suffering from major constraints to agricultural production*

| Region | Total area | Area with major constraints[a] | % of total area |
|---|---|---|---|
| Asia and Pacific | 28 989 | 22 246 | 77 |
| Europe | 6 843 | 4 739 | 69 |
| North Africa and Near East | 12 379 | 11 298 | 91 |
| North America | 19 237 | 14 120 | 73 |
| North Asia, east of Urals | 21 033 | 12 573 | 60 |
| South and Central America | 20 498 | 16 526 | 81 |
| Sub-Saharan Africa | 23 754 | 19 419 | 82 |
| World total | 132 733 | 100 921 | 76 |

[a]Constraints include salinity, low CEC, steepness, shallowness, hydromorphy, erosion risk, Al toxicity, vertic properties and high P fixation capacity.

Data compiled by FAO (2006).

There is likely to be a substantial geographical bias associated with the nature and extent of potential issues that is related to the type of stresses involved (Table 18.1). These may include salinization, erosion, contamination with persistent substances such as pesticides or heavy metals, and nutrient enrichment. The impact of such stressors on soil may be cumulative, additive and strongly soil property dependent. Although commonly used, a certain degree of ambiguity can be associated with the terms soil 'quality' or 'health' (Anderson, 2003). These are often measured using some aspect of soil functioning or more indirectly through the monitoring of other ecosystem components, e.g. drainage water quality. For this to be achieved, and in order to provide the necessary underpinning rationale, it has been important to better define specific roles and functioning of soil within a general ecosystem functioning and food production context. The following quotation provides some context and summary of the problem.

> Every day soil is irreversibly damaged or lost. According to the European Environmental Agency, over the past 20 years, built-up areas have been steadily increasing, leading to 6–9% of Europe's land area being completely or partially sealed by urbanisation, which has increased flood and drought risks. 17% of Europe's total land is affected by soil erosion, leading to losses of about 53 Euro $ha^{-1}$ $year^{-1}$ in agricultural areas and between 2 and 35 Euro per year per capita are spent in clean-up costs for soils polluted by mixed and toxic waste disposal and industrial activities. The potential to increase the effectiveness of environmental measures, once mandatory soil protection and monitoring standards are set, is immense. Climate, air, water and nature protection, pesticide controls and regulation of industrial emissions could be improved and guidance for environmentally sound land use management and farming practices provided.
>
> *(EEB, 2004)*

Increasingly soils are required and expected to act as receivers, transformers and retainers of the various anthropogenically derived materials and waste products in order to reduce the risk of potential impacts on the wider environment. One intrinsic, but soil specific, property is a capacity to buffer against change, which provides greater stability of local conditions. This buffering ability of soil has long been utilized for soil fertility and food production purposes to regulate water and nutrient supply to growing crops. However, like all buffered systems each soil has a variable but finite capacity, which once exceeded loses many of its important properties.

The emphasis of this final chapter is therefore placed upon the need for greater process-based understanding of soils that can be used for modelling purposes as a way of helping to preserve their multiple functionality capabilities in the future. The case studies used are heavily studied and widely reported in the literature and provide examples of where a multidisciplined research approach towards a commonly agreed and clearly stated goal(s) is the primary objective for future research.

## 18.1 Soils and their buffering capacities

The value placed upon soil as a buffering mechanism is inherent within many aspects of environmental science. For example, risk assessments developed for predicting the likelihood of the downward leaching of potentially polluting substances contaminating groundwater often incorporates soil depth as a major factor (e.g. Nolan *et al.*, 1998). Soils are now exposed (both accidentally and on purpose) to multiple sources of potential contamination, raising concern with regard to the continued sustainability of soil functionality and the processes they help mediate. The likely extent of regional and global consequences that this situation, if allowed to continue, represents is difficult to predict. There are a number of well-documented examples where environmental degradation at either local or global geographical scales is occurring. The extent to which this degradation represents an irreversible situation varies and this worrying overall situation raises a number of general questions in relation to soil processes:

- How can the continued intensive utilization of the global soil resource be optimized for its and our long-term sustainability?
- How can soil processes be utilized to help in the continual provision of long-term protection of the wider environment?

And some more specific issues:

- What is the extent of spatial and temporal variability associated with individual soil processes across different pedoclimatic environments?

- What is the range of fundamental soil processes and what are their sensitivities to changing environmental conditions?
- What are the likely consequences for individual processes of accumulated exposure to multiple environmental stressors?

## 18.2 The soil resource

As described previously in this book, soil consists of an intimate mixture of solid, solution and gaseous phases. Understanding of how a soil will respond to changing conditions when *in situ* increasingly requires a fully integrated and interdisciplinary approach to soils-related research. Many soils have been heavily modified through a combination of changes to either their biological, chemical and/or physical properties. The consequences that these modifications are likely to have upon the wider environment are becoming increasingly apparent.

Soil mediates most of the substance cycling and fluxes that occur within terrestrial ecosystems. Soil can act as a short or longer-term store for substances; however, the storage capability of any particular soil is finite and heavily dependent upon prevailing environmental conditions. This inherent capacity to retain and/or transform substances such as water and chemicals has both a historical and future significance with respect to its continued long-term functioning. As with all buffers they are finite and only continue to operate efficiently within certain constraining boundaries. There is increasing evidence that under certain circumstances these boundaries are being exceeded and the soil's buffering capacity greatly diminished or lost all together. It is necessary to consider the extent to which these buffers operate, how they vary with soil properties and the ease with which they can be restored in those situations where they are already compromised. This is discussed in relation to phosphorus.

## 18.3 Soil phosphorus

The example of phosphorus accumulation within agricultural soil usefully highlights some of these particular issues and demonstrates how susceptible substantial land areas are to inappropriate management. Importantly it is necessary to remember that P itself represents a non-renewable resource (Steen, 1998). Annual application rates of P as either fertilizer and/or livestock manures have continuously exceeded crop requirement over wide geographical areas. This is especially noticeable in areas having intensive animal production (e.g., Chesapeake Bay area, Eastern USA (Chesapeake Bay Program, 2002 and United States Geological Survey, 1995) and a region of north-west Europe comprising parts of the Netherlands, Denmark and Germany). Here the accumulated result of large P surpluses (where total P input is greater than the total P output) has created situations where serious and long-lasting impacts on both surface water and groundwater environments have occurred. In the case of north-west Europe this has been aggravated by sandy textured, low P retentive soils.

This represents an extreme example, but demonstrates that it is possible to completely overwhelm a soil's retentive capacity so that it becomes essentially 'saturated' with respect to a particular contaminant. The accumulation of P experienced by these soils is additive, taking place over a number of years (but in the examples used here represents a comparatively instantaneous timescale with respect to those involved with soil development). The precise relationship between P adsorbed to soil with that which is present in soil solution while being soil property dependent (Breeuwsma *et al.*, 1986), is generally positive but non-linear. It is likely that the soil system will show strong hysteretic properties in response to reduced application of P.

It is perhaps worth considering a number of questions relating to how and why situations such as that briefly outlined above for P actually develop. The following list is certainly not exhaustive and produced with hindsight, which is a wonderful thing. Its value may lie instead with asking if current approaches to carrying out and interpreting of soils-related research are effective.

- Why wasn't the likelihood of saturation of soil by excessive P applications ever considered to represent an operation that carried a 'high risk' for downgradient ecosystems?
- How regular were soil samples taken for advisory/extension purposes and how were the actual test results interpreted and used for recommendations?
- Was any consideration given to the significance of soil type (through its influence upon P retention capacity) in these recommendation systems?
- How was the advice from extension services interpreted and used by the agricultural industry?
- Was there sufficient dialogue between 'research' and 'extension' practitioners of soil science?
- Was the advice available from environment protection agencies regarding 'possible risks' sufficient and accurate?
- What is the range of management options available to help remedy the situation?
- What usable new knowledge has been learnt from the recent substantial investment in soils-related P research?
- How is this, often highly location, time and treatment specific, information going to be compiled and extrapolated to the wider situation?
- Are there other comparable situations that are developing which require urgent consideration?

## 18.4 Soil processes

As a medium for supporting many of the life sustaining processes, soil functioning is susceptible to anthropogenic factors that may impose either a direct or indirect influence over the dominance of an individual process. Although not explicitly mentioned in the titles, many of the preceding chapters deal in some direct or indirect way with processes. There is a continuing need for developing an understanding of all aspects relating to the kinetics of soil processes. The requirement to understand, quantify and represent soil processes within a spatial and temporal modelling framework will increasingly provide the underpinning and linking component for future soils and related research. A focus on soil processes provides one framework that would encourage:

- the development of multidisciplined research that involves all soil science disciplines,
- the development of sampling techniques that allow an adequate quantification of spatial and temporal variability over all geographical scales,
- extrapolation of pedological and site specific influences on processes expressed at the landscape scale,
- development of robust transport models,
- provision of increased capacity for scenario testing.

As a central component of most natural and managed terrestrial ecosystems soil and the processes it supports exert a significant influence over biological productivity, ecosystem diversity, drainage water quantity and quality and gaseous exchange. Despite the apparent organizational complexity and visual differences that exist between ecosystems, the range of fundamental soil mediated processes is probably not wide. If we assume (for purpose of the present discussion) that all soil has the intrinsic capability to perform the majority of fundamental processes, then the extent to which individual processes are expressed and dominate depend upon a range of rate limiting factors. These may include general 'environmental' factors, such as temperature and moisture content, which influence the kinetics of most processes and are likely to impart a strong seasonal variability, while other attributes such as soil acidity and specific substrate availability might be expected to show a greater sensitivity to soil properties and land use/management. Soil moisture content and pH have been subjected to continuous and extensive manipulation in order to optimize local soil conditions for plant growth. Regular liming of intensively cropped, non-calcareous soils has been used to maintain soil pH well above (often one order of magnitude) their 'natural' value. The instillation of artificial land drainage has completely altered soil hydrological properties and indirectly shifted the balance of redox reactions. These commonly accepted and widely practised land management options have dramatically influenced the relative significance of key soil processes with global consequences. In extreme situations this may lead to a shift in the balance between organic matter accumulation and oxidation resulting in reduced soil structural stability and increased erosion risk. It is this combination of site, pedological and management factors that influence the relative capacity of a soil to allow individual processes to proceed that contribute to defining and determining the wider ecosystem properties. The situation involving N deserves particular attention as it demonstrates many of the attributes discussed previously.

## 18.5 Nitrogen cycling

The issues that involve N are numerous and diverse, having impacts at local, regional and global scales. Nitrogen is distributed ubiquitously throughout the natural environment (Table 18.2). Its central role in enhancing primary biological productivity has been a principal reason for the extensive anthropogenic modification that has taken place to the N cycle – most obvious being the increased availability of N within agriculture as a result of the application of artificial fertilizers and increased use of N 'fixing' crops. The benefits that have been associated with increased capacity for food production are obvious, however; so are the number of environmental and health issues related to the increased transport and availability of N within the wider environment. Even remote 'pristine' environments do not escape being impacted by the growing global atmospheric N cycle and the possibility of long-range N transport thought to be responsible for the nutrient imbalances implicated by the N saturation hypothesis (Galloway, 1998; Green *et al.*, 2004). The combustion of fossil fuels may contribute substantially to this atmospheric component. Environmental consequences include ecosystem level changes in species diversity, structure and functioning, rising $NO_3$ concentrations in surface water and groundwater and eutrophication.

Table 18.2. *Distribution of N within the main global components. The figures in brackets represent the percentage of total 'available' N (which excludes that present in rock and sediments)*

| Location | N concentration (Pg) |
|---|---|
| Rock and sediments | 190 400 120 |
| Atmosphere | 3 900 000 (99.4) |
| Ocean | 23 348 (0.59) |
| Soils | 460 (0.01) |
| Terrestrial plants | 14 (< 0.001) |
| Terrestrial animals | 0.2 (< 0.0001) |

Data from Robarts and Wetzel (2000).

Nitrate for example, now represents the major form of soluble N present in soil solution and lost from soil. Concentrations of nitrate in surface water and groundwater have increased dramatically in certain areas (> 10-fold).

Many intensive agricultural production systems are currently experiencing a surplus of nutrients as a consequence of excessive use of fertilizers and the inefficient utilization and recycling of the resulting animal waste products (Steele, 1995; Smaling *et al.*, 1999). Primary sources of N into agricultural ecosystems include fertilizer and biological fixation. Substantial amounts of N are transported in the form of food products, as nitrate in drainage water, and gaseous compounds (ammonium and $NO_x$). Not only has the quantity of N within the cycle increased but also the form of N is changing. Unfortunately terrestrial nutrient cycles are inevitably 'leaky', which is particularly so for $NO_3$, while point sources are often major contributors of P. Even when apparently achievable objectives have been decided upon, the buffering mechanisms operating within soil–plant nutrient cycles introduce a strong hysteresis into the recovery. An example of the difficulties that this can introduce has become evident for the Baltic Sea where responses to management action plans were slower than anticipated (Grimvall *et al.*, 2000).

Nitrogen is an abundant element present in atmospheric, terrestrial and aquatic ecosystems and involved in a truly global cycle. The amount of N present in biologically active or available forms is increasing globally. Specific ecological impacts of increased N availability are associated with stimulated primary production under conditions of N limitation or the perceived human health risks associated with excessive dietary intake of nitrate (Table 18.3).

Table 18.3. *The range of issues involving nitrogen and some environmental consequences*

| Issue | Impact | Consequences |
|---|---|---|
| N enrichment of drainage waters | Increased primary productivity of receiving aquatic ecosystems | Reduced biodiversity |
| N enrichment of natural ecosystems | Increased primary productivity | Reduced biodiversity and change in community structure |
| Acidification | Increased soil and/or water acidity | Change in community structure, mobilization of potentially toxic elements |
| Increased nitrate in potable waters | Increased dietary intake of nitrate | Perceived change in human health |
| Increased nitrous oxide emissions | Greenhouse gas | Heat balance of the Earth |

This case study raises a number of general and specific issues that require addressing. Importantly from a potential transport context, N has a readily soluble, mobile species and gaseous component. There are local and regional issues. A single element is involved, which is considered an essential input for high productivity agriculture, and is a by-product of the internal combustion engine.

It is impossible to visualize a situation where artificially produced N will not continue to be utilized within agriculture, thus there must be improvements to the level of efficiency. This really does require a multidisciplined approach to thinking and research.

The dominance of the atmospheric component is immediately apparent where $\sim 99.5\%$ is present as $N_2$. The terrestrial component contains $< 0.02\%$ of the total available N, which is completely unrepresentative of its significance within the human N cycle. Currently the annual human production of reactive N is approximately 0.17 Pg, much of which is used in fertilizer manufacture (Galloway *et al.*, 2003). It is perhaps not immediately apparent why so much attention and concern has therefore been focused on the issue of N but can be explained through the need for a comprehensive understanding of any particular element cycle rather than just absolute amounts. When the annual N cycle is considered, the anthropogenic components become much more significant. Recently the amount of N fixed annually using industrial processes has for the first time exceeded that fixed through natural processes (Vitousek *et al.*, 1997b). The N cycle can now be realistically considered at two organizational levels or scales: (a) the local 'ecosystem' level and (b) a regional and global level. Although closely related and interlinked, the main drivers and issues that arise from the two cycles differ.

## 18.6 The continued investigation of soil processes

A useful distinction between soil processes can be made depending upon whether they predominantly involve abiotic or biotic reactions. For many soil profiles this translates to a spatial separation into a lower 'geochemical' zone dominated by weathering related reactions which transgresses into surface horizons where roots and 'biochemical' processes involving the transformation of organic material dominate (Ugolini *et al.*, 1977). The latter include the anabolic and catabolic processes that directly control carbon and soil organic matter dynamics. The potential significance that processes mediated by individual enzymes can make to the soil carbon balance has been demonstrated by Freeman *et al.* (2001).

The two zones are strongly interlinked within a single soil profile through the movement of soil solution and transfer of substances it contains. One particularly important concept worthy of much wider consideration in this respect is the simple statement that 'the reaction products at one depth in the soil may be the reactants for the next depth' and 'percolating water reaching each soil horizon has a composition determined by the previous path' (Ugolini *et al.*, 1977).

The value of making measurements on soil solution composition for interpreting the significance of individual processes has been discussed in this book. The logistical and analytical issues that still surround obtaining actual samples of soil solution were outlined. This continuing, but rather basic, problem highlights the difficulty of working with a complex, multiphase material such as soil. Despite the tremendous advances (with respect to detection limits and sample volumes required) in end-of-line analytical techniques (such as induced coupled plasma-mass spectroscopy, ICP-MS) these are of little consequence if matching advances in the procedures used for obtaining samples are not forthcoming. The development of novel sampling techniques that offer advantages over existing procedures remains critical, and one promising recently developed technique is diffusive gradients in thin films (DGT; Davison and Zhang, 2005).

Many scientific observations/measurements that are made on soil solution reflect and describe the net balance between opposing processes. One important consideration is the extent to which the current state reflects a balance; that is, is a given soil acting as a source or sink over short or longer timescales? How have various human activities acted to modify this balance? Do these changes have any long-term significance, do they continue and can they be reversed? Within the context for most soils-related research it is important to continue developing an understanding of the extent to which individual processes operate, together with the factors that influence their overall expression. Equally desirable is that the decision makers, and humankind in general, become fully aware that biologically productive soils are a finite resource, so as to arrest what now is appearing as an unstoppable profligacy of this resource.

# Appendix: Naming soils and soil horizons

*Stanley W. Buol*
*Giacomo Certini*
*Riccardo Scalenghe*

## Naming genetic horizons

The genetic horizons are not the equivalent of the diagnostic horizons. Genetic horizons are used in profile descriptions to identify observations of soil profiles *in situ*. Diagnostic horizons are quantitatively defined features that are used to differentiate between taxa in soil classification systems. A diagnostic horizon may encompass several genetic horizons, and the pedogenic features implied by genetic horizon designations may not be sufficient to justify recognition of different diagnostic horizons. In genetic horizon nomenclature capital letters are used to represent the master horizons (Table I).

Where a substantial thickness is present between two master horizons a transitional or combination horizon may be described. Transitional horizons dominated by properties of one master horizon while having subordinate properties of an adjacent master horizon are designated by two master horizon capital letters. The first letter indicates the dominant master horizon characteristics and the second letter indicates the subordinate horizon characteristics. For example an AB horizon indicates a transitional horizon between the A and B horizon, but more like the A horizon than the B horizon. Also a BA horizon is a transitional horizon between the A and B master horizons but more like the B horizon than the A horizon. Other commonly designated transitional horizons include AE, EA, EB, BE, BC, CB, and AC.

Combination horizons are recognized where separate components of two master horizons are recognizable in the horizon and one of the components surrounds the other. Such combination horizons are designated as A/B, B/A, E/B, B/E or B/C. The first symbol designates the material of greatest volume in the horizon.

*Soils: Basic Concepts and Future Challenges*, ed. Giacomo Certini and Riccardo Scalenghe.
Published by Cambridge University Press. 

Table I. *Master horizon designations*

| | |
|---|---|
| O | Organic material dominated; some are saturated with water for long periods, some have never been saturated, and some are artificially drained. Many O horizons consist of undecomposed and partially decomposed litter (leaves, needles, twigs, moss and lichens) deposited on the surface of mineral or organic soils; the mineral fraction is only a minor percentage of the volume, generally less than half. [Soil Taxonomy uses O for any organic soil material while World Reference Base uses O for organic soil material not saturated with water.] |
| H | Organic material dominated; both on top of mineral soils and any depth beneath the surface if it is buried and saturated for long periods. [World Reference Base, only.] |
| L | Limnic horizons or layers include both organic and mineral materials that were either (a) deposited in water by precipitation or through the actions of aquatic organisms, such as algae and diatoms, or (b) derived from underwater and floating aquatic plants and subsequently modified by aquatic animals (include coprogenous earth, sedimentary peat, diatomaceous earth, and marl). [Used only in organic soils; use the following subordinate distinctions co, di, ma; do not use any of the subordinate distinctions of the other master horizons and layers.] [Soil Taxonomy, only.] |
| A | Mineral material dominated; formed at the surface or below an O horizon; accumulated humified organic matter closely mixed with the mineral fraction (not dominated by properties characteristic of E or B horizons) or with clear inheritance from cultivation or pasturing. [Recent alluvial or aeolian deposits that retain fine stratification are not considered to be A horizons unless cultivated.] |
| E | Mineral material dominated; featuring loss of clay, Fe, Al, leaving a concentration of sand and silt particles; differentiated from the underlying B or C horizon by a paler colour (higher value[a] or lower chroma[b] according to the ®Munsell system; Landa, 2004) and/or a coarser texture; contains less organic matter than an overlying A or O horizon. |
| B | Mineral material dominated below an A, E, O or H horizon; featuring illuvial concentration of silicate clay (clay films), Si, Fe, Al, organic matter, or either accumulation or removal of $CaCO_3$ and/or $CaSO_4 \cdot 2H_2O$ or weathering that forms silicate clay or liberates oxides from primary minerals; B horizons are usually redder in hue or higher in chroma according to the ®Munsell system), with granular, blocky or prismatic structure; cemented or not. |
| C | Mineral material dominated (excluding strongly cemented and harder bedrock) underlying an A, O, H, E or B horizon. C horizons are little altered by soil forming processes although they can be penetrated by plant roots; C horizon materials include sediment, saprolite[c], moderately cemented bedrock; either like or unlike the material from which the soil has formed. |
| R | Bedrock (granite, basalt, quartzite, limestone, sandstone etc.) or strongly cemented horizon so hard it is impractical to dig with hand tools. Roots penetrate only in cracks. |
| M | Nearly continuous root-limiting manufactured layers. [Soil Taxonomy, only.] |
| W | Water dominated; Wf if permanently frozen. [Not used for shallow water, ice, or snow above the soil surface.] [Soil Taxonomy, only.] |

[a]Value, lightness or darkness of a colour.
[b]Chroma, the purity of a colour, sometimes called saturation.
[c]Saprolite, soft, friable, isovolumetrically weathered bedrock that retains the fabric and structure of the parent rock, exhibiting extensive intercrystal and intracrystal weathering; any unconsolidated residual material underlying the soil and grading to hard bedrock below.

Table II. *Modifiers of the master horizons in the Soil Taxonomy*

| | |
|---|---|
| a | Highly decomposed organic material (fibre content by volume of less than 17% after rubbing). [Used only with O horizons.] |
| b | Buried horizon with major genetic features developed before burial. [Not used in organic soils, nor is it used to separate an organic layer from a mineral layer.] |
| c | Concretions/nodules of Fe, Al, Mn, Ti. [No silica, dolomite, calcite, or more soluble salts.] |
| co | *Coprogenous earth.* [Used only with L, indicates a limnic layer of coprogenous earth (or sedimentary peat).] |
| d | Physical root restriction, e.g. dense basal till, plough-induced pans, or other mechanically compacted zones. |
| di | *Diatomaceous earth.* [Used only with L, indicates a limnic layer of diatomaceous remnants.] |
| e | Intermediately decomposed organic material (fibre content by volume 17–40% after rubbing). [Used only with O horizons.] |
| f | Permanently frozen soil, rock or water. [Not used for dry or seasonally frozen layers.] |
| ff | Dry permafrost (continually colder than 0 °C without enough ice to be cemented). |
| g | Gleying (reduced Fe and removed or preserved as reduced), greyish colours of 2 or less chroma according to the ®Munsell system). [Not used for C horizons of natural low chroma without history of wetness or E horizons.] |
| h | Illuvial accumulation of amorphous complexes of organic matter and Fe- Al-(hydr)oxides in B horizons. |
| i | Slightly decomposed organic material (fibre content by volume exceeds 40% after rubbing). [Used only with O horizons.] |
| j | Jarosite presence, yellowish colour (yellow hue[a] and high chroma according to the ®Munsell system). |
| jj | Evidence of cryoturbation (irregular broken horizon boundaries, sorted rock fragments, or undecomposed organic material). |
| k | Carbonates accumulation in B and C horizons. |
| m | Cementation or induration (continuous in more than 90% by volume). [Most often used in combination to identify cementing material, i.e km = carbonates, sm =oxides, etc.] |
| ma | *Marl.* [Used only with L, indicates a limnic layer of marl.] |
| n | Accumulation of sodium. |
| o | Residual Fe and Al (hydr-)oxides (sesquioxides) accumulation as a result of silicate weathering. |
| p | Anthropogenic disturbances (agriculture, pasturing, or similar uses). [In A horizons.] |
| q | Silica accumulation. |
| r | Soft bedrock or saprolite (moderately cemented). |
| s | Illuvial Fe- Al-(hydr)oxides accumulations. |
| ss | Slickensides (shear failure, wedge-shaped, polished and grooved ped surfaces produced by one soil mass sliding past another). |
| t | Clay accumulation (ped surface coatings, in pores, as *lamellae*, or bridges). [Used in B and C horizons.] |
| u | Presence of manufactured materials (artefacts) |
| v | Plinthite, firm iron-rich, humus-poor reddish material that hardens in place upon repeated wetting and drying. |
| w | Colour or structure development in B horizons (little or no illuvial accumulation of material). |
| x | Fragipan (firmness and brittleness combination, high bulk density). [In B and C horizons.] |
| y | Gypsum accumulation in B and C horizons. |
| z | Accumulation of salts more soluble than carbonates and gypsum. |

[a] Hue, dominant wavelength (colour) of light. Munsell classified colours into five major hues (red, R, yellow, Y, green, G, blue, B and purple, P), five half hues (e.g. yellow-red, YR) and four gradations of each.

Arabic numerals are used as prefixes to master or transitional horizon designations, i.e. 2B, 3B, etc. to indicate discontinuities (e.g. change in particle-size distribution or mineralogy resulting from geologic processes). Where more consecutive horizons have formed in the same kind of material, the same prefix number is applied (numbering is used only on the second layer and consecutive deeper layers). The prime symbol (′) is added after the master-horizon designation if two or more horizons of the same kind are separated by one or more horizons of a different kind in a pedon. The caret symbol (^) is used as a prefix to master-horizon designation to indicate layers of human-transported material.

Lowercase letters are used as suffixes to modify and designate specific properties within the master horizons (Table II).

It is often desirable to subdivide horizons or layers when sampling or describing a soil. For this purpose, arabic numerals are added to the letters of the horizon designation. For example, Bt1, Bt2, Bt3 identify three subdivisions with slight differences in a master B horizon with clay accumulation. These conventions apply whatever the purpose of the subdivision.

## Diagnostic horizons

Both the World Reference Base (WRB) and Soil Taxonomy (ST) have named and defined soil properties as diagnostic horizons for use in their classification systems. Diagnostic horizons are defined by specific physical, chemical or mineralogical properties, and depth and thickness of occurrence in a soil profile and do not correspond to genetic horizon nomenclature used in profile descriptions. Several diagnostic horizons require sampling and laboratory analyses for identification. Each diagnostic horizon may include properties from only a portion of a horizon identified by a horizon nomenclature symbol or more than one horizon identified by horizon symbols.

## Naming soil: classification systems

Soil classification systems have been developed to correlate experiences on similar soils all over the world. The classification process involves formation of classes by grouping the objects on the basis of their common properties. In general, environments that share comparable soil forming factors produce similar types of soils. Numerous classification systems are in use worldwide. Developments in soil classification have accompanied parallel progress in our understanding of the soil system. Theories behind the classification systems and the purposes for which they were created have changed over time. Most early soil classification systems were based on concepts of soil forming processes. Experience has proven that soil properties are more easily quantifiable than soil processes. Modern systems classify soils based on quantitative characteristics defined as diagnostic horizons, properties and materials. Although Soil Science was born in Russia, where the development of pedology was characterized by intensive accumulation of empirical soil-mapping knowledge and clarification of fundamental issues (genesis, diagnostics, classification and cartography), at present the most used classification systems worldwide are the World Reference Base for Soil Resources (WRB) and Soil Taxonomy.

In 1998, the International Union of Soil Sciences (IUSS) officially adopted the WRB (FAO/ISRIC/ISSS, 1998), strongly influenced by the FAO-Unesco Soil Classification System. Recently, a second edition of the WRB has been published (IUSS Working Group WRB, 2006). To classify a soil correctly the user must follow the Key to Reference Soil Groups (RSG) starting at the beginning of the key and eliminate, one by one, all classes that include criteria that do not fit the soil in question. The soil belongs to the first RSG listed for which it meets all the required criteria (Table III).

Each Reference Soil Group of WRB is provided with a set of prefixes and suffixes, referring to the diagnostic horizons, properties and soil materials of Table IV, as unique qualifiers in a priority sequence. These enable the user to classify individual soil profiles at a second lower categoric level.

Curtis F. Marbut introduced the first formal system of soil classification in the United States in the 1930s. The first version of Soil Taxonomy was introduced to the world in 1960 as the 7th Approximation (Soil Survey Staff, 1960). The current system of Soil Taxonomy has six levels or categories of classification structured hierarchically (Soil Survey Staff, 2006). These levels or categories, from general to specific, are: Orders, Suborders, Great Groups, Subgroups, Families, and Series (Table V).

Like the WRB the key to Soil Orders (Table VI) must be followed sequentially to classify a soil correctly. Similar keys are used for each lower category. Attempting to classify a soil by referring to descriptions at any category level will often lead to frustration and erroneous placement.

Somewhat different diagnostic horizon definitions and the sequences in which the soils key out in the WRB and Soil Taxonomy classification systems make it impossible reliably to cross reference soils in the two systems. However, there are major areas of agreement between the two systems and Table VII is an approximate harmonization of the WRB Reference Soil Groups and the Soil Taxonomy Orders.

Table III. *Abbreviated key to WRB Reference Soil Groups*

| Abbreviated key criteria to Reference Soil Groups[a] | Reference Soil Group |
|---|---|
| Soils with a histic or folic (organic-rich) horizon 0.4 m or more thick | Histosols |
| Other soils profoundly modified by agricultural management | Anthrosols |
| Other soils with 20% or more human made materials | Technosols |
| Other soils having a cryic (perennially frozen) horizon within 1 m | Cryosols |
| Other soils less than 0.25 m to continuous hard rock | Leptosols |
| Other soils with a vertic (expandable clay minerals rich) horizon and more than 30% clay in all parts after mixing top 0.2 m; more than 0.5 m thick and developing cracks during most years | Vertisols |
| Other soils with fluvic material within 0.25 m and continuing to 0.5 m | Fluvisols |
| Other soils with a natric (illuvial clay and exchangeable Na rich) horizon within 1 m | Solonetz |
| Other soils with a salic (soluble salts rich) horizon within 0.5 m and no thionic (extremely acid) horizon | Solonchaks |
| Other soils with a gleyic colour pattern within 0.5 m | Gleysols |
| Other soils with an andic or vitric horizon within 0.25 m | Andosols |
| Other soils with a spodic (illuvial Al-humus complexes rich) horizon within 2 m | Podzols |
| Other soils with a plinthic (Fe-rich, humus-poor highly weathered) horizon within 0.5 m | Plinthosols |
| Other soils with a ferralic horizon at some depth within 1.5 m | Ferralsols |
| Other soils with an abrupt textural change with associated reducing conditions within 1 m | Planosols |
| Other soils with reducing conditions | Stagnosols |
| Other soils with a mollic (dark-coloured, highly base saturated) horizon with moist chroma 2 to a depth of 0.2 m or more and secondary carbonates within 0.5 m below the mollic horizon | Chernozems |
| Other soils with a mollic horizon and secondary carbonate within 0.5 m below the mollic horizon | Kastanozems |
| Other soils with a mollic horizon and base saturation of 50% or higher throughout upper 1 m, but no secondary carbonates | Phaeozems |
| Other soils with a gypsic or petrogypsic (gypsum-rich) horizon within 1 m | Gypsisols |
| Other soils with a duric or petroduric (silica cemented) horizon within 1 m | Durisols |
| Other soils with a calcic or petrocalcic (secondary calcium carbonate rich) horizon within 1 m | Calcisols |
| Other soils with an argic (directly or indirectly enriched in clay) horizon within 1 m and albeluvic tonguing | Albeluvisols |
| Other soils with an argic horizon with a $CEC_7$ of 24 $cmol_c^+ kg^{-1}$ clay or more starting within 1 m and a base saturationof less than 50% in the major part between 0.25 and 0.5 m | Alisols |
| Other soils with a nitic horizon starting within 1 m | Nitisols |
| Other soils with an argic horizon with a $CEC_7$ less than 24 $cmol_c^+ kg^{-1}$ clay and a base saturation less than 50% in the major part between 0.5 and 1m | Acrisols |
| Other soils with an argic horizon with a $CEC_7$ more than 24 $cmol_c^+ kg^{-1}$ clay | Luvisols |
| Other soils with an argic horizon within 1 m | Lixisols |
| Other soils with an umbric (dark-coloured, base-depleted) or mollic horizon | Umbrisols |
| Other soils coarser than sandy loam to at least 1 m | Arenosols |
| Other soils with a cambic horizon starting within 0.5 m and ending beyond 0.25 m | Cambisols |
| Other soils | Regosols |

[a]Surface and subsurface horizons are described in Chapter 2 of this book.

Table IV. *Diagnostic horizons*[a], *properties and soil materials of World Reference Base and approximate Soil Taxonomy equivalent or description*

| World Reference Base diagnostic horizons, properties and materials | Approximate Soil Taxonomy equivalent, or description in parentheses |
|---|---|
| Abrupt textural change | Abrupt textural change |
| Albeluvic tonguing | Glossic horizon and interfingering of albic materials |
| Albic horizon | Albic horizon and albic materials |
| Andic properties | Andic soil properties |
| Anthraquic horizon | Anthric saturation |
| Anthric horizon | (Surface horizons resulting from long-term cultivation) |
| Argic horizon | Argillic horizon |
| Aridic properties | (Surface features resulting from wind) |
| Artefacts | Human-made, -modified or -excavated materials |
| Calcaric material | (More than 2% calcium carbonate equivalent; strong effervescence with 10% HCl in most of the fine earth) |
| Calcic horizon | Calcic horizon |
| Cambic horizon | Cambic horizon |
| Colluvic material | Sediments from human-induced erosion |
| Continuous rock | Lithic contact |
| Cryic horizon | Permafrost (with cryoturbation) |
| Duric horizon | (10% or more silica cemented durinodes) |
| Ferralic horizon | Oxic and kandic horizons |
| Ferralic properties | ($CEC_7$ <24 $cmol_c^+\ kg^{-1}$ clay) |
| Ferric horizon | (Coarse red mottles) |
| Fluvic material | (Stratified flood deposits) |
| Folic horizon | Folistic epipedon |
| Fragic horizon | Fragipan (non-cemented pedogenic horizon, that restricts roots and water movement to cracks) |
| Fulvic horizon | (Dark brown, highly organic andic surface horizon that fails to be a melanic horizon because of colour or melanic index) |
| Geric properties | (ECEC of 1.5 $cmol_c^+\ kg^{-1}$ clay or less) |
| Gleyic colour pattern | Aquic conditions |
| Gypsic horizon | Gypsic horizon |
| Gypsiric material | (5% or more gypsum) |
| Histic horizon | Histic epipedon |
| Hortic horizon | Anthropic epipedon |
| Hydragric horizon | (Redox features resulting from wet cultivation) |
| Irragric horizon | (Unstratified sediments from continous irrigation) |
| Limnic material | Limnic materials |
| Lithological discontinuity | Lithological discontinuities |
| Melanic horizon | Melanic epipedon |
| Mineral material | Mineral soil material |
| Mollic horizon | Mollic epipedon |
| Natric horizon | Natric horizon |
| Nitic horizon | (Argillic horizon with shiny ped faces) |
| Organic material | Organic soil material |
| Ornithogenic material | (Excrements and remnants of birds) |
| Petrocalcic horizon | Petrocalcic horizon |
| Petroduric horizon | Duripan |
| Petrogypsic horizon | Petrogypsic horizon |

Table IV. (Cont.)

| World Reference Base diagnostic horizons, properties and materials | Approximate Soil Taxonomy equivalent, or description in parentheses |
|---|---|
| Petroplinthic horizon | Petroferric contact |
| Pisoplinthic horizon | (Nodules strongly cemented with Fe) |
| Plaggic horizon | Plaggen epipedon |
| Plinthic horizon | Plinthite |
| Reducing conditions | (Features due to long-term lack of oxygen) |
| Salic horizon | Salic horizon |
| Secondary carbonates | Identifiable secondary carbonates |
| Sombric horizon | Sombric horizon |
| Spodic horizon | Spodic horizon and spodic materials |
| Stagnic colour pattern | (pronounced mottling induced by water saturation) |
| Sulfidic material | Sulfidic materials |
| Takyric horizon | (Clayey surface crust in flooded arid soils) |
| Technic hard rock | (consolidated material from industrial processes) |
| Tephric material | (Slightly weathered volcanic tephra) |
| Terric horizon | (Mineral material applied by humans) |
| Thionic horizon | Sulfuric horizon |
| Umbric horizon | Umbric epipedon |
| Vertic horizon and properties | (Slickensides and parallelepiped aggregates) |
| Vitric properties | (5% or more volcanic glass) |
| Voronic horizon | Mollic epipedon |
| Yermic horizon | (Surface layer of gravel, desert pavement) |

[a]Surface and subsurface horizons are described in Chapter 2.

Table V. *Categories of Soil Taxonomy*

| Category | Number of taxa | Nature of differentiating soil properties |
|---|---|---|
| Order | 12 | Presence or absence of major diagnostic horizons, mineralogical properties and extremes of soil temperature and moisture regime |
| Suborder | 64 | Soil moisture regimes and diagnostic horizons |
| Great Group | 319 | Degree of diagnostic horizon expression within each suborder taxa |
| Subgroup | 2468 | Properties that intergrade to taxa in other Orders, Suborders and Great Groups or extragrade to non-soil material |
| Family | | Soil temperature regimes, particle-size and mineralogy of the pedon control section |
| Series | (22 000[a]) | Any consistently identifiable soil property not specifically identified as criteria in a higher category[b] |

[a]Approximate number in USA.
[b]Profile descriptions of each Series at http://soils.usda.gov/ (verified on 23 March 2006).

Table VI. *Simplified key to Soil Orders in Soil Taxonomy*[a]

| Abreviated criteria for each Order | Order |
|---|---|
| Soils with permafrost or gelic material[b] within 1 m | Gelisols |
| Other soils with more than 30% organic matter content to a depth of 0.4 m or more | Histosols |
| Other soils with a spodic horizon within a depth of 2 m | Spodosols |
| Other soils with andic soil properties[c] in one-half or more of the upper 0.6 m | Andisols |
| Other soils with an oxic horizon, or containing more than 40% clay in the surface 0.18 m and a kandic horizon with less than 10% weatherable minerals | Oxisols |
| Other soils containing more than 30% clay in all horizons and cracks that open and close periodically | Vertisols |
| Other soils with some diagnostic subsoil horizon(s) and an aridic soil moisture regime | Aridisols |
| Other soils with an argillic or kandic horizon and a base saturation percentage (at pH 8.2) less than 35 at a depth of 1.8 m | Ultisols |
| Other soils with a mollic epipedon and a base saturation percentage (at pH 7) of 50 or more in all depths above 1.8 m | Mollisols |
| Other soils with an argillic, kandic or natric horizon | Alfisols |
| Other soils with an umbric, mollic, or plaggen epipedon, or a cambic horizon | Inceptisols |
| Other soils | Entisols |

[a]Epipedons and subsurface horizons are described in Chapter 2.
[b]Gelic materials are either mineral or organic and affected by cryoturbation (frost churning, irregular and broken horizons, involutions, organic matter accumulation on top of and within the permafrost, oriented coarse fragments, and silt-enriched layers), ice segregation (ice crystals, ice lenses, vein ice, ice wedges), and/or thermal contraction cracking (filled, or not, with ice or soil material); occur in any horizons of the active (seasonally frozen) layer and/or the upper permafrost. More details in Bockheim *et al.*, 1997.
[c]Andic soil properties result mainly from the presence of significant amounts of Al-humus complexes or short-range-order Al- and Fe-minerals, such as allophane, imogolite, and ferrihydrite.

Table VII. *Approximated cross-harmonization of WRB Reference Soil Groups (rows) and Soil Taxonomy Orders (columns)*
☺ *indicates all or almost all of soils in this WRB Group are in ST Order or vice versa,* 😐 *indicates some soils in this WRB Group are in ST Order or vice versa, while NO SYMBOL indicates soils in this WRB Group not likely to be in ST Order or vice versa.*

| WRB/ST | *Alfisols* | *Andisols* | *Aridisols* | *Entisols* | *Gelisols* | *Histosols* | *Inceptisols* | *Mollisols* | *Spodosols* | *Oxisols* | *Ultisols* | *Vertisols* | ST/WRB |
|---|---|---|---|---|---|---|---|---|---|---|---|---|---|
| ACRISOLS | 😐 | | | | | | | | | 😐 | ☺ | | ACRISOLS |
| ALBELUVISOLS | ☺ | | | | | | | 😐 | | | 😐 | | ALBELUVISOLS |
| ALISOLS | 😐 | | | | | | | | | | ☺ | | ALISOLS |
| ANDOSOLS | | ☺ | | | | | | | | | | | ANDOSOLS |
| ANTHROSOLS | | 😐 | 😐 | ☺ | 😐 | | 😐 | | | 😐 | | | ANTHROSOLS |
| ARENOSOLS | | | | ☺ | | | | | | | | | ARENOSOLS |
| CALCISOLS | 😐 | | ☺ | | | | 😐 | | | | | | CALCISOLS |
| CAMBISOLS | | | 😐 | | | | ☺ | | | | | | CAMBISOLS |
| CHERNOZEMS | | | | | | | | ☺ | | | | | CHERNOZEMS |
| CRYOSOLS | | | | | ☺ | | | | | | | | CRYOSOLS |
| DURISOLS | 😐 | | ☺ | | | | 😐 | | | | 😐 | | DURISOLS |
| FERRALSOLS | 😐 | | | | | | | | | ☺ | 😐 | | FERRALSOLS |
| FLUVISOLS | | | | ☺ | | | | | | | | | FLUVISOLS |
| GLEYSOLS | 😐 | 😐 | | 😐 | | | 😐 | 😐 | | 😐 | 😐 | | GLEYSOLS |
| GYPSISOLS | 😐 | | ☺ | | | | 😐 | | | | | 😐 | GYPSISOLS |
| HISTOSOLS | | | | | 😐 | ☺ | | | | | | | HISTOSOLS |
| KASTANOZEMS | | | | | | | | ☺ | | | | | KASTANOZEMS |
| LEPTOSOLS | | | 😐 | ☺ | | 😐 | | 😐 | | | | | LEPTOSOLS |

| | | | | | | | | | | | | | |
|---|---|---|---|---|---|---|---|---|---|---|---|---|---|
| LIXISOLS | ☺ | | | | | | | | | | 😐 | | LIXISOLS |
| LUVISOLS | ☺ | | | | | | | | | | 😐 | | LUVISOLS |
| NITISOLS | 😐 | | | | | | | | | 😐 | ☺ | | NITISOLS |
| PHAEOZEMS | | | | | | | | ☺ | | | | | PHAEOZEMS |
| PLANOSOLS | 😐 | | | | | | 😐 | 😐 | | | 😐 | | PLANOSOLS |
| PLINTHOSOLS | 😐 | | | | | | | | | 😐 | 😐 | | PLINTHOSOLS |
| PODZOLS | | | | | | | | | ☺ | | | | PODZOLS |
| REGOSOLS | | | | ☺ | | | | | | | | | REGOSOLS |
| SOLONCHAKS | | | ☺ | | | | 😐 | | | | | | SOLONCHAKS |
| SOLONETZ | 😐 | | 😐 | | | | | | | | | | SOLONETZ |
| STAGNOSOLS | 😐 | | | 😐 | | | 😐 | 😐 | | | 😐 | | STAGNOSOLS |
| TECHNOSOLS | | 😐 | 😐 | ☺ | 😐 | | 😐 | | | | | 😐 | TECHNOSOLS |
| UMBRISOLS | | | | | | | ☺ | | | | | | UMBRISOLS |
| VERTISOLS | | | | | | | | | | | | ☺ | VERTISOLS |
| ☝WRB/ST☞ | *Alfisols* | *Andisols* | *Aridisols* | *Entisols* | *Gelisols* | *Histosols* | *Inceptisols* | *Mollisols* | *Spodosols* | *Oxisols* | *Ultisols* | *Vertisols* | ☜ST/WRB☝ |

# References

Agnelli, A., Ugolini, F. C., Corti, G. and Pietramellara, G. (2001). Microbial biomass C and basal respiration of fine earth and highly altered rock fragments of two forest soils. *Soil Biology and Biochemistry*, **33**:613–20.

Alakukku, L., Weisskopf, P., Chamen, W. C. T. *et al.* (2003). Prevention strategies for field traffic-induced subsoil compaction: a review. Part 1. Machine/soil interactions. *Soil and Tillage Research*, **73**:145–60.

Amali, S., Rolston, D. E. and Yamaguchi, T. (1996). Transient multicomponent gas-phase transport of volatile organic chemicals in porous media. *Journal of Environmental Quality*, **158**:106–14.

Amundson, R. (2005a). Eugene Woldemar Hilgard. In Hillel, D. (ed.) *Encyclopedia of Soils in the Environment*, pp. 182–7. Oxford: Elsevier.

Amundson, R. (2005b). Hans Jenny. In Hillel, D. (ed.) *Encyclopedia of Soils in the Environment*, pp. 293–300. Oxford: Elsevier.

Amundson, R. and Jenny, H. (1991). The place of humans in the state factor theory of ecosystems and their soils. *Soil Science*, **151**:99–109.

Amundson, R. and Jenny, H. (1997). On a state factor model of ecosystems. *Bioscience*, **47**:536–43.

Amundson, R. and Yaalon, D. H. (1995). Eugene Woldemar Hilgard and John Wesley Powell: efforts for a joint agricultural and geological survey. *Soil Science Society of America Journal*, **59**:4–13.

Amundson, R., Wang, Y., Chadwick, O. *et al.* (1994). Factors and processes governing the $^{14}C$ content of carbonate in desert soils. *Earth and Planetary Science Letters*, **125**:385–405.

Amundson, R., Guo, Y., and Gong, P. (2003). Soil diversity and land use in the United States. *Ecosystems*, **6**:470–82.

Anderson, J. M. (1975). Succession, diversity and trophic relationships of some soil animals in decomposing leaf litter. *Journal of Animal Ecology*, **44**:475–95.

Anderson, T.-H. (2003). Microbial eco-physiological indicators to assess soil quality. *Agriculture Ecosystems and Environment*, **98**:285–93.

André, H.-M., Noti, M. I. and Lebrun, P. (1994). The soil fauna: the other last biotic frontier. *Biodiversity and Conservation*, **3**:45–56.

Angers, D. A., Recous, S. and Aita, C. (1997). Fate of carbon and nitrogen in water-stable aggregates during decomposition of (CN)-C-13-N-15-labelled wheat straw in situ. *European Journal of Soil Science*, **48**:295–300.

Arnold, R. W. (1983). Concepts of soils and pedology. In Wilding, L. P., Smeck, N. E. and Hall, G. F. (eds.) *Pedogenesis and Soil Taxonomy. I: Concepts and Interactions*, pp. 1–21. Amsterdam: Elsevier.

Arnold, R. W. and Eswaran, H. (2003). Conceptual basis for soil classification: lessons from the past. In Eswaran, H., Ahrens, R. J., Rice, T. J. and Stewart, B. A. (eds.) *Soil Classification: A Global Desk Reference*, pp. 27–41. Boca Raton, FL: CRC Press.

Arnold, R. W., Szabolcs, I. and Targulian, V. O. (eds.) (1990). *Global Soil Change*. Report of an IIASA-UNEP task force on the role of soils in global change, CP-90–2. Laxenburg, Austria: IIASA.

Austin, M. P., Cunningham, R. B., and Fleming, P. M. (1984). New approaches to direct gradient analysis using environmental scalars and statistical curve-fitting procedures. *Vegetatio*, **55**:11–27.

Bailey, S. W. (1980). Structures of layer silicates. In Brindley, G. W. and Brown G. (eds.) *Crystal Structures of Clay Minerals and their X-ray Identification*, pp. 1–124. London: Mineralogical Society.

Baldock, J. A. (2001). Interactions of organic materials and microorganisms with minerals in the stabilization of soil structure. In Huang, P. M., Bollag, J. M. and Senesi, N. (eds.), *Interactions between Soil Particles and Microorganisms: Impact on the Terrestrial Ecosystem*, pp. 85–131. New York: John Wiley.

Baldock, J. A. and Skjemstad, J. O. (2000). Role of the soil matrix and minerals in protecting natural organic materials against biological attack. *Organic Geochemistry*, **31**:697–710.

Balesdent, J. (1996). The significance of organic separates to carbon dynamics and its modelling in some cultivated soils. *European Journal of Soil Science*, **47**:485–94.

Bandibas, J., Vermoesen, A., De Croot, C. J. and Van Cleemput, O. (1994). The effect of different moisture regimes and soil characteristics on nitrous oxide emission and consumption by different soils. *Soil Science*, **158**:106–14.

Belnap, J. (2003). Biological soil crusts in deserts: a short review of their role in soil fertility, stabilization, and water relation. *Algological Studies*, **148**:113–26.

Benoit, P., Barriuso, E., Bergheaud, V. and Etiévant, V. (2000). Binding capacities of different soil size fractions in the formation of herbicide-bound residues. *Agronomie*, **20**:505–12.

Besnard, E., Chenu, C. and Robert, M. (2001). Influence of organic amendments on copper distribution among particle size and density fractions in Champagne vineyard soils. *Environmental Pollution*, **112**:329–37.

Birkeland, P. W. (1984). *Soils and Geomorphology*, 2nd edition. New York: Oxford University Press.

Birkeland, P. W. (1999). *Soils and Geomorphology*. 3rd edition. New York: Oxford University Press.

Blackwood, C. B., Marsh, T., Kim, S. H. and Paul, E. A. (2003). Terminal restriction fragment length polymorphism data analysis for quantitative comparison of microbial communities. *Applied Environmental Microbiology*, **69**:926–32.

Blaser, P., Zysset, M., Zimmermann, S. and Luster, J. (1999). Soil acidification in southern Switzerland between 1987 and 1997: a case study based on the critical load concept. *Environmental Science and Technology*, **33**:2383–9.

Blaser, P., Zimmermann, S., Luster, J. and Shotyk, W. (2000). Critical examination of trace element enrichments and depletions in soils: As, Cr, Cu, Ni, Pb, and Zn in Swiss forest soils. *The Science of the Total Environment*, **249**:257–80.

Bloem, J., Hopkins, D. W. and Benedetti, A. (eds.) (2005). *Microbiological Methods for Assessing Soil Quality*. Wallingford: CABI Publishing.

Blum, W. E. H. (1988). *Problems of Soil Conservation*. Nature and Environment, 40. Strasbourg: Council of Europe.

Blum, W. E. H. (1997). Soil as an open, complex system of biotic and abiotic interactions – energy concept. Extended abstracts of the international symposium on soil system behaviour in time and space. *Communications of the Austrian Soil Science Society*, **55**:13–15.

Bockheim, J. G. and Gennadiyev, A. N. (2000). The role of soil-forming processes in the definition of taxa in Soil Taxonomy and the World Soil Reference Base. *Geoderma*, **95**:53–72.

Bockheim, J. G., Tarnocai, C., Kimble, J. M. and Smith, C. A. S. (1997). The concept of gelic materials in the new Gelisol order for permafrost-affected soils. *Soil Science*, **162**:927–39.

Bockheim, J. G., Gennadiyev, A. N., Hammer, R. D. and Tandarich, J. P. (2005). Historical development of key concepts in pedology. *Geoderma*, **124**:23–36.

Boesten, J. J. T. I. and Van der Linden, A. M. A. (1991). Modeling the influence of sorption and transformation on pesticide leaching and persistence. *Journal of Environmental Quality,* **20**:425–35.

Boettinger, J. L. and Richardson, J. L. (2001). Saline and wet soils of wetlands in dry climates. In Richardson, J. L. and Vepraskas, M. J. (eds.) *Wetland Soils: Genesis, Hydrology, Landscapes, and Classification*, pp. 383–90. Boca Raton, FL: Lewis Publishers.

Bond-Lamberty, B., Wang, C. and Gower, S. T. (2004). A global relationship between the heterotrophic and autotrophic components of soil respiration? *Global Change Biology*, **10**:1756–66.

Booltink, H. W. G. and Bouma, J. (2002). Suction crust infiltrometer and bypass flow. In Dane, J. H. and Topp, G. C. (eds.) *Methods of Soil Analysis. Part 4: Physical Methods*, Book Series 5, pp. 926–37. Madison, WI: Soil Science Society of America.

Boulaine, J. (1989). *Historie des Pedologues et de la Science des Sols*. Paris: Institut Internationale de la Recherche Agronomique.

Bouma, J. (1979). Subsurface applications of sewage effluent. In Beatty M. T., Petersen, G. W. and Swindale, L. D. (eds.) *Planning the Uses and Management of Land*, Agronomy 21, pp. 665–703. Madison, WI: ASA-CSSA-SSSA.

Bouma, J. (1989). Using soil survey data for quantitative land evaluation. In Stewart, B. A. (ed.) *Advances in Soil Science*, **9**:177–211. Berlin: Springer Verlag.

Bouma, J. (1991). Influence of soil macroporosity on environmental quality. *Advances in Agronomy*, **46**:1–37.

Bouma, J. (2000). Land evaluation for landscape units. In Summer, M. E. (ed.) *Handbook of Soil Science*, pp. E393–E412. Boca Raton, FL: CRC Press.

Bouma, J. (2001a). The role of soil science in the land negotiation process. *Soil Use and Management*, **17**:1–6.

Bouma, J. (2001b). The new role of soil science in a network society. *Soil Science*, **166**:874–9.

Bouma, J. and Droogers, P. (1999). Comparing different methods for estimating the soil moisture supply capacity of a soil series subjected to different types of management. *Geoderma*, **92**:185–97.

Bouma, J. and Jones, J. W. (2001). An international collaborative network for agricultural systems applications (ICASA). *Agricultural Systems*, **70**:355–68.

Bouma, J., Stoorvogel, J., van Alphen, B. J. and Booltink, H. W. G. (1999). Pedology, precision agriculture and the changing paradigms of agricultural research. *Soil Science Society of America Journal*, **63**:343–8.

Bouma, J., van Alphen B. J. and Stoorvogel, J. J. (2002). Fine tuning water quality regulations in agriculture to soil differences. *Environmental Science and Policy*, **5**:113–20.

Bowden, W. A. (1986). Gaseous nitrogen emissions from undisturbed terrestial ecosystems. *Biogeochemistry*, **12**:249–79.

Brady, N. C. and Weil, R. R. (eds.) (1996). *The Nature and properties of Soils*. 11th edition. Upper Saddle River, NJ: Prentice-Hall.

Breeuwsma, A., Wosten, J. H. M., Vleeshouwer, J. J., Vanslobbe, A. M. and Bouma, J. (1986). Derivation of land qualities to assess environmental problems from soil surveys. *Soil Science Society of America Journal*. **50**:186–90.

Breland, T. A. and Hansen, S. (1996). Nitrogen mineralization and microbial biomass as affected by soil compaction. *Soil Biology and Biochemistry*, **28**:655–63.

Bridges, E. M. and Batjes, N. H. (1996). Soil gaseous emissions and global climate change. *Geography*, **81**:155–69.

Brimhall, G. H., Chadwick, O. A., Lewis, C. J. *et al.* (1991). Deformational mass transport and invasive processes in soil evolution. *Science*, **255**:695–702.

Brinkman, R. (1970). Ferrolysis, a hydromorphic soil forming process. *Geoderma*, **3**:199–206.

Bryant, R. B. and Macedo, J. (1990). Differential chemoreductive dissolution of iron oxides in a Brazilian Oxisol. *Soil Science Society of America Journal*, **54**:819–21.

Bundt, M., Albrecht, A., Froidevaux, P., Blaser, P. and Flühler, H. (2000). Impact of preferential flow on radionuclide distribution in soil. *Environmental Science and Technology*, **34**:3895–9.

Bundt, M., Krauss, M., Blaser, P. and Wilcke, W. (2001). Forest fertilization with wood ash: effect on the distribution and storage of polycyclic aromatic hydrocarbons (PAHs) and polychlorinated biphenyls (PCBs). *Journal of Environmental Quality*, **30**:1296–304.

Buol, S. W. and Eswaran, H. (2000). Oxisols. *Advances in Agronomy*, **68**:151–95.

Buol, S. W., Hole, F. D., McCracken, R. J. and Southard, R. J. (1997). *Soil Genesis and Classification.* 4th edition. Ames: Iowa State University Press.

Buol, S. W., Southard, R. J., Graham, R. C. and McDaniel, P. A. (2003). *Soil Genesis and Classification.* 5th edition. Ames, IA: Iowa State Press, Blackwell Publishing.

Burdon, J. (2001). Are the traditional concepts on the structures of humic substances realistic? *Soil Science*, **166**:752–69.

Burt, R., Wilson, M. A., Kanyanda, C. W., Spurway, J. K. R. and Metzler, J. D. (2001). Properties and effects of management on selected granitic soils in Zimbabwe. *Geoderma*, **101**:119–41.

Butler, B. E. (1959). *Periodic Phenomena in Landscape as a Basis for Soil Studies.* Australia Soil Publications No. 14. Victoria, Australia: CSIRO Publishing.

Byzov, B. A., Chernjakovskaya, G. M., Zenova, G. M. and Dobrovolskaya, T. G. (1996). Bacterial communities associated with soil diplopods. *Pedobiologia*, **40**:67–79.

Campbell, G. S. (1985). *Soil Physics with BASIC.* Amsterdam: Elsevier.

Canny, M. J. (1981). A universe comes into being when a space is severed: some properties of boundaries in open systems. *Proceedings of Ecological Society of Australia*, **11**:1–11.

Carballas, M., Carballas, T. and Jacquin, F. (1979). Biodegradation and humification of organic matter in humiferous Atlantic soils. *Anales de Edafologa y Agrobiologa*, **38**:1699–717.

Carballas, M., Cabaneiro, A., Guitian Ribeira F., and Carballas, T. (1980). Organo-metallic complexes in Atlantic humiferous soils. *Anales de Edafologa y Agrobiologa*, **39**:1033–43.

Certini, G. (2005). Effects of fire on properties of forest soils. A review. *Oecologia*, **143**:1–10.

Certini, G., Ugolini, F. C., Corti, G. and Agnelli, F. (1998). Early stages of podzolization under Corsican pine (*Pinus nigra* Arn. ssp. *laricio*). *Geoderma*, **83**:103–25.

Certini, G., Corti, G., Agnelli, A. and Sanesi, G. (2003). Carbon dioxide efflux and concentrations in two soils under temperate forests. *Biology and Fertility of Soils*, **37**:39–46.

Chadwick, O. A., Derry, L. A., Vitousek, P. M., Heubert, B. J. and Hedin, L. O. (1999). Changing sources of nutrients during four million years of ecosystem development. *Nature*, **397**:491–7.

Chadwick, O. A., and Graham, R. C. (2000). Pedogenic processes. In Sumner, M. E. (ed.) *Handbook of Soil Science*, pp. E41–E75. Boca Raton, FL: CRC Press.

Chamran, F., Gessler, P. E. and Chadwick, O. A. (2002). Spatially explicit treatment of soil-water dynamics along a semiarid catena. *Soil Science Society of America Journal*, **66**:1571–83.

Chenu, C. (1995). Extracellular polysaccharides: an interface between microorganisms and soil constituents. In Huang, P. M., Berthelin, J., Bollag, J. M., McGill, W. B. and Page, A. L. (eds.), *Environmental Impact of Soil Component Interactions. I: Natural and Anthropogenic Organics*, pp. 217–33. Boca Raton, FL: Lewis.

Chenu, C. and Guérif, J. (1991). Mechanical strength of clay minerals as influenced by an adsorbed polysaccharide. *Soil Science Society of America Journal*, **55**:1076–80.

Chenu, C., Le Bissonnais, Y. and Arrouays, D. (2000). Organic matter influence on clay wettability and soil aggregate stability. *Soil Science Society of America Journal*, **64**:1479–86.

Chesapeake Bay Program (2002). *Chesapeake Bay*. Annapolis, MD. www.chesapeakebay.net. [verified on 15 May, 2005].

Chesworth, W. (1973). The parent rock effect in the genesis of soil. *Geoderma*, **10**:215–25.

Cho, C. M., Burton, D. L. and Chang, C. (1997). Denitrification and fluxes of nitrogenous gases from soil under steady oxygen distribution. *Canadian Journal of Soil Science*, **77**:261–9.

Christensen, B. T. (1996). Carbon in primary and secondary organomineral complexes. *Advances in Soil Science*, **27**:97–165.

Churchman, G. J. (2000). The alteration and formation of soil minerals by weathering. In Sumner, M. E. (ed.) *Handbook of Soil Science*, pp. F3–F76. Boca Raton, FL: CRC Press.

Churchman, G. J., and Burke, C. M. (1991). Properties of subsoils in relation to various measures of surface area and water content. *Journal of Soil Science*, **42**:463–78.

Churchman, G. J. and Tate, K. R. (1986). Aggregation of clays in six New Zealand soil types as measured by disaggregation procedures. *Geoderma*, **37**:207–20.

Churchman, G. J., Skjemstad, J. O. and Oades, J. M. (1993). Influence of clay minerals and organic matter on effect of sodicity on soils. *Australian Journal of Soil Research*, **31**:779–800.

Churchman, G. J., Slade, P. G., Self, P. G. and Janik, L. J. (1994). Nature of interstratified kaolin-smectites in some Australian soils. *Australian Journal of Soil Research*, **32**:805–22.

CIA (2006). *The World Factbook 2005*. Pittsburgh, PA: Central Intelligence Agency. www.cia.gov/cia/publications/factbook/index.html [verified on 20 March 2006].

CIESIN (Center for International Earth Science Information Network)/IFPRI (International Food Policy Research Institute)/World Bank/CIAT (Centro Internacional de Agricultura Tropical) (2004). *Global Rural-Urban Mapping Project (GRUMP)*. Palisades, NY: CIESIN, Columbia University. http://sedac.ciesin.columbia.edu/gpw [verified on 20 March 2006].

Ciolkosz, E. J., Cronce, R. C., Cunningham, R. L. and Petersen, G. W. (1985). Characteristics, genesis, and classification of Pennsylvania minesoils. *Soil Science*, **139**:232–8.

CMS (1984). Report of the Clay Minerals Society Nomenclature Committee for 1982 and 1983. *Clays and Clay Minerals*, **32**:239–40.

Coleman, D. and Fry, B. (eds.) (1991). *Carbon Isotope Techniques*. Isotopic Techniques in Plant, Soil, and Aquatic Biology Series, Vol. 1. New York: Academic Press.

Courchesne, F., Séguin, V. and Dufresne, A. (2000). Solid-phase fractionation of metals in the rhizosphere of forest soils. In Gobran, G. R., Wenzel, W. W. and Lombi, E. (eds.) *Trace Elements in the Rhizosphere*, pp. 189–206. Boca Raton, FL: CRC Press.

Courchesne, F., Kruyts, N. and Legrand, P. (2006). Labile zinc and copper ion activity in the rhizosphere of forest soils. *Environmental Toxicology and Chemistry*, **25**, 635–42.

Crescimanno, G. and De Santis, A. (2004). Bypass flow, salinization and sodication in a cracking clay soil. *Geoderma*, **121**:307–21.

Cronan, C. and Grigal, D. F. (1995). Use of calcium/aluminum ratios as indicators of stress in forest ecosystems. *Journal of Environmental Quality*, **24**:209–26.

Crutzen, P. J. (2002). Geology of mankind: the Anthropocene. *Nature*, **415**:23.

Curry, J. P. (2004). Factors affecting the abundance of earthworms in soils. In Edwards, C. A. (ed.) *Earthworm Ecology*, 2nd Edition, pp. 91–113. Boca Raton, FL: CRC Press.

Dahlgren, R. A. (1993). Comparison of soil solution extraction procedures: effect on solute chemistry. *Communications in Soil Science and Plant Analysis*, **24**:1783–94.

Dahlgren, R. A. (1994). Soil acidification and nitrogen saturation from weathering of ammonium-bearing rock. *Nature*, **368**:838–41.

Dahlgren, R. A. and Marrett, D. J. (1991). Organic carbon sorption in arctic and sub-alpine Spodosol B horizons. *Soil Science Society of America Journal*, **55**:1382–90.

Dahlgren, R. A. and Ugolini, F. C. (1989a). Aluminum fractionation of soil solutions from unperturbed and tephra-treated Spodosols, Cascade Range, Washington, USA. *Soil Science Society of America Journal*, **53**:559–66.

Dahlgren, R. A. and Ugolini, F. C. (1989b). Effects of tephra addition on soil processes in Spodosols in the Cascade Range, Washington, U.S.A. *Geoderma*, **45**:331–55.

Dahlgren, R. A. and Ugolini, F. C. (1991). Distribution and characterization of short-range-order minerals in Spodosols from the Washington Cascades. *Geoderma*, **48**:391–413.

Dahlgren, R. A., Dragoo, J. P. and Ugolini, F. C. (1997a). Weathering of Mt. St. Helens tephra under a cryic-udic climatic regime. *Soil Science Society of America Journal*, **61**:1519–25.

Dahlgren, R. A., Percival, H. J. and Parfitt, R. L. (1997b). Carbon dioxide degassing effects on soil solutions collected by centrifugation. *Soil Science*, **162**:648–55.

Dahlgren, R. A., Singer, M. J. and Huang, X. (1997c). Oak tree and grazing impacts on soil properties and nutrients in a California oak woodland. *Biogeochemistry*, **39**:45–64.

Dahlgren, R. A., Ugolini, F. C. and Casey, W. H. (1999). Field weathering rates of Mt. St. Helens tephra. *Geochimica et Cosmochimica Acta*, **63**:587–98.

Dahlgren, R. A., Saigusa, M. Ugolini, F. C. (2004). The nature, properties and management of volcanic soils. *Advances in Agronomy*, **82**:113–82.

Dalrymple, J. B., Blong, R. J. and Conacher, A. J. (1968). An hypothetical nine-unit landsurface model. *Zeitschrift für Geomorphologie*, **12**:60–76.

Daniels, R. B. and Gamble, E. E. (1967). The edge effect in some Ultisols in the North Carolina Coastal Plain. *Geoderma,* **1**:117–24.

Darwin, C. (1985). *The Origin of Species by Means of Natural Selection or the Preservation of Favoured Races in the Struggle for Life*. London: Penguin Books.

Davison, B. and Zhang, H. (2005). *Diffusion gradients in thin-films (DGT)*. www.dgtresearch.com [verified on 15 May, 2005].

Delvaux, B. and Herbillon, A. J. (1995). Pathways of mixed-layer kaolin-smectite formation in soils. In Churchman, G. J., Fitzpatrick, R. W. and Eggleton, R. A. (eds.) *Clays Controlling the Environment*, pp. 457–61. Melbourne: CSIRO Publishing.

Deppenmeier, U., Müller, V. and Gottschalk, G. (1996). Pathways of energy conservation in methanogenic archaea. *Archive of Microbiology*, **165**:149–63.

Dieffenbach, A. and Matzner, E. (2000). *In situ* soil solution chemistry in the rhizosphere of mature Norway spruce (*Picea abies* [L.] Karst.) trees. *Plant and Soil*, **222**:149–61.

Dilly, O. (2005). Microbial energetics in soil. In Buscot, F. and Varma, A. (eds.) *Microorganisms in soils: roles in genesis and functions*, pp. 123–38. Berlin: Springer Verlag.

Dilly, O. and Blume, H.-P. (1998). Indicators to assess sustainable land use with reference to soil microbiology. *Advances in GeoEcology*, **31**:29–36.

Dilly, O. and Munch, J. C. (1995). Microbial biomass and activities in partly hydromorphic agricultural and forest soils in the Bornhöved Lake region of Northern Germany. *Biology and Fertility of Soils*, **19**:343–47.

Dilly, O., Bartsch, S., Rosenbrock, P., Buscot, F. and Munch, J. C. (2001). Shifts in physiological capabilities of the microbiota during the decomposition of leaf litter in a black alder (*Alnus glutinosa* (Gaertn.) L.) forest. *Soil Biology and Biochemistry*, **33**:921–30.

Dilly, O., Blume, H.-P., Sehy, U., Jimenez, M. and Munch, J. C. (2003). Variation of stabilised, microbial and biologically active carbon and nitrogen in soil under contrasting land use and agricultural management practices. *Chemosphere*, **52**:557–69.

Dilly, O., Gnaß, A. and Pfeiffer, E.-M. (2005). Humus accumulation and microbial activities in Calcari-Epigleyic Fluvisols under grassland and forest diked in for 30 years. *Soil Biology and Biochemistry*, **37**:2163–66.

Dixon, J. B., and Schulze, D. G. (eds.) (2002). *Soil Mineralogy with Environmental Applications*. Madison, WI: Soil Science Society of America.

Dixon, J. B. and Weed, S. B. (eds.) (1989). *Minerals in Soil Environments*, 2nd edition. Madison, WI: Soil Science Society of America.

Dokuchaev, V. (1883). *Russian Chernozem*. Selected Works of V. V. Dokuchaev, Vol 1. Translated by the Israel Program for Scientific Translations (1967), Jerusalem.

Doran, J. W. and Parkin, T. B. (1996). Quantitative indicators of soil quality: a minimum data set. In Doran, J. W. and Jones, A. J. (eds.) *Methods for Assessing Soil Quality*, pp. 25–37. SSSA Spec. Pub. 49. Madison, WI: SSSA.

Driscoll, C. T., Lawrence, G. B., Bulger, A. J. *et al.* (2001). Acidic deposition in the northeastern United States: sources and inputs, ecosystem effects, and management strategies. *BioScience*, **51**:180–98.

EEA (1999). *Environment in the European Union at the Turn of the Century*, chapter 3.6. Environmental Assessment Report 2. Copenhagen: European Environmental Agency. reports.eea.eu.int/92-9157-202-0/en/3.6.pdf [verified on 21 December, 2005].

EEA (2003). *Europe's Environment: The Third Assessment*. Copenhagen: European Environment Agency.

EEA/UNEP (2000). *Down to Earth: Soil degradation and Sustainable Development in Europe – a Challenge for the 21st Century.* Environmental Issue Report 16. Copenhagen: EEA and United Nations Environment Programme Regional Office for Europe.

EEB (2004). *European Soil Protection Policy: time to move from neglect to protection.* Brussels: European Environmental Bureau. www.eeb.org [verified on 20 March 2006].

Egashira, K., Tsuda, T. and Takuma, K. (1985). Relationship between soil properties and the erodibility of Masa soils (granitic soils). *Soil Science and Plant Nutrition*, **31**:105–11.

Eldridge, D. J. (2003). Exploring some relationships between biological soil crusts, soil aggregation and wind erosion. *Journal of Arid Environments*, **3**:457–66.

Ellis, J. R. (1998). Post flood syndrome and vesicular-arbuscular mycorrhizal fungi. *Journal of Production Agriculture*, **11**:200–4.

Emerson, W. W. (1995). Water retention, organic C and soil texture. *Australian Journal of Soil Research*, **33**:241–351.

Eswaran, H., Bin, W. C. (1978). A study of a deep weathering profile on granite in Peninsular Malaysia. Parts I, II and III. *Soil Science Society of America Journal*, **42**:144–58.

Eurogas (2005). *Statistics 2004.* www.eurogas.org/preview/frameset.asp?page=11 [verified on 20 March 2006].

European Commission (2002). Towards a thematic strategy for soil protection. COM (2002)179. Brussels: European Parliament, Economic and Social Committee, and Committee of the Regions.

Fairbridge, R. W. (1968). Mountain and hilly terrain. In Fairbridge, R. W. (ed.) *The Encyclopedia of Geomorphology*, pp. 745–747. New York: Reinhold Book Corporation.

Falloon, P., Smith, P., Coleman, K. and Marshall, S. (1998). Estimating the size of the inert organic matter pool from total soil organic carbon content for use in the Rothamsted carbon model. *Soil Biology and Biochemistry*, **30**:1207–11.

Fanning, D. S. and Fanning, M. C. B. (1989). *Soil Morphology, Genesis, and Classification.* New York: John Wiley.

FAO (2006). *TERRASTAT: Global Land Resources GIS Models and Databases for Poverty and Food Insecurity.* Rome: Food and Agriculture Organization of the United Nations. www.fao.org/ag/agl/agll/terrastat/ [verified on March 20, 2006].

FAO/ISRIC/ISSS (1998). *World Reference Base for Soil Resources.* World Soil Resources, Report 84. Rome: Food and Agriculture Organization of the United Nations, International Soil Reference and Information Centre, and International Society of Soil Science. www.fao.org/ag/agl/agll/wrb/ [verified on 9 March, 2005].

Feller, C. (1997). La matière organique des sols. Questions, concepts et méthodologies. Quelques aspects historiques et état actuel. *Comptes Rendus de l'Académie d'Agriculture de France*, **83**:85–98.

FitzPatrick, E. A. (1956). An indurated soil horizon formed by permafrost. *Journal of Soil Science*, **7**:248–54.

FitzPatrick, E. A. (1980). *Soils: Their Formation, Classification and Distribution.* Reprinted in 1983 and 1991. London: Longman.

FitzPatrick, E. A. (2002). Horizon identification using Excel. In *Proceedings of the 17th World Congress of Soil Science, Bangkok, Thailand, 1983*, pp. 1–5. www.ldd.go.th/Wcss 2002 /papers/1983.pdf [verified on 21 December 2005].

Foissner, W. (1987). Soil Protozoa: fundamental problems, ecology significance, adaptations in ciliates and testaceans, bioindicators, and guide to literature. *Progress in Protistology*, **2**:69–212.

Folland, C. K., Karl, T. R. and Vinnikov, K. Y. A. (1990). Observed climate variations and change. In Houghton, J. T., Jenkins, G. J. and Ephraums, J. J. (eds.) *Climate Change: The IPCC Scientific Assessment*, pp. 195–238. Cambridge: Cambridge University Press.

Frazier, C. S. and Graham, R. C. (2000). Pedogenic transformation of fractured granitic bedrock, southern California. *Soil Science Society of America Journal*, **64**:2057–69.

Freeman, C., Ostle, N. and Kang, H. (2001). An enzymic 'latch' on a global carbon store: a shortage of oxygen locks up carbon in peatlands by restraining a single enzyme. *Nature*, **409**:149.

Fridland, V. M. (1976). *Pattern of the Soil Cover*. Translation of 1972 Russian edition. Jerusalem: Keter Publishing House.

Fukui, Y. and Doskey, P. V. (1996). Technique for measuring nonmethane organic compound emissions from grassland. *Journal of Environmental Quality*, **25**:601–10.

GACGC (1995). *World in Transition: The Threat to Soils*. German Advisory Council on Global Change 1994 Annual Report. Bonn: Economia Verlag.

Galloway, J. N. (1998). The global nitrogen cycle: changes and consequences. *Environmental Pollution*, **102**:15–24.

Galloway, J. N., Aber, J. D., Erisman, J. W. *et al.* (2003). The nitrogen cascade. *BioScience*, **53**:341–56.

Garcia, R. and Millan, E. (1998). Assessment of Cd, Pb and Zn contamination in roadside soils and grasses from Gipuzkoa (Spain). *Chemosphere*, **37**:1615–25.

Garcia Rodeja, E., Macias, Vazquez, F. and Guitian Ojea, F. (1984). Reacción con FNa del suelos de Galicia. II: Suelos sobre rocas granticas. *Anales de Edafologa y Agrobiologia*, **43**:787–807.

Garcia-Talegon, J., Molina, E. and Vicente, M. A. (1994). Nature and characteristics of 1:1 phyllosilicates from weathered granite, central Spain. *Clay Minerals*, **29**:727–34.

Garnier, P., Neel, C., Mary, B. and Lafolie, F. (2001). Evaluation of a nitrogen transport and transformation model in a bare soil. *European Journal of Soil Science*, **52**:253–68.

Gellert, R., Rieder, R., Anderson, R. C. *et al.* (2004). Chemistry of rocks and soils in Gusev Crater from the alpha particle X-ray spectrometer. *Science*, **305**:829–32.

Gerasimov, I. P. and Glazovskaya, M. A. (1965). *Fundamentals of Soil Science and Soil Geography*. Translation of 1960 Russian edition. Jerusalem: Israel Program of Scientific Translations.

German Advisor Council on Global Change (2003). *World in Transition: Towards Sustainable Energy Systems*. London: Earthscan.

Gessler, P. E., Chadwick, O. A., Chamran, F., Althouse, L. and Holmes, K. (2000). Modeling soil-landscape and ecosystem properties using terrain attributes. *Soil Science Society of America Journal*, **64**:2046–56.

Gilkes, R. J., Scholz, G. and Dimmock, G. M. (1973). Lateritic deep weathering of granite. *Journal of Soil Science*, **24**:523–36.

Gillman, G. P. and Bell, L. C. (1976). Surface charge characteristics of six weathered soils from tropical North Queensland. *Australian Journal of Soil Research*, **14**:351–60.

Gillman, G. P. and Sumpter, E. A. (1986). Surface charge characteristics and lime requirements of soils derived from basaltic, granitic and metamorphic rocks in high-rainfall tropical Queensland. *Australian Journal of Soil Research*, **24**:173–92.

Glentworth, R. and Muir, J. W. (1963). *The Soils of the Country Round Aberdeen, Inverurie and Fraserburgh.* Memoirs of the Soil Survey of Great Britain. Edinburgh: HMSO.

Golchin, A. (1994). Soil structure and carbon cycling. *Australian Journal of Soil Research*, **32**:1043–68.

Golchin, A., Baldock, J. A. and Oades, J. M. (1998). A model linking organic matter decomposition, chemistry, and aggregate dynamics. In Lal, R., Kimble, J. M., Follett, R. F. and Stewart, B. A. (eds.), *Soil Processes and the Carbon Cycle*, pp. 245–66. Advances in Soil Science. Boca Raton, FL: CRC Press.

Goss, J. C. and Phillips, F. M. (2001). Terrestrial in situ cosmogenic nuclides: theory and application. *Quaternary Science Reviews*, **20**:1475–560.

Gosz, J. R., Likens, G. E. and Bormann, F. H. (1973). Nutrient release from decomposing leaf and branch litter in the Hubbard Brook Forest, New Hampshire. *Ecological Monograph*, **43**:173–91.

Graham, R. C. and Buol, S. W. (1990). Soil-geomorphic relations on the Blue Ridge Front. II: Soil characteristics and pedogenesis. *Soil Science Society of America Journal*, **54**:1367–77.

Graham, R. C., Schoenberger, P. J., Anderson, M. A., Sternberg, P. D. and Tice, K. R. (1997). Morphology, porosity and hydraulic conductivity of weathered granitic bedrock and overlying soils. *Soil Science Society of America Journal*, **61**:516–22.

Green, P. A., Vörösmarty, C. J., Meybeck, M., Galloway, J. N., Peterson, B. J. and Boyer, E. W. (2004). Pre-industrial and contemporary fluxes of nitrogen through rivers: a global assessment based on typology. *Biogeochemistry*, **68**:71–105.

Green, T. R., Ahuja, L. R. and Benjamin, J. G. (2003). Advances and challenges in predicting agricultural management effects on soil hydraulic properties. *Geoderma*, **116**:3–27.

Gregorich, E. G., Carter, M. R., Angers, D. A., Monreal, C. M. and Ellert, B. H. (1995). Towards a minimum data set to assess soil organic matter quality in agricultural soils. *Canadian Journal of Soil Science*, **74**:367–85.

Grimvall, A., Stålnacke, P. and Tonderski, A. (2000). Time scales of nutrient losses from land to sea: a European perspective. *Ecological Engineering*, **14**:363–71.

Hagedorn, F., Schleppi, P., Mohn, J., Bucher, J. B. and Flühler, H. (2001). Retention and leaching of elevated N deposition in a forest ecosystem with Gleysols. *Water, Air, and Soil Pollution*, **129**:119–42.

Hagedorn, F., Blaser, P. and Siegwolf, R. (2002). Elevated atmospheric $CO_2$ and increased N deposition effects on dissolved organic carbon-clues from $\delta^{13}C$ signature. *Soil Biology and Biochemistry*, **34**:355–66.

Hagedorn, F., Spinnler, D., Bundt, M., Blaser, P. and Siegwolf, R. (2003). The input and fate of new C in two forest soils under elevated $CO_2$. *Global Change Biology*, **9**:862–72.

Hahn, F. (2004). *Künstliche Beschneiung im Alpenraum: Ein Hintergrundbericht.* Schaan, Liechtenstein: CIPRA International. http://www.alpmedia.net/pdf/Dossier_Kunstschnee_D.pdf [verified on 24 March 2006].

Hajabbasi, M. A., Jalalian, A. and Karimzadeh, H. R. (1997). Deforestation effects on soil physical and chemical properties, Lordegan, Iran. *Plant and Soil*, **190**:301–8.

Hamdan, J., Ruhana, B. and McRae, S. G. (2000). Characteristics of a regolith developed on basalt in Pahang, Malaysia. *Communications in Soil Science and Plant Analysis*, **31**:981–93.

Hanson, R. S. and Hanson, T. E. (1996). Methanotrophic bacteria. *Microbiological Reviews*, **60**:439–71.

Hattori, T. (1992). Distribution and movement of Protozoa within and among soil aggregates. *Bulletin of Japanese Society of Microbial Ecology*, **7**:69–74.

Heimsath, A. M., Dietrich, W. E., Nishiizumi, I. and Finkel, R. C. (1997). The soil production function and landscape equilibrium. *Nature*, **388**:358–61.

Heimsath, A. M., Dietrich, W. E., Nishiizumi, K. and Finkel, R. C. (1999). Cosmogenic nuclides, topography, and the spatial variation of soil depth. *Geomorphology*, **27**:151–72.

Helms, D. (2002). Early Leaders of the Soil Survey. In Helms, D., Effland, A. B. W. and Durana, P. J. (eds.), *Profiles in the History of the U.S. Soil Survey*, pp. 19–64. Ames: Iowa State Press.

Hillel, D. (1998). *Environmental Soil Physics*. San Diego, CA: Academic Press.

Hinkelman, E. G. (2005). *Dictionary of International Trade*. Novato, CA: World Trade Press.

Hinsinger, P. (1998). How do plant roots acquire mineral nutrients? Chemical processes involved in the rhizosphere. *Advances in Agronomy*, **64**:225–65.

Hirano, Y., Zimmermann, S., Graf Pannatier, E. and Brunner, I. (2004). Induction of callose in roots of Norway spruce seedlings after short-term exposure to Al. *Tree Physiology*, **24**:1270–83.

Holmgren, G. G. S. (1988). The point representation of soil. *Soil Science Society of America Journal*, **52**:712–16.

Houghton, R. A., Callander, B. A. and Varney, S. K. (1992). *Climate Change*. Cambridge: Cambridge University Press.

Hunckler, R. V. and Schaetzl, R. J. (1997). Spodosol development as affected by geomorphic aspect, Baraga County, Michigan. *Soil Science Society of America Journal*, **61**:1105–15.

Hutchinson, G. E. (ed.) (1970). *Ianula: An Account of the History and Development of the Lago di Monterosi, Latium, Italy*. Transactions of the American Philosophical Society, 60. Philadelphia PA: APS.

Hyvönen, R. and Huhta, V. (1989). Effects of lime, ash and nitrogen fertilizers on nematode populations in Scots pine forest soils. *Pedobiologia*, **33**:129–43.

Ingersoll, R. B., Inman, R. E. and Fisher, W. R. (1974). Soils potential as a sink for atmospheric carbon monoxide. *Tellus*, **26**:151–8.

IRF (2003). *World Road Statistics*. Geneva: International Road Federation.

Irmler, U. (2000). Changes in the fauna and its contribution to mass loss and N release during leaf litter decomposition in two deciduous forests. *Pedobiologia*, **44**:105–18.

Irmler, U., Wachendorf, C. and Blume, H.-P. (1997). Landscape pattern of humus forms and their soil biology. *Recent Research Developments in Soil Biology and Biochemistry*, **1**:93–108.

Isbell, R. F., Stephenson, P. J., Murtha, G. G. and Gillman, G. P. (1976). *Red Basaltic Soils in North Queensland*. I: *Environment, Morphology, Particle Size Characteristics and Clay Mineralogy*. II: *Chemistry*. Division of Soils, Technical Paper 28. Adelaide: CSIRO.

Isbell, R. F., Gillman, G. P., Murtha, G. G. and Jones, P. N. (1977). *Brown Basaltic Soils in North Queensland*. Division of Soils, Technical Paper 34. Adelaide: CSIRO.

ISRIC (1990). *GLASOD. Global Assessment of Human Induced Soil Degradation*. Wageningen: International Soil Reference and Information Center.

IUSS Working Group WRB (2006). *World Reference base for Soil Resources* (2006). 2nd edition. World Soil Resources Reports No.103. Rome: FAO.

Jamagne, M. and King, D. (2003). The current French approach to a soilscape typology. In Eswaran, H., Rice, T., Ahrens, R. and Stewart, B. A. (eds.) *Soil Classification. A Global Desk Reference*, pp. 158–78. Boca Raton, FL: CRC Press.

Jaworski, N. A., Howarth, R. W. and Hetling, L. J. (1997). Atmospheric deposition of nitrogen oxides onto the landscape contributes to coastal eutrophication in the Northeast United States. *Environmental Science and Technology*, **31**:1995–2005.

Jaynes, W. F., Bigham, J. M., Smeck, N. E. and Shipitalo, M. J. (1989). Interstratified 1:1–2:1 mineral formation in a polygenetic soil from southern Ohio. *Soil Science Society of America Journal*, **53**:1888–94.

Jenkinson, D. S. and Rayner, J. H. (1977). The turnover of soil organic matter in some of the Rothamsted classical experiments. *Soil Science*, **123**:298–305.

Jenny, H. (1941). *Factors of Soil Formation. A System of Quantitative Pedology*. New York: McGraw Hill.

Jenny, H. (1958). Role of the plant factor in the pedogenic functions. *Ecology*, **39**:5–16.

Jenny, H. (1961). *E.W. Hilgard and the Birth of Modern Soil Science*. Pisa: Collana della Rivista 'Agrochimica'.

Jenny, H. (1980). *The Soil Resource*. New York: Springer-Verlag.

Jenny, H. (1989). *Hans Jenny: Soil Scientist, Teacher, and Scholar. Interviews by D. Haher and K. Stuart*. University History Series. Berkeley, CA: Regional Oral History Office, The Bancroft Library, University of California.

Jobbagy, E. G. and Jackson, R. B. (2004). The uplift of soil nutrients by plants: biogeochemical consequences across scales. *Ecology*, **85**:2380–9.

Johnson, D. L. and Watson-Stegner, D. (1987). Evolution model of pedogenesis. *Soil Science*, **143**:349–66.

Johnson, D. W. and Curtis, P. S. (2001). Effects of forest management on soil C and N storage: meta analysis. *Forest Ecology and Management*, **140**:227–38.

Kay, G. F. (1916). Gumbotil, a new term for Pleistocene geology. *Science*, **44**:637–8.

Keller, L. P. and McKay, D. S. (1997). The nature and origin of rims on lunar soil grains. *Geochimica et Cosmochimica Acta*, **61**:2331–41.

Kelly, E. F., Chadwick, O. A. and Hilinski, T. E. (1998). The effects of plants on mineral weathering. *Biogeochemistry*, **42**:21–53.

Khalil, M. A. K., Rasmussen, R. A. and Shearer, M. J. (1998). Effects of production and oxidation processes on methane emissions from rice fields. *Journal of Geophysical Research*, **103**:233–9.

Kittrick, J. A. (1967). Gibbsite-kaolinite equilibria. *Soil Science Society of America Proceedings*, **31**:314–16.

Kristufek, V., Ravasz, K. and Pizl, V. (1992). Changes in densities of bacteria and microfungi during gut transit in *Lumbricus rubellus* and *Aporrectodea caliginosa* (Oligochaeta: Lumbricidae). *Soil Biology and Biochemistry*, **24**:1499–500.

Kropff, M., Jones, J. and van Laar, G. (2001). Advances in systems approaches for agricultural development. *Agricultural Systems*, **70**:353–4.

Krupenikov, I. A. (1992). *History of Soil Science*. Translation of 1981 Russian edition, TT89-7-0184. New Delhi: Amerind Publishing Corporation.

Kruse, C. W., Moldrup, P. and Iversen, N. (1996). Atmosperic methane diffusion and consumption in a forest soil. *Soil Science*, **161**:355–65.

Kuhn, T. S. (1962). *The Structure of scientific Revolutions. Chicago, IL: University of Chicago Press.*

Kuyvenhoven, A., Bouma, J. and van Keulen, H. (1998). Policy analysis for sustainable land use and food security. *Agricultural Systems*, **58**:281–481.

Lal, R. (2004). Soil carbon sequestration impacts on global climate change and food security. *Science*, **304**:1623–7.

Landa, E. R. (2004). Albert H. Munsell: a sense of color at the interface of art and science. *Soil Science*, **169**:83–9.

Lauren, A. (1997). Physical properties of the mor layer in a Scots pine stand. II: Air permeability. *Canadian Journal of Soil Science*, **77**:635–42.

Lay, M. G. (1992).*Ways of the World.* New Brunswick, NJ: Rutgers University Press.

Lee, B. D., Graham, R. C., Laurent, T. E. and Amrhein, C. (2004). Pedogenesis in a wetland meadow and surrounding serpentinitic landslide terrain, northern California, USA. *Geoderma*, **118**:303–20.

Legrand, P., Turmel, M.-C., Sauvé, S. and Courchesne, F. (2005). Speciation and bioavailability of trace metals (Cd, Cu, Ni, Pb, Zn) in the rhizosphere of contaminated soils. In Huang, P. M. and Gobran, G. R. (eds.) *Biogeochemistry of Trace Elements in the Rhizosphere*, pp. 261–99. New York: Elsevier.

Letey, J. (1985). Relationship between soil physical properties and crop production. *Advances in Soil Science*, **1**:277–94.

Lin, H., Bouma, J., Wilding, L., Richardson, J., Kutilek, M. and Nielsen, D. R. (2004). Advances in hydropedology. *Advances in Agronomy*, **85**:1–89.

Lindsay, W. L. (1979). *Chemical Equilibria in Soils*. New York: John Wiley.

Lipiec, J., Arvidsson, J. and Murer, E. (2003). Review of modeling crop growth, movement of water and chemicals in relation to topsoil and subsoil compaction. *Soil and Tillage Research*, **73**:15–29.

Logsdon, S. D. and Karlen, D. L. (2004). Bulk density as a soil quality indicator during conversion to no-tillage. *Soil and Tillage Research*, **78**:143–9.

Mamilov, A. S. and Dilly, O. (2002). Soil microbial eco-physiology as affected by short-term variations in environmental conditions. *Soil Biology and Biochemistry*, **34**:1283–90.

Manna, S., Courchesne, F., Roy, A. G., Turmel, M.-C. and Côté, B. 2002. Trace metals in the forest floor, litterfall patterns and topography. Proceedings of the 17th World Congress of Soil Science. Bangkok, Thailand, **1306**:1–9. www.ldd.go.th/Wcss 2002/ papers/1306.pdf [verified on 21 December 2005].

Maraun, M., Alphei, J., Bonkowski, M. *et al.* (1999). Middens of the earthworm *Lumbricus terrestris* (Lumbricidae): microhabitats for micro- and mesofauna in forest soil. *Pedobiologia*, **43**:276–86.

Markewitz, D. (1997). Soil without life? *Nature*, **389**:435.

Marschner, H. (1995). *The Mineral Nutrition in Higher Plants*. London: Academic Press.

Marschner, H. and Römheld, V. (1996). Root-induced changes in the availability of micronutrients in the rhizosphere. In Waisel, Y., Eshel, A. and Kafkafi, U. (eds.) *Plant Roots: the Hidden Half*, pp. 557–79. New York: Marcel Dekker.

Martin, R. R., Naftel, S., Macfie, S., Skinner, W., Courchesne, F. and Séguin, V. (2004). Time of flight secondary ion mass spectrometry studies on the distribution of metals between the soil, rhizosphere and roots of *Populus tremuloïdes* Minchx growing in forest soil. *Chemosphere*, **54**:1121–5.

Mautner, M. (1997). Biological potential of extraterrestrial materials. 1: Nutrients in carbonaceous meteorites and effects on biological growth. *Planetary and Space Science*, **45**:653–64.

McCarthy, P. (2001). The principles of humic substances. *Soil Science*, **166**:738–51.

McCoy, B. J., and Rolston, D. E. (1992). Convective transport of gases in moist porous media. *Environmental Science and Technology*, **26**:2468–76.

McDaniel, P. A. and Buol, S. W. (1991). Manganese distributions in acid soils of the North Carolina Piedmont. *Soil Science Society of America Journal*, **55**:152–8.

McMartin, I., Henderson, P. J., Plouffe, A. and Knight, R. D. (2002). Comparison of Cu-Hg-Ni-Pb concentration in soils adjacent to anthropogenic point sources: examples from four Canadian sites. *Geochemistry: Exploration, Environment, Analysis*, **2**:57–73.

McNabb, D. H., Stratsev, A. D. and Nguyen, H. (2001). Soil wetness and traffic level effects on bulk density and air-filled porosity of compacted boreal forest soils. *Soil Science Society of America Journal*, **65**:1238–47.

Milne, G. (1935). Some suggested units of classification and mapping, particularly for East African Soils. *Soil Research*, **4**:183–198.

Ming, D. W. and Henninger, D. L. (eds.) (1989). *Lunar Base Agriculture: Soils for Plant Growth.* Madison, WI: ASA, CSSA, and SSSA.

Mitchell, B. D. and Jarvis, R. A. (1956). *The Soils of the Country Round Kilmarnock.* Memoirs of the Soil Survey of Great Britain, Scotland. Edinburgh: HMSO.

Mokma, D. L. and Evans, C. V. (2000). Spodosols. In Sumner, M. E. (ed.) *Handbook of Soil Science*, pp. E307–E321. Boca Raton, FL: CRC Press.

Moody, L. E. and Graham, R. C. (1997). Silica-cemented terrace edges, central California coast. *Soil Science Society of America Journal*, **61**:1723–1729.

Moore, I. D., Gessler, P. E., Nielsen, G. A. and Peterson, G. A. (1993). Soil attribute prediction using terrain analysis. *Soil Science Society of America Journal*, **57**:443–452.

Moore, J. N. and Luoma, S. N. (1990). Hazardous wastes from large-scale metal extraction: a case study. *Environmental Scicnce and Technology*, **24**: 1278–85.

Morris, R. V., Klingelhöfer, G., Bernhardt, B. *et al.* (2004). Mineralogy at Gusev Crater from the Mössbauer Spectrometer on the Spirit Rover. *Science*, **305**:833–6.

Murase, J. and Kimura, M. (1996). Methane production and its fate in paddy fields. *Soil Science and Plant Nutrition*, **42**:187–90.

Murty, D., Kirschbaum, M. U. F., McMurtrie, R. E. and McGilvray, H. (2002). Does conversion of forest to agricultural land change soil carbon and nitrogen? A review of the literature. *Global Change Biology*, **8**:105–23.

Nevo, E., Travleev, A. P., Belova, N. A. *et al.* (1998). Edaphic interslope and valley bottom differences at 'Evolution Canyon', Lower Nahal Oren, Mount Carmel, Israel. *Catena*, **33**:241–54.

Newnham, R. M., Lowe, D. J. and Williams, P. W. (1999). Quaternary environmental change in New Zealand: a review. *Progress in Physical Geography*, **23**:567–610.

Nockolds, S. R. (1954). Average chemical composition of some igneous rocks. *Bulletin of the Geological Society of America*, **65**:1007–32.

Nolan, B. T., Ruddy, B. C., Hitt, K. J. and Helsel, D. R. (1998). *Nitrate in Ground Waters of the United States.* Proceedings of the National Water Quality Monitoring Council. Reno, NV: NWQMC.

Nordt, L. C., Wilding, L. P., Hallmark, C. T. and Jacob, J. S. (1996). Stable carbon isotope composition of pedogenic carbonates and their use in studying pedogeneis. In Boutton, T. W. and Yamasaki, S. (eds.) *Mass Spectrometry of Soils*, pp. 133–54. New York: Marcel-Dekker.

Norman, J. M., Kucharik, C. J., Gower, S. T. *et al.* (1997). A comparison of six methods for measuring soil-surface carbon dioxide fluxes. *Journal of Geophysical Research*, **102**:28771–7.

Norrish, K. and Rosser, H. (1983). Mineral phosphate. In *Soils: An Australian Viewpoint*, pp. 335–61. Melbourne: CSIRO and London: Academic Press.

Norrish, K. and Pickering, J. G. (1983). Clay minerals. In *Soils: An Australian Viewpoint*, pp. 281–308. Melbourne: CSIRO and London: Academic Press.

NRCS (1996). *Soil Survey Laboratory Methods Manual.* Natural Resources Conservation Service Soil Survey Investigations, Report No. 42. Washington, DC: U.S. Govt. Printing Office.

Nyborg, M., Laidlaw, J. W., Solberg, E. D. and Malhi, S. S. (1997). Denitrification and nitrous oxide emissions from a black chernozemic soil during spring thaw in Alberta. *Canadian Journal of Soil Science*, **77**:153–60.

Olowolafe, E. A. (2002). Soil parent materials and properties in two separate catchment areas on the Jos Plateau, Nigeria. *GeoJournal*, **56**:210–12.

Olson, C. G., Thompson, M. L. and Wilson, M. A. (2000). Phyllosilicates. In Sumner, M. E. (ed.) *Handbook of Soil Science*, pp. F77–F123. Boca Raton, FL: CRC Press.

Oxford Dictionary (1966). *The Oxford Dictionary of English Etymology*. Friedrichsen, G. W. S. and Bunchfield, R. W. (eds.). London: Oxford University Press.

Ozima, M., Seki, K., Terada, N., Miura, Y. N., Podosek, F. A. and Shinagawa, H. (2005). Terrestrial nitrogen and noble gases in lunar soils. *Nature*, **436**:655–9.

Packer, I. J. and Hamilton, G. J. (1993). Soil physical and chemical changes due to tillage and their implications for erosion and productivity. *Soil and Tillage Research*, **27**:327–39.

Palmer, S. M. and Driscoll, C. T. (2002). Acidic deposition: decline in mobilization of toxic aluminium. *Nature*, **417**:242–3.

Paquet, H. and Millot, G. (1972). Geochemical evolution of clay minerals in the weathered products in soils of Mediterranean climate. In Serratosa, J. M. (ed.) *Proceedings of the International Clay Conference*, pp. 199–206. Madrid: Tipografia Artistica.

Parfitt, R. L., Russell, M. and Orbell, G. E. (1983). Weathering sequence of soils from volcanic ash involving allophane and halloysite, New Zealand. *Geoderma*, **29**:41–57.

Paton, T. R., Humphreys, G. S. and Mitchell, P. B. (1995). *Soils: A New Global View*. London: Yale University Press.

Paul, E. A. and Clark, F. E. (1989). *Soil Microbiology and Biochemistry*. New York: Academic Press.

Petersen, L. V., El-Farhan, Y. H. Moldrup, P. *et al.* (1996). Transient diffusion, adsorption and emission of volatile organic vapors in soils with fluctuating low water contents. *Journal of Environmental Quality*, **25**:1054–63.

Piccolo, A. (2002). The supramolecular structure of humic substances: a novel understanding of humus chemistry and implications in soil science. *Advances in Agronomy*, **75**:57–134.

Pierzynski, G. M., Sims, T. and Vance, G. F. (2000). *Soils and Environmental Quality*, 2nd edition. Boca Raton, FL: CRC Press.

Pillans, B. (1998). *Regolith Dating Methods. A Guide to Numerical Dating Techniques.* Perth, Australia: Cooperative Research Centre for Landscape Evolution and Mineral Exploration.

Pillans, B. (2004). Geochronology of the Australian Regolith. *Geochronology*, **168**:117–30.

Pimentel, D., Harvey, C., Resosudarmo, P. *et al.* (1995). Environmental and economic costs of soil erosion and conservation benefits. *Science*, **267**:1117–23.

Plante, A. F. and McGill, W. B. (2002). Soil aggregate dynamics and the retention of organic matter in laboratory-incubated soil with differing simulated tillage frequencies. *Soil and Tillage Research*, **66**:79–92.

Poole, D. K. and Miller, P. C. (1975). Water relations of selected species of chaparral and coastal sage communities. *Ecology*, **56**:1118–28.

Post, W. M., Emmanuel, W. R., Zinke, P. J. and Stangenberger, A. G. (1982). Soil carbon pools and world life zones. *Nature*, **298**:156–9.

Post, W. M., Pastor, J., Zinke, P. J. and Stangenberger, A. G. (1985). Global patterns of soil nitrogen storage. *Nature*, **317**:613–16.

Postel, S. L., Daily, G. C. and Ehrlich, P. R. (1996). Human appropriation of renewable freshwater. *Science*, **271**:785–8.

Poulet, F., Bibring, J.-P., Mustard, J. F. *et al.* (2005). Phyllosilicates on Mars and implications for early Martian climate. *Nature*, **438**:623–7.

Preston, C., Newman, R. H. and Rother, P. (1994). Using $^{13}C$ CPMAS NMR to assess effects of cultivation on the organic matter of particle size fractions in a grassland soil. *Soil Science*, **157**:20–35.

Puget, P., Chenu, C. and Balesdent, J. (1995). Total and young organic carbon distributions in aggregates of silty cultivated soils. *European Journal of Soil Science*, **46**:449–59.

Pulleman, M. M., Bouma, J., van Essen, E. A. and Meijles, E. W. (2000). Soil organic matter content as a function of different land use history. *Soil Science Society of America Journal*, **64**:689–94.

Rebertus, R. A. and Buol, S. W. (1985). Intermittency of illuviation in Dystrochrepts and Hapludults from the Piedmont and Blue Ridge provinces of North Carolina. *Geoderma*, **36**:277–91.

Reid-Soukup, D. A. and Ulery, A. L. (2002). Smectites. In Dixon, J. B. and Schulze, D. G. (eds.) *Soil Mineralogy with Environmental Applications*, pp. 467–99. Madison, WI: Soil Science Society of America.

Reimann, C., Banks, D. and Kashulina, G. (2000). Processes influencing the chemical composition of the O-horizon of podzols along a 500-km north-south profile from the coast of the Barents Sea to the Arctic Circle. *Geoderma*, **95**:113–39.

Rengasamy, P. and Sumner, M. E. (1998). Processes involved in sodic behaviour. In Sumner M. E. and Naidu, R. (eds.) *Sodic Soils*, pp. 35–50. New York: Oxford University Press.

Riesen, T. K., Zimmermann, S. and Blaser, P. (1999). Spatial distribution of $^{137}Cs$ in forest soils of Switzerland. *Water, Air, and Soil Pollution*, **114**:277–85.

Righi, D., Bravard, S., Chauvel, A., Ranger, J. and Robert, M. (1990). In situ study of soil processes in an Oxisol-Spodosol sequence of Amazonia (Brazil). *Soil Science*, **150**:438–45.

Ritter, E., Vesterdal, L. and Gundersen, P. (2003). Changes in soil properties after afforestation of former intensively managed soils with oak and Norway spruce. *Plant and Soil*, **249**:319–30.

Robarts, R. and Wetzel, R. (2000). *The Global Water and Nitrogen Cycles.* www.globalchange.umich.edu/globalchange1/current/lectures/kling/water_nitro/water_and_nitrogen_cycles.htm [verified on 17 May 2005].

Robertson, I. D. M. and Eggleton, R. A. (1991). Weathering of granitic muscovite to kaolinite and halloysite and of plagioclase-derived kaolinite to halloysite. *Clays and Clay Minerals*, **39**:113–26.

Rojstaczer, S., Sterling, S. M. and Moore, N. J. (2001). Human appropriation of photosynthesis products. *Science*, **294**:2549–52.

Rudolf, J., Rothfuss, F. and Conrad, R. (1996). Flux between soil and atmosphere, vertical concentration profiles in soil, and turnover of nitric oxide. *Journal of Atmospheric Chemistry*, **23**:253–73.

Ruf, M. and Brunner, I. (2003). Vitality of tree fine roots: reevaluation of the tetrazolium test. *Tree Physiology*, **23**:257–63.

Ruhe, R. V. (1975). *Geomorphology: Geomorphic Processes and Surficial Geology.* Boston, MA: Houghton Mifflin Company.

Ryszkowski, L. (1975). Energy and matter economy of ecosystems. In Van Dobben, W. H. and Lowe-McConnel, R. H. (eds.) *Unifying Concepts in Ecology*. The Hague: Junk.

Sandaa, R. A., Torsvik, V. and Enger, O. (2001). Influence of long-term heavy-metal contamination on microbial communities in soil. *Soil Biology and Biochemistry*, **33**:287–95.

Sanderson, E. W., Jaiteh, M., Levy, M. A., Redford, K. H., Wannebo, A. V. and Woolmer, G. (2002). The human footprint and the last of the wild. *Bioscience*, **52**:891–904.

Sase, T. and Hosono, M. (1996). Vegetation histories of Holocene volcanic ash soils in Japan and New Zealand: relationship between genesis of melanic volcanic ash soils and human impact. *Earth Science (Chikyu Kagaku)*, **50**:466–82.

Scalenghe, R., Certini, G., Corti, G., Zanini, E. and Ugolini, F. C. (2004). Segregated ice and liquefaction effects on compaction of fragipans. *Soil Science Society of America Journal*, **68**:204–14.

Schaefer, M. and Schauermann, J. (1990). The soil fauna of beech forests: comparison between a mull and a moder soil. *Pedobiologia*, **34**:299–314.

Scharpenseel, H. W. (1993). Major carbon reservoirs of the pedosphere: source-sink relation, potential of 14C and 13C as supporting methodologies. *Water, Air, and Soil Pollution*, **70**:431–42.

Scheinost, A. C. and Schwertmann, U. (1999). Color identification of iron oxides and hydroxysulfates: use and limitations. *Soil Science Society of America Journal*, **63**:1463–71.

Scheu, S. (1987). The role of substrate feeding earthworms (Lumbricidae) for bioturbation in a beechwood soil. *Oecologia*, **72**:192–6.

Schoeneberger, P. J., Wysocki, D. A., Benham, E. C. and Broderson, W. D. (1998). *Fieldbook for Describing and Sampling Soils.* Lincoln, NE: Natural Resources Conservation Service, USDA, National Soil Survey Center.

Schwertmann, U. (1985). The effect of pedogenic environments on iron oxide minerals. *Advances in Soil Science*, **1**:172–99.

Scow, K. M. and Johnson, C. R. (1997). Effect of sorption on biodegradation of soil pollutants. *Advances in Agronomy*, **58**:1–56.

Séguin, V., Gagnon, C. and Courchesne, F. (2004). Changes in water extractable metals, pH and organic carbon concentrations at the soil-root interface of forested soils. *Plant and Soil*, **260**:1–17.

Seiler, W. (1974). The cycle of atmospheric CO. *Tellus*, **26**:116–35.

Seligman, B. J. (2000). Long-term variability of pipeline: permafrost interactions in North-West Siberia. *Permafrost and Periglacial Processes*, **11**:5–22.

Shainberg, I. (1992). Chemical and mineralogical components of crusting. In Sumner, M. E. and Stewart, B. A. (eds.) *Soil Crusting Chemical and Physical Processes*, pp. 33–53. Advances in Soil Science. Boca Raton, FL: Lewis Publishers.

Sharp, W. D., Ludwig, K. R., Chadwick, O. A., Amundson, R. and Glaser, L. L. (2003). Dating fluvial terraces by $^{230}$Th/U on pedogenic carbonate, Wind River Basin, Wyoming. *Quaternary Research*, **59**:139–50.

Shiklomanov, I. A. (2000). Appraisal and assessment of world water resources. *Water International*, **25**:11–32.

Shishov, L. L., Tonkonogov, V. D., Lebedeva, I. I. and Gerasimova, M. I. (2001). *Russian Soil Classification System*. English translation. Moscow: Dokuchaev Soil Science Institute.

Shoji, S. (1986). Mineral characteristics. I: Primary minerals. In Wada, K. (ed.) *Ando Soils in Japan*, pp. 21–40. Fukuoka, Japan: Kyushu University Press.

Shoji, S. and Masui, J. (1971). Opaline silica of recent volcanic ash soils in Japan. *Journal of Soil Science*, **22**:101–12.

Shoji, S. and Takahashi, T. (2002). Environmental and agricultural significance of volcanic ash soils. *Global Environmental Research*, **6**:113–35.

Shoji, S., Nanzyo, M. and Dahlgren, R. (1993a). *Volcanic Ash Soils: Genesis, Properties and Utilization*. Amsterdam: Elsevier.

Shoji, S., Nanzyo, M., Shirato, Y. and Ito, T. (1993b). Chemical kinetics of weathering in young Andisols from northeastern Japan using soil age normalized to 10 °C. *Soil Science*, **155**:53–60.

Shoji, S., Nanzyo, M., Dahlgren, R. A. and Quantin, P. (1996). Evaluation and proposed revisions of criteria for Andosols in the World Reference Base for Soil Resources. *Soil Science*, **161**:604–15.

Simonson, R. W. (1995). Airborne dust and its significance to soils. *Geoderma*, **65**:1–43.

Singer, A. (1977). Extractable sesquioxides in six Mediterranean soils developed on basalt and scoria. *Journal of Soil Science*, **28**:125–35.

Singer, A. (1987). Land evaluation of basaltic terrain under semi-arid to Mediterranean conditions in the Golan Heights. *Soil Use and Management*, **3**:155–62.

Singer, C. (1959). *A History of Scientific Ideas*. New York: Dorsett Press.

Singer, M. J. and Le, Bissonnais, Y. (1998). Importance of surface sealing in the erosion of some soils from a Mediterranean climate. *Geomorphology*, **24**:79–85.

Singer, M. J. and Shainberg, I. (2004). Mineral soil surface crusts and wind and water erosion. *Earth Surface Processes and Landforms*, **29**:1065–75.

Singer, M. J. and Ugolini, F. C. (1974). Genetic history of two well-drained subalpine soils formed on complex parent material. *Canadian Journal Soil Science*, **54**:475–89.

Singer, M. J. and Warrington, D. N. (1992). Crusting in the Western United States. In Sumner, M. E. and Stewart, B. A. (eds.) *Soil Crusting Chemical and Physical Processes*, pp. 179–204. Advances in Soil Science. Boca Raton, FL: Lewis Publishers.

Singh, B. and Gilkes, R. J. (1992). The electron-optical investigation of the alteration of kaolinite to halloysite. *Clays and Clay Minerals*, **40**:212–29.

Sitaula, B. K., Warner, W. S., Bakken, L. R. *et al.* (1995). An interdisciplinary approach for studying greenhouse gases at the landscape scale. *Norwegian Journal of Agricultural Science*, **9**: 189–209.

Six, J. and Jastrow, J. (2002). Organic matter turnover. In Lal, R. (ed.) *Encyclopedia of Soil Science*, pp. 936–42. New York: Marcel Dekker.

Six, J., Elliott, E. T. and Paustian, K. (2000). Soil macroaggregate turnover and microaggregate formation: a mechanism for C sequestration under no-tillage agriculture. *Soil Biology and Biochemistry*, **32**:2099–2103.

Six, J., Connant, R. T., Paul, E. A. and Paustian, J. (2002). Stabilisation mechanisms of soil organic matter: implications for C-saturation of soils. *Plant and Soil*, **241**:155–76.

Smagin, A. V. (1994). Theory of soil stability. *Eurasian Soil Science*, **27**:17–32.

Smagin, A. V. (1999). Functioning regimes of bio-abiotic systems. *Eurasian Soil Science*, **32**:1277–90.

Smagin, A. V. (2000). The gas function of soils. *Eurasian Soil Science*, **33**:1061–71.

Smagin, A. V. (2003). Theory and methods of evaluating the physical status of soils. *Eurasian Soil Science*, **36**:301–12.

Smagin, A. V., Smagina, M. V., Vomperskii, S. E. and Glukhova, T. V. (2000). Generation and emission of greenhouse gases in bogs. *Eurasian Soil Science*, **33**:959–66.

Smaling, E. M. A., Oenema, O. and Fresco, L. O. (eds.) (1999). *Nutrient Disequilibria in Agroecosystems: Concepts and Case Studies*. Wallingford, UK: CAB International.

Smith, D. (2003). *The Atlas of War and Peace*. London: Earthscan.

Smith, W. H. (1981). *Air Pollution and Forests*. New York: Springer-Verlag.

Snakin, V. V., Prisyazhnaya, A. A. and Kovacs-Lang, E. (2001). *Soil Liquid Phase Composition*. Amsterdam: Elsevier.

Soil Survey Staff. (1960). *Soil Classification, a Comprehensive System:—7th Approximation*. USDA-SCS. Washington, DC: United States Government Printing Office.

Soil Survey Staff (1975). *Soil Taxonomy. A Basic System of Soil Classification for Making and Interpreting Soil Surveys*. Agricultural Handbook No. 436. Washington, DC: USDA.

Soil Survey Staff (1998). *Keys to Soil Taxonomy*. 8th edition. USDA, NRCS. Washington, DC: United States Government Printing Office.

Soil Survey Staff (1999). *Soil Taxonomy. A Basic System of Soil Classification for Making and Interpreting Soil Surveys*. USDA, Handbook 436, 2nd edition. Washington, DC: United States Government Printing Office.

Soil Survey Staff (2006). *Keys to Soil Taxonomy*. 10th edition. USDA, NRCS. Washington, DC: United States Government Printing Office.

Soil Survey Staff (2005). *National Soil Survey Characterization Data*. Lincoln, NE: Soil Survey Laboratory, National Soil Survey Center, USDA, NRCS.

Sokolov, I. A. (1996). The paradigm of pedology from Dokuchaev to the present day. *Eurasian Soil Science*, **29**:250–63.

Sonneveld, M. P. W. and Bouma, J. (2003). Methodological considerations for nitrogen policies in the Netherlands including a new role of research. *Environmental Science and Policy*, **6**:501–11.

Sonneveld, M. P. W., Bouma, J. and Veldkamp, A. (2002). Refining soil survey information for a Dutch soil series using land use history. *Soil Use and Management*, **18**:157–63.

Southard, R. J. and Buol, S. W. (1988). Subsoil blocky structure formation in some North Carolina Paleudults and Paleaquults. *Soil Science Society of America Journal*, **52**:1069–76.

Sposito, G. (1981). *The Thermodynamics of Soil Solutions*. Oxford/New York: Clarendon Press.

Sposito, G., Schultz, A., Gersper, P. and Amundson, R. (1992). Hans Jenny. In *In Memoriam*, pp. 79–81. Berkeley, CA: Academic Senate, The University of California.

State Committee of the Russian Federation on Environmental Protection (1997). *National Report on the State of the Environment in the Russian Federation in 1996*. Moscow (in Russian).

Steele, K. (ed.) (1995). *Animal Waste and the Land-Water Interface*. Boca Raton, FL: Lewis Publishers.

Steen, I. (1998). Phosphorus availability in the 21st century: management of a non-renewable resource. *Phosphorus and Potassium*, **217**:25–31. www.nhm.ac.uk/mineralogy/phos/p&k217/steen.htm [verified on 15 May 2005].

Stevenson, F. J. (1994). *Humus Chemistry: Genesis, Composition, Reactions*. New York: John Wiley.

Stigliani, W. M. (1991). *Chemical Time Bombs: Definitions, Concepts and Examples*. Laxenburg, Austria: International Institute for Applied Systems Analysis.

Stremski, M. (1975). *Ideas Underlying Soil Systematics*. Translation of 1971 Polish edition. TT73–54013. Warsaw: Foreign Scientific Publisher.

Suarez, D. L. (1987). Prediction of pH errors in soil-water extractors due to degassing. *Soil Science Society of America Journal*, **51**:64–7.

Sundquist, B. (2004). *The Earth's Carrying Capacity. Some Literature Reviews*. Edition 1. http://home.alltel.net/bsundquist1/ [verified on 7 April 2005].

Suzuki, T., Kondo, H., Yaguchi, K., Maki, T. and Suga, T. (1998). Estimation of leachability and persistence of pesticides at golf courses from point-source monitoring and model to predict pesticide leaching to groundwater. *Environmental Science and Technology*, **32**:920–9.

Takahashi, T. and Shoji, S. (1996). Active aluminum status in surface horizons showing continuous climosequence of volcanic ash-derived soils in Towada district, northeastern Japan. *Soil Science and Plant Nutrition*, **42**:113–20.

Takahashi, T., Nanzyo, M. and Shoji, S. (2004). Proposed revisions to the diagnostic criteria for andic and vitric horizons and qualifiers of Andosols in the World Reference Base for Soil Resources. *Soil Science and Plant Nutrition*, **50**:431–7.

Tandarich, J. P. and Sprecher, S. W. (1994). The intellectual background for the factors of soil formation. In Amundson, R., Harden, J. and Singer, M. (eds.) *Factors of Soil Formation: A Fiftieth Anniversary Retrospective*, pp. 1–13. Special Publication No. 33. Madison, WI: Soil Science Society of America.

Tandarich, J. P., Darmody, R. G., Follmer, L. R. and Johnson, D. L. (2002). Historical development of soil and weathering profile concepts from Europe to the United States of America. *Soil Science Society of America Journal*, **66**:335–46.

Tansley, A. G. (1935). The use and abuse of vegetational concepts and terms. *Ecology*, **16**:284–307.

Tardy, Y., Bocquier, G., Paquet, H. and Millot, G. (1973). Formation of clay from granite and its distribution in relation to climate and topography. *Geoderma*, **10**:271–84.

Thompson, M., Zhang, H., Kazemi, M. and Sandor, J. (1989). Contribution of organic matter to cation exchange capacity and specific surface area of fractionated soil materials. *Soil Science*, **148**:250–7.

Tiller, K. (1983). Micronutrients. In *Soils: An Australian Viewpoint*, pp. 365–87. Melbourne: CSIRO and London: Academic Press.

Tilman, D. (1987). Secondary succession and the pattern of plant dominance along experimental nitrogen gradients. *Ecological Monographs*, **57**:189–214.

Tilman, D., Cassman, K. G., Matson, P. A., Naylor, R. and Polasky, S. (2002). Agricultural sustainability and intensive production practices. *Nature*, **418**:671–7.

Toop, E. and Pattey, E. (1997). Soils as sources and sinks for atmospheric methane. *Canadian Journal of Soil Science*, **77**:167–78.

Trumbore, S. E., Chadwick, O. A. and Amundson, R. (1996). Rapid exchange between soil carbon and atmospheric carbon dioxide driven by temperature change. *Science*, **272**:393–6.

Tutzing Project 'Time Ecology' (1998). *Preserving Soils for Life. Proposal for a Convention on Sustainable use of Soils (Soil Convention)*. Munich: Schriftenreihe zur politischen Ökologie.

Ugolini, F. C. (2005). Dynamic pedology. In Hillel, D. (ed.) *Encyclopedia of Soils in the Environment*, Vol. 1, pp. 156–65. Oxford: Elsevier.

Ugolini, F. C. and Anderson, D. M. (1973). Ionic migration and weathering in frozen Antarctic soils. *Soil Science*, **115**: 461–70.

Ugolini, F. C. and Dahlgren, R. A. (1987). The mechanism of podzolization as revealed by soil solution studies. In Righi, D. and Chauvel, A. (eds.) *Podzols et Podzolisation*, pp. 195–203. Paris: AFES – INRA.

Ugolini, F. C. and Dahlgren, R. A. (2002). Soil development in volcanic ash. *Global Environmental Research*, **6**:69–81.

Ugolini, F. C. and Edmonds, R. L. (1983). Soil biology. In Wilding, L. P., Smeck, N. E. and Small, G. F. (eds.) *Pedogenesis and Soil Taxonomy. I: Concepts and Interactions*, pp. 193–231. Amsterdam: Elsevier.

Ugolini, F.C. and Spaltenstein, H. (1992). The pedosphere. In Charlson, R., Orions, G., Butcher, S. and Wolf, G. (eds.) *Global Biogeochemical Cycles*, pp. 85–153. San Diego, CA: Academic Press.

Ugolini, F.C., Dawson, H.C. and Zachara, J. (1977). Direct evidence of particle migration in the soil solution of a Podzol. *Science*, **198**:603–605.

Ugolini, F.C., Dahlgren, R.A., Shoji, S. and Ito, T. (1988). Andosolization and podzolization as revealed by soil solution studies, South-Hakkoda, Northeastern Japan. *Soil Science*, **145**:111–25.

UN-ECE (1998). *Convention on Long-range Transboundary Air Pollution. Protocol on Heavy Metals*. United Nations Economic Commission for Europe. www.unece.org/env/lrtap/welcome.html [verified on 21 December 2005].

United States Geological Survey (1995). *Chesapeake Bay: Measuring Pollution Reduction*. www.water.usgs.gov/wid/html/chesbay.html [verified on 15 May 2005].

U.S. Census Bureau (2005a). *Historical Estimates of World Population*. www.census.gov/ipc/www/worldhis.html [verified on 24 March 2006].

U.S. Census Bureau (2005b). *Total Midyear Population for the World: 1950–2050*. www.census.gov/ipc/www/worldpop.html [verified on 24 March 2006].

Van den Akker, J.J.H., Arviddsson, J. and Horn, R. (2003). Introduction to the special issue on experiences with the impact and prevention of subsoil compaction in the European Union. *Soil and Tillage Research*, **73**:1–8.

Van Straalen, N.M. and Verhoef, H.A. (1997). The development of bioindicator system for soil acidity based on arthropod pH preferences. *Journal of Applied Ecology*, **34**:217–32.

Vegter, J.J., De Bie, P. and Dop, H. (1988). Distributional ecology of forest floor Collembola (Entomobryidae) in the Netherlands. *Pedobiologia*, **31**:65–73.

Veldkamp, A. and Lambin, E. (2001). Predicting land-use change. *Agriculture Ecosystems and Environment*, **85**:1–292.

Vepraskas, M.J. (1988). Bulk density values diagnostic of restricted root growth in coarse textured soils. *Soil Science Society of America Journal*, **52**:1117–21.

Vepraskas, M.J. (2001). Morphological features of seasonally reduced soils. In Richardson, J.L. and Vepraskas, M.J. (eds.) *Wetland Soils: Genesis, Hydrology, Landscapes, and Classification*, pp. 163–82. Boca Raton, FL: Lewis Publishers.

Vitousek, P. M. (2004). *Nutrient Cycling and Limitation. Hawai'i as a Model System*. Princeton, NJ: Princeton University Press.

Vitousek, P.M., Mooney, H.A., Lubchenco, J. and Mellilo, J.M. (1997a). Human domination of Earth's ecosystems. *Science*, **277**:494–9.

Vitousek, P.M., Aber, J., Howarth, R.W. *et al.* (1997b). Human alteration of the global nitrogen cycle: causes and consequences. *Ecological Applications*, **7**:737–50.

Volkoff, B., Melfi, A.J. and Cerri, C.C. (1979). Les sols sur roches cristallines formés sous climat sub-tropical humide au Brésil. *Cahiers ORSTOM, Pedologie*, **17**:163–83.

Wada, K., Kakuto, Y., Wilson, M.A. and Hanna, J.V. (1991). The chemical composition and structure of a 14 Å intergradient mineral in a Korean Ultisol. *Clay Minerals*, **26**:449–61.

Wagner, D., Pfeiffer, E.-M. and Bock, E. (1999). Methane production in aerated marshland and model soils: effects of microflora and soil texture. *Soil Biology and Biochemistry*, **31**:999–1006.

Wardle, D. A., Bardgett, R. D., Klironomos, J. N. *et al.* (2004). Ecological linkages between aboveground and belowground biota. *Science*, **304**:1629–33.

Warscheid, Th., Oelting, M. and Krumbein, W. E. (1991). Physico-chemical aspects of biodeterioration processes on rocks with special regard to organic pollutants. *International Biodeterioration*, **28**:37–48.

Watanabe, T., Sawada, Y., Russell, J. D., McHardy, W. J. and Wilson, M. J. (1992). The conversion of montmorillonite to interstratified halloysite-smectite by weathering in the Omi acid clay deposit. *Clay Minerals*, **27**:159–73.

Wauthy, G., Mundon-Izay, N. and Dufrene, M. (1989). Geographic ecology of soil oribatid mites in deciduous forest. *Pedobiologia*, **33**:399–416.

Weaver, C. E. (1989). *Clays, Muds and Shales.* Developments in Sedimentology 44. Amsterdam: Elsevier.

Wedepohl, K. H. (ed.) (1978). *Handbook of Geochemistry.* Berlin: Springer Verlag.

Weitkamp, W. A., Graham, R. C., Anderson, M. A. and Amrhein, C. (1996). Pedogenesis of a vernal pool Entisol-Alfisol-Vertisol catena in southern California. *Soil Science Society of America Journal*, **60**:316–23.

Wenzel, W. W., Sletten, R. S., Brandstetter, A., Wieshammer, A. and Stingeder, G. (1997). Adsorption of trace metals by tension lysimeters: nylon membrane vs. porous ceramic cup. *Journal of Environmental Quality*, **26**:1430–4.

White, G. N. and Dixon, J. B. (2002). Kaolin-Serpentine minerals. In Dixon, J. B. and Schulze, D. G. (eds.) *Soil Mineralogy with Environmental Applications*, pp. 389–414. Madison, WI: Soil Science Society of America.

Whittaker, R. H. (1967). Gradient analysis of vegetation. *Biological Reviews*, **42**:207–64.

Williams, E. G. (1959). Influence of parent materials and drainage conditions on soil phosphorus relationships. *Agrochimica*, **3**:279–309.

Wilson, M. J. (1976). Exchange properties and mineralogy of some soils derived from lavas of Lower Old Red Sandstone (Devonian) age.II: Mineralogy. *Geoderma*, **15**:289–304.

Wilson, M. J. and Logan, J. (1976). Exchange properties and mineralogy of some soils derived from lavas of Lower Old Red Sandstone (Devonian) age. I: Exchangeable cations. *Geoderma*, **15**:273–88.

Winteringham, F. P. W. (1989). Radioactive fallout in soils, crops and food. FAO Soils Bulletin 61. Rome: Food and Agriculture Organization of the United Nations.

Wipf, S., Rixen, C., Fisher, M., Schmid, B. and Stoeckli, V. (2005). Effects of ski piste preparation on alpine vegetation. *Journal of Applied Ecology*, **42**:306–16.

Wolt, J. (1994). *Soil Solution Chemistry: Applications to Environmental Science and Agriculture.* New York: John Wiley.

Wood, Y. A., Graham, R. C. and Wells, S. G. (2005). Surface control of desert pavement pedologic process and landscape function, Cima Volcanic Field. *Catena*, **59**:205–30.

World Watch (1994). *Back on Track: The Global Rail Revival.* Worldwatch Paper 118. Washington, DC: WorldWatch.

Wosten, J. H. M., Pachepsky, Ya. A. and Rawls, W. J. (2001). Pedotransfer functions: bridging the gap between available basic soil data and missing soil hydraulic characteristics. *Journal of Hydrology*, **251**:123–50. www.nwqmc.org/98proceedings/Papers/63-NOLAN.htm [verified on 15 May 2005].

WRF (2000). *World Resources 2000–2001.* Washington, DC: World Resources Institute.

Yaalon, D. H. and Berkowicz, S. (eds.) (1997). *History of Soil Science: International Perspectives.* Advances in Geoecology 29. Reiskirchen: Catena Verlag.

Yang, S. Y. N., Connell, D. W., and Hawker, D. W. and Kayal, S. I. (1991). Polycyclic aromatic hydrocarbons in air, soil and vegetation in the vicinity of an urban roadway. *The Science of the Total Environment*, **102**:229–40.

Zabowski, D. and Ugolini, F. C. (1990). Lysimeter and centrifuge soil solutions: seasonal differences between methods. *Soil Science Society of America Journal*, **54**:1130–5.

Zabowski, D. and Ugolini, F. C. (1992). Seasonality in the mineral stability of a subalpine Spodosol. *Soil Science*, **154**:497–507.

Zimmer, M. and Topp, W. (1998). Microorganisms and cellulose digestion in the gut of the woodlouse *Porcellio scaber. Journal of Chemical Ecology*, **24**:1397–408.

# Index


Abrazem 10
acid soil 71, 78
acidic deposition 69
acidification 202, 236–8
acidity 68
Acrisol 147
actinomycete 167
adhesion 53
adsorption 53, 59
age 110, 190
aggregate 51, 53, 62
aggregation 53
agricultural revolution 194
agriculture 195–8
Agrobrazem 10
agronomy 212
Agrozem 10
air entry potential 84
air permeability 83
airport 199
albic 17
albite 33
Alfisol 134, 136, 147, 158
algae 166, 167
Al-humus complex 67, 121, 127, 137, 141, 143
alic property 144
alkaline soil 78
allophane 18, 37, 40–2, 66, 134, 137, 141
Al-rich allophane 41
aluminium (Al) 18, 31, 37, 66, 71, 134, 137, 139, 159, 186, 237–8
aluminium silicate clay 16
amino acid 50
amphibian 167
analcime 29
anatase 42
Andisol 66, 137
Andosol 121, 138–47
andosolization 66–8
anion exchange capacity (AEC) 120
anthraquic 15
Anthropocene 9, 193
anti-allophanic effect 144
antimony (Sb) 241
apatite 126
*Aporrectodea caliginosa* 96
aquic conditions 19
Archae 97
Archimedes 104
Arenosol 100
argic 17
argillic 17, 18
argon (Ar) 77
arid 39
arid climate 17, 19
arid region 36, 160
aridic 132
aridic environment 146
Aridisol 135
Aristotle 2
arsenate ($AsO_4^{3}$) 32
arsenic (As) 241
artefact 44, 63, 65
arthropod acidity index (AAI) 100
Artifabricat 10
ash 70
aspect 151, 156
aspen 170, 176
automobile exhaust 80

back slope 153
bacteria 50, 53, 95, 165, 167
basaltic soil 128
base saturation 120, 124, 237, 238
basin 151–3
bauxite 184
beech 145, 147, 176
beidellite 36
bentonite 41
bicarbonate 17, 62
bioavailability 62
biocide 246
biocorrosive activity 91
biocycling 12, 14
biogenic activity 82

biomarker 242
biomolecule 50
biota 50
biotic factor 109–10
biotite 19, 33, 114
birch 171, 176
bird 167
blocky structure 21
boehmite 38, 42, 188
bog 78, 136
bond 32, 53
boreal forest 134
boulder clay (glacial drift) 185
brackish 19
Braun-Blanquet 104
bromium (Br) 210
bubbles formation 83
buffer 236, 258
buffer function 211, 218
bulk density 59, 117, 119, 224
by-pass flow 216

$^{13}C$ 54
$^{14}C$ 54
C4-derived carbon 55
cable 200
cadmium (Cd) 169, 179, 241
caesium ($^{134}Cs$, $^{137}Cs$) 203, 250
calcareous 29
calcaric 176
calcic 17–18, 146
Calcisol 135
calcite 23, 33, 160, 189
calcium (Ca) 16, 18, 71, 238
calcium carbonate ($CaCO_3$) 16
calcrete (calcite cemented horizon) 189
callose 239
cambic 18
Cambisol 134, 147, 187
canal 199
Canny 106
canopy 72, 166
capillarity 17, 59
capillary action 16
capillary tension 21
capillary water 61, 62, 63
carbohydrate 47, 50
carbon (C) 205
carbon dioxide ($CO_2$) 77, 78
carbon monoxide (CO) 89
carbonate ($CO_3^{2}$) 17, 18, 36, 113, 132, 135, 146, 237
carbonic acid ($H_2CO_3$)16, 66, 71, 72, 78, 134, 135
carrier function 211, 217
casts 166
catena 110, 156–7, 158
cation exchange capacity (CEC) 18, 31, 45, 53, 120
Cato 3, 4
cell 238
cemented horizon 230
chamber 85
channel 152
Chemdegrazem 10
chemical protection 53
'chemical time bomb' 243
chemo-organotrophic bacteria 91
chernic 14
Chernozem 4, 95, 147
chinochlore 29
chlorite 29, 35, 36
chromium (Cr) 241
chrysotile 29
classification 3
clay 17, 18
clay mineral 29, 53
clay skin (or film or tonhatchen) 17, 18
claypan 227
clearcutting 69, 194
climate 109, 160, 168, 236, 251
climate change 205, 252–3
climatic regime 6
climosequence 109
closed drainage 152
cobalt (Co) 123
cohesion 53
cold climate 36
cold desert 132
colloid 61
colonizer 94
colour 19, 20, 39, 148, 159, 173, 179
Columella 3–4, 4
communism 214
compacted soil 117
compactibility 59
compaction 197, 217, 220, 223, 224–7, 229–30, 231
compost 15
compound attributes 162
concavity 153
co-neogenesis 40
conflict 204
coniferous 68, 134, 142, 145
conservation tillage 229
consistency 59
contamination 236
Convention on Long-range Transboundary Air Pollution 245
Convention on Soil 253
convexity 153
copper (Cu) 123, 169, 170, 241
co-production 220
corundum 42
crack 21
creep movement 16
Cretaceous 184, 186, 188
crusting 223, 227–9, 230, 231, 232
cryic 132
Cryosol 134
cultural component 111
cultural heritage 213
cultural, function 212, 219
cyanobacteria 95

Darwin 105
dating 182
De Agricultura 4
decomposition 80
deficiency 127, 241, 249
degradation 247
deposition 12, 157
depositional crust 228
diffusion coefficient 82
digital elevation model (DEM) 162
dioxin 246
displacement 63
dissolved organic carbon (DOC) 62, 64, 171
dissolved organic matter (DOM) 252
dissolved solid 59
divide 152
DNA 242
Dokuchaev 4–5, 103
dolomite 33
drainage 152
dry area 39
dry region 65
dump 201
dune 91
dung 3
duration 205
duripan 62, 146, 227
dynamic pedology 13

earthquake 32, 207
earthworm 99, 167, 237
ecosystem 108, 167
edaphos 3
eddy correlation 86
effect 242
Eh 65
electrical conductivity (EC) 61
electrical resistance block 60
electrochemical technique 65
electrode 65
eluviation 16, 37
emission 85
emission of $CO_2$ 80–1, 88
endellite 29
enrichment factor (EF) 244–5
Entisol 132, 146, 158, 162
enzyme 50
epipedon 14
epsomite 160
erosion 12, 157, 195, 231, 196
eucalypt 123
'Eureka' moment 112
eutrophication 248–50, 255
evaporite mineral 160–1
expansive soil 32
exposure 242
exudate 21

Fabricat 10
faeces 94, 95, 96
Fallou 103
FAO Legend 8
faunal activity 101
Fe-hydroxide 132, 134, 135
feldspar 38, 114, 115, 121, 127, 183
Fe-oxide 207
ferralic 17
Ferralsol 135, 147, 182
ferric 42
ferrihydrite 37, 42
ferrolysis 42
ferrous 34, 42
fertilization 211, 217
fertilizer 89, 102, 195, 202, 221, 236, 243, 248, 258, 262
field capacity 60, 61
filter function 211, 218, 237, 241, 249
fir 171
fire 42, 205–7
flint 201
flood 32
fluid 223
food chain 239, 241
foot slope 153
forest 69, 71, 78, 99, 176, 194, 226, 236, 238
Fourier transform infrared spectroscopy (FITR) 46
fragipan 207, 227
frigid 132
fulvic acid (FA) 48, 50, 66, 134
fungi 21, 50, 53, 95, 166
fungicide 245
furan 246

gamma-ray absorption 59
garbage 201
gas diffusivity 82, 84
gas emission 80
Gelisol 134
genome 93
Gibbs' free energy 34
gibbsite 18, 34, 38, 41, 42, 115, 116, 188
gilgai 32
glacial drift (boulder clay) 185
glacial till 115, 116, 118, 123, 187
glass 71
gleization 20
gley (glei) 20
gleyic properties 19
global change 109
globalization 213
glomalin 21
goethite 19, 42, 116, 126, 135, 159
golf 203
gopher 157
gradient analyses 111
granitic soil 128
granular structure 18, 21
grass 14, 166
grassland 134–5
gravitational potential 60
gravitational water 60, 61, 62, 65

gravity 59
grazing 194–5
gumbotil 182
gypsic 18
Gypsisol 135
gypsum ($CaSO_4$ $2H_2O$) 18, 42, 160, 230

habitat function 212, 218–19
haematite 19, 34, 42, 116, 126, 135, 159
halide 36
halite 33, 160
halloysite 38, 40–1, 115, 147
halloysite-smectite 40
haploidization 14
head slope 152
heavy metal 49, 50, 51, 240–2
Henry's constant 77
herbicide 245
Herodotus 3
Hilgard 103
histic 15
historic function 212, 219
Histosol 45, 136, 147
Holocene 1, 110, 147, 185, 187
horizon 8, 9, 13
horizonation 14
hortic 15
hot desert 135
human activity 10, 102, 185
human impact 220
human production 263
human-impacted soil 195
humankind 193, 208, 219
humans 110, 148
humic acid (HA) 48, 143, 148
humic substance 50–1
humid climate 38
humid (udic) environment 142
humid region 160, 161
humin 48
humus 41, 47, 48, 99
humus complex 68
husbandry 249
hydraulic conductivity 61, 117, 118
hydrogen sulphide ($H_2S$) 89
hydromorphic soil 77, 88, 97
hydroxide 32
hydroxy-interlayered smectite (HIS) 37
hydroxy-interlayered vermiculite (HIV) 37
hygroscopic water 60, 61
hyperthermic 132

ice 31, 186, 209
illite 31, 35, 39, 115
illuviation 16, 17
ilmenite 42, 209
imogolite 18, 37, 40, 66, 134, 137, 141
Inceptisol 134, 147
incongruent dissolution 34, 71–2
independent variable 107
induced convection 83
industrial emission 88
industrial process 263
industry 218
infiltration 15, 217, 226, 227, 230, 251
insect 167
insecticide 245
intensity 205
intensity factor 62
interfluve 152
ions 16
iridium (Ir) 208
iron (Fe) 18, 19, 20, 31, 66, 37, 71, 125, 134, 135, 159, 172, 231, 241, 186
iron oxide 17, 18, 19, 33, 39, 42, 129, 198, 210
ironstone 21
irragric 15
irrigation 1, 19, 197, 211, 251
isothermal diffusion 81
isotope 110

jarosite 42
Jenny 103, 106–8
Jenny equation 35
Jenny's approach 132

kandic 17
kaolinite 18, 31, 33, 34, 38, 39, 43, 115, 135, 161, 189
kaolin-smectite 40
Kozeni-Karman's theory 83
Krasnozems 126
krotovina 166
Kuhn 111

lacustrine 39
land use 211, 212, 215, 220, 236, 255
landfill 83, 201
landscape 6, 9, 151, 219
laterite 21, 39
lateritic 42
lava 207
leaching 12, 16, 17, 19, 36, 37, 38, 39, 40, 50, 54, 64, 68, 135, 137, 160, 161
lead (Pb) 169, 241, 245
lenticular structure 186
lepidocrocite 42
lessivage 17, 18, 160
lichen 166
lignin 47
limonite 29
liquefaction 207
litter 87, 92, 99, 156, 169, 176, 237
Little Ice Age 147
livestock manure 248
Lixisol 147
low-molecular-weight organic acid (LMWOA) 66
*Lumbricus rubellus* 95
*Lumbricus terrestris* 96
lunar regolith 209
Luvisol 134, 147
lysimeter 13, 18, 64, 65, 247

macrofauna 95
maghaemite 42, 116
magma 33, 207
magnesium (Mg) 18, 31, 39
magnetic field 200
magnetite 42, 116
maize 220
mammal 166
manganese (Mn) 19, 20, 159, 241
manganese oxide 42
mangrove 43
manure 15
maple 171, 176
Marbut 104
Mars 209–10
marshes 97
mass transfer 81
mass wetness 59
matric force 59
matric potential 60, 61
mean residence time (MRT) 48, 54
melanic epipedon 145
melanization 134
membrane diffusion technique 77
mercury (Hg) 241, 242
mesic 132
mesofauna 95
metahalloysite 29
metal-organo complex 66
meteorite 208
methane ($CH_4$) 77, 83, 88–9, 97
methyl bromide ($CH_3Br$) 90
mica 23, 32, 35, 36, 39, 69, 115, 121, 134
Michaelis-Menten's dependency 80
microaggregate 52, 53
microbe 19
microbial activity 53
microbial biomass 50
microclimate 109
microfauna 95
micrometeorological method 86
micromorphology 188
micro-organism 20, 50, 92, 242
microtopographic feature 155
microtopography 156
midden 96
Middle Age 194
mineral 18, 33
mineralization 54
minimum tillage 229
mining 102, 201
Miocene 184
model 216, 220
moder 99
mollic 14, 15
Mollisol 135, 147
monolith 64, 65
montmorillonite 162
Moon 209
moss 166
mottled pattern 20
mucilage 170
mulching 230
mull 99
muscovite 33, 114
mycorrhizosphere 170

$^{15}N$ 54
natric 18
natrojarosite 42
natural convection 82–3
Naturfabricat 10
nematocide 246
neoformation 37
Neolithic 198
net primary productivity (NPP) 193, 197
neutron scattering 59
Ni (nickel) 169, 170
nickel (Ni)
nitrate ($NO_3$) 19, 20, 65, 69, 221
nitrate pollution 217
nitric acid ($HNO_3$) 69
nitrogen (N) 69, 77, 89, 92, 121, 125, 134, 205, 248–50, 261–3
non-liquid, water 61
non-polar solvent 63
non-soil 10, 105
non-solvent water 61
noosphere 10
nose slope 152
nuclear magnetic resonance spectroscopy (NMR) 46
'nutrient pollution' 248

oak 72
ochric 15
olivine 33, 209
opal 189
Operational Taxonomic Unit (OTU) 242
optimum water content 224
organic acid 18, 50
organic contaminant 245–7
organic farming 98
organic matter 45
organomineral association 51–3
osmotic force 59
osmotic potential 60, 61
oxic 18
oxidation 34
oxide 32
Oxisol 126, 135, 136, 147, 161
oxygen (O) 19, 20, 77
oxygen starvation 84–5
oxyhydroxide 32

paddy soil 83
palaeoclimate 147
palaeosol 43, 147
palygorskite 33, 39
pampas grass 142, 145
parent material 110

particulate organic matter (POM) 48, 49
P-E index 142
peat 78, 79, 80, 83, 84, 97, 136, 211
pedogenic chlorite 37–8
pedolandscape 9
pedology 103, 105
pedon 6, 7–8
pedotransferfunction 216
pedoturbation 16
penetrability 59
penguin 165
Penman equation 82
periglacial 184
permafrost 134, 186, 200
permanent charge 31
permanent wilting point 60, 61
permeability 251
persistent organic pollutant (POP) 246
pesticide 49, 64, 245–7
petrocalcic 18, 146, 227
petroferric 21, 227
petrogypsic 18, 227
petroplinthic 21
pH 63, 66, 69
phenol 50
phosphate ($PO_4^{3}$) 32
phosphorus (P) 122, 126, 248–50, 258–9
phyllosilicate 29, 207
physical comminution 35
physical protection 53
phytolith 166, 167
phytosiderophore 170
pine 165
pipeline 200
plaggen 15, 219
plaggic 15
plagioclase 33, 71, 114, 115, 209
plagioclimax 185
plant debris 49
plant-available water 60, 118
plasticity 59
Pleistocene 184–5
plinthic 21
plinthite 20
Pliny, the elder 4
Pliocene 184
plough pan 225
ploughing (subsoiling) 196, 230
plutonium ($^{238}$Pu, $^{239}$Pu, $^{240}$Pu, $^{241}$Pu, $^{242}$Pu) 203
podzol 37, 41, 42, 142, 145, 147, 176, 186, 187, 134
podzolic soil 78
podzolization 37, 50, 66–8
point of zero charge (PZC) 31
polar desert 132
pollution 202, 240
polyacrylamide 230
polychlorinated biphenyl (PCB) 246
polycyclic aromatic hydrocarbon (PAH) 246
polygenetic soil 110
polypedon 6, 8
polysaccharide 50, 52
pore 16, 59, 60, 223
porosity 53, 224
potassium (K) 36, 122, 126
preferential flow 247
primary attributes 162
primary mineral 23
primary production 54
processes 17
Proctor curve 224
production function 211, 216–17
profile 6
protein 47, 50, 51
proto-imogolite 66
proton donor 66, 71
psychrometer 60
pyrite 33, 42, 102, 211, 218
pyrolysis 46
pyrophyllite 29, 32
pyroxene 209

$Q_{10}$ 80
quantity factor 62
quarries 201
quart 2, 23, 34, 114, 115, 127, 183, 190
Quasizem 10

radioactivity 202
radionuclide 203, 250–1
railway 199
rainfall 155, 228
rainforest 135
rainwater 79
rate of decomposition 54
reactor function 211, 218
recommendation 259
redox 42, 62, 65, 159
reduction 20, 34
regionalization 213
Regosol 146, 187
relative humidity (RH) 59, 76
relict soil 110
relief 151
reptile 167
residence time 247
resilience 100, 253
resource function 212, 218
retention time 218
rhizosphere 169, 171
rice 15
ripping 230
roads 198–9
rodent 157
rodenticide 246
Roman road 199
root 50, 166
Rothamsted model 54
rubbish 201
run-off 232
ruthenium ($^{106}$Ru) 203
rutile 29, 42

salic 19
salinization 197, 211, 251–2
salt 132, 160, 197, 204, 251
salt accumulation 36
salt-affected soil 39
sampling volume 7–8
sandstone 91
saponite 115, 116
saprolite 117, 119
saturated conditions 15
saturation 259, 261
saturation index 59, 68
savanna 135
sealing 223, 227–9, 230, 231
secondary mineral 29
semi-arid (ustic) environment 145
sequence 108
serpentine 115, 161
severity 205
sewage sludge 249
shocked quartz ('stishovite') 208
short-range order 18
shoulder 153
shrink-swell 21
side slope 152
siderophore 172
silica 18, 41, 141, 142, 162, 189
silicate 16, 29, 209
silicic acid, 'aqueous silica' ($H_4Sio_4$) 33, 68, 135
silicon, silicium (Si) 31, 39, 71, 137, 138
silt capping 186
silver (Ag) 241
Simonson 165
Si-rich allophane 41
skiing 203
slaking 229
slope 152–3
smectite 31, 32, 36, 39, 39–40, 51, 116, 134, 135, 147, 161
smectite-protein complex 51
smectitic soil 32
snow 155, 204
snowpack (snowcover) 72
soda 160
sodic soil 251
sodium (Na) 18, 32, 71, 231
soil 'quality' 256
soil aggregate 49, 175
soil aggregation 50
soil atmosphere 75, 76
Soil Conservation Service 5
soil function 10
Soil Group 8
soil inorganic carbon (SIC) 45
soil moisture 7, 132
soil organic carbon (SOC) 45
soil organic matter (SOM) 45, 241
soil resilience 221
soil respiration 86
soil series 8
soil solution 35, 57–8, 63
soil structure 55, 251
soil survey 212, 219
Soil Survey Manual 7
Soil Taxonomy (ST) 7, 8, 14, 15, 17, 19, 132, 134, 229
soil temperature 7, 132, 155
soil:solution ratio 63
soilscape 9
soil-water potential 59–60
solar radiation 82
Solonchak 135
Solonet 135
solubility of gases 78
soluble organic matter 50
soluble salt 19, 36
solvation 36
sorption reaction 31
spodic 18
Spodosol 66, 68, 70, 134, 136, 147, 156, 161
sport 85, 203
spruce 239
stability diagram 34
stability field 34
stabilization 55
stable isotope 89
stable isotope tracing 46
stable pool 54
state factors 107
static pedology 13
steppe 134
stickiness 59
stishovite 208
stones 187
Strahler method 187
strength 59
strontium ($^{90}Sr$) 203, 250
structural crust 228
student 112
sulphate ($SO_4^{2}$19, 36, 70
sulphide mineral 42
sulphur 89
sulphur dioxide ($SO_2$) 89
sulphuric 19
sulphuric acid ($H_2SO_4$) 42
summit 153
sun 156
supramolecular association 51
surface area 30
surface diffusion 82
susceptibility 242
suspension effect 65
swamp 43, 97
swelling 32
syngenesis 40

talc 29, 32
Telford and McAdam 199
temperate deciduous forest 134
temperature 131, 138, 252
temperature coefficient 80
tensiometer 60

tephra 41, 71–2, 137, 207
Terminal Restriction Fragment Length Polymorphism (T-RFLP) 242
termitaries 88
terric 15
Tertiary 182
Theophrastos 3
thermal sliding 82
thermic 132
thermodiffusion 82
thermodynamic approach 34–5
thermodynamic potential 60
threshold value 217
tillage 53, 101, 225, 196
time 110, 114, 182, 235
time-domain reflectometry (TDR) 59
titanium (Ti) 244
titanium oxide 42
toe slope 153
topographic attributes 162–3
topography 110, 151
toposequence 9, 110, 161, 161
Toxifabricat 10
trace element 122, 127, 244, 251,
trace metal 169, 178
translocation 12, 68
transmission electron microscopy (TEM) 44
transport of gas 83
tree vegetation 15, 17
tree removal 72
tree-line 132, 134
tropical environment 39
tundra 97, 132–4, 186
tyre 229

udic 132
Ultisol 147, 161
umbric 14–15
uranium (U) 204
urban area 198
urban waste 83
ustic 132

vadose zone 62
vapour 61, 76
vapour pressure 59
variable charge 31
Varro 3, 4
vascular ‘vascular’ transport 84
vermiculite 29, 32, 36, 39, 121, 147, 161,
Vertisol 119, 147, 158, 162
viscosity 83
void ratio 225
volcanic ash 40, 69–70, 131, 137
volcanic soil 66, 113
volume wetness 59

Waksman 104
war 204
waste 201, 205, 218
water ($H_2O$) 16, 59, 73
water content 59
water quality 249
water reservoir 198
water table 20
water-holding capacity (WHC) 61
waterlogging 197
water-saturated soil 82
watershed 151–3
weapon 250
weathering 12, 23, 33, 35, 43, 50, 66, 71, 116, 117, 118, 129, 132, 135, 189, 240, 243
wetland 77, 83, 84, 88, 161
wettability 53
Wiegner 104
wind 156
World Reference Base for Soil Resources (WRB) 8, 19, 14, 134
World Trade Organization (WTO) 213

xeric 132
X-ray diffraction (XRD) 36, 44

zeolite 32
zero-tension 64, 65
zinc (Zn) 123, 169, 241
zircon 183
zirconium (Zr) 244, 245